Fascinating Life Sciences

This interdisciplinary series brings together the most essential and captivating topics in the life sciences. They range from the plant sciences to zoology, from the microbiome to macrobiome, and from basic biology to biotechnology. The series not only highlights fascinating research; it also discusses major challenges associated with the life sciences and related disciplines and outlines future research directions. Individual volumes provide in-depth information, are richly illustrated with photographs, illustrations, and maps, and feature suggestions for further reading or glossaries where appropriate.

Interested researchers in all areas of the life sciences, as well as biology enthusiasts, will find the series' interdisciplinary focus and highly readable volumes especially appealing.

More information about this series at http://www.springer.com/series/15408

Christian Foth • Oliver W. M. Rauhut
Editors

The Evolution of Feathers

From Their Origin to the Present

Springer

Editors
Christian Foth
Department of Geosciences
University of Fribourg
Fribourg, Switzerland

Oliver W. M. Rauhut
Department for Earth and Environmental Sciences
Ludwig-Maximilians-Universität, Palaeontology and Geobiology, GeoBioCenter
München, Germany

ISSN 2509-6745 ISSN 2509-6753 (electronic)
Fascinating Life Sciences
ISBN 978-3-030-27222-7 ISBN 978-3-030-27223-4 (eBook)
https://doi.org/10.1007/978-3-030-27223-4

This Springer imprint is published by the registered company Springer Nature Switzerland AG.
The registered company address is: Gewerbestrasse 11, 6330 Cham, Switzerland

Preface

Birds, the last living dinosaurs, have fascinated humans since ancient times, primarily due to their ability to fly and their colourful appearance. Both characteristics are closely related to their unique epidermal integument, the feather, which represents the morphologically most complex and diverse skin derivate within vertebrates. Due to their high diversity in shape that varies between different body regions, feathers can fulfil a large number of biological roles, including flight, body insulation, display and sensory function. But how did this aberrant integumentary structure evolve, and what was its initial biological role?

With the discovery of the famous Urvogel *Archaeopteryx* in the lithographic limestones of southern Germany just two years after the publication of the Darwin's seminal book *The Origin of Species*, it became evident that the presence of modern feather types, including flight feathers, extended back to at least the Upper Jurassic, preceding the origin of toothless beaks, clawless hands with fused fingers or a strongly reduced tail. For decades, it was considered to be common sense that the origin of feathers was primarily driven by the evolution of flight, using *Archaeopteryx* as key witness from the fossil record. Since the description of *Sinosauropteryx* from the Early Cretaceous of China in 1998, numerous new discoveries of non-avian dinosaurs covered with various types of feathers have challenged this idea fundamentally and led to new evolutionary scenarios for the origin of feathers, their changing functional significance and a new understanding of dinosaurs in general. This book is devoted to the origin and evolution of feathers and highlights the crucial impact of palaeontology on this field of research, documenting the successive increase of morphological complexity along the line towards modern birds.

However, this book would not have been possible without the help of many colleagues and friends who need to be acknowledged. First, we thank Verena Penning for inviting us to edit a book on the evolution of feather. Lars Körner and Rajeswari Balachandran are thanked for their patience and editorial support during the production of this book. We further thank all authors for their contributions as well as Richard Butler, Tom Holtz, Jingmai O'Connor, Michael Pittman and Randall Widelitz for their careful reviews. Finally, the senior editor, Christian Foth, would like to thank Ragnar Kinzelbach and Stefan Richter for their contemplative mentoring during his studies at the University of Rostock, when he started exploring the fascinating world of feathers.

Fribourg, Switzerland — Christian Foth
München, Germany — Oliver W. M. Rauhut

Contents

Introduction to the Morphology, Development, and Ecology of Feathers

1

Christian Foth

1.1 Introduction

Feathers are a characteristic of modern birds that differentiate them from all other extant non-avian reptiles. The origin of feathers goes back deep into the Mesozoic, preceding the origin of flight (Norell and Xu 2005; Xu and Guo 2009; Rauhut et al. 2012), and early protofeathers were probably present in the ancestral Tetanurae (Barrett et al. 2015), Dinosauria, or even Ornithodira (Rauhut et al. 2012; Godefroit et al. 2014). Among extant vertebrates, the feathers of modern birds are morphologically the most complex integumentary structure (Fig. 1.1) with enormous shape diversity (Fig. 1.2b–i) resulting from a hierarchical organization of repetitive morphological and developmental modules (Prum and Williamson 2001; Prum and Dyck 2003). In this chapter, the morphological ground patterns of modern feathers, their underlying developmental processes, and the biological roles of different feather types are reviewed.

1.2 Main Morphological Organization

The skin of birds is relatively thin compared to non-avian reptiles, but can form different kinds of derivates, including scales (in the tarsal region), glands (uropygial gland), feathers, and other epidermal outgrowths (e.g., the turkey beard, combs) (Lucas and Stettenheim 1972). Except among secondary flightless birds (Busching 2005), the skin of birds is organized in pterylae and apteria (Fig. 1.2a), skin portions that grow feathers or remain naked, respectively (Lucas and Stettenheim 1972). The distribution of pterylae and apteria varies between different bird species (Burckhardt 1954; Wetherbee 1957; Lucas and Stettenheim 1972), but remains constant in every individual throughout its ontogeny after initial formation during embryogenesis (Burckhardt 1954, see below).

Feathers itself contain several morphological units, which vary morphologically between different feather types (see below). The main units are the calamus, barbs, barbules, and the rhachis and hyporhachis (Fig. 1.1a; Lucas and Stettenheim 1972; Prum and Brush 2002). The calamus is the most proximal portion of the feather, tubular in shape, and anchors the feather into the skin in a so-called follicle. The follicle is associated with a complex mesh of muscles, which connects nearby follicles with each other, and thereby allows for a synchronized movement of the feather within each pteryla. The calamus contains two openings, the distally located inferior umbilicus and the proximal superior umbilicus. The calamus is hollow and contains a number of horizontally orientated pulp caps, which represents serial epidermal overgrowths of the dermal pulp (pulpa) that form during

C. Foth (✉)
Department of Geosciences, University of Fribourg, Fribourg, Switzerland

C. Foth, O. W. M. Rauhut (eds.), *The Evolution of Feathers*, Fascinating Life Sciences,
https://doi.org/10.1007/978-3-030-27223-4_1

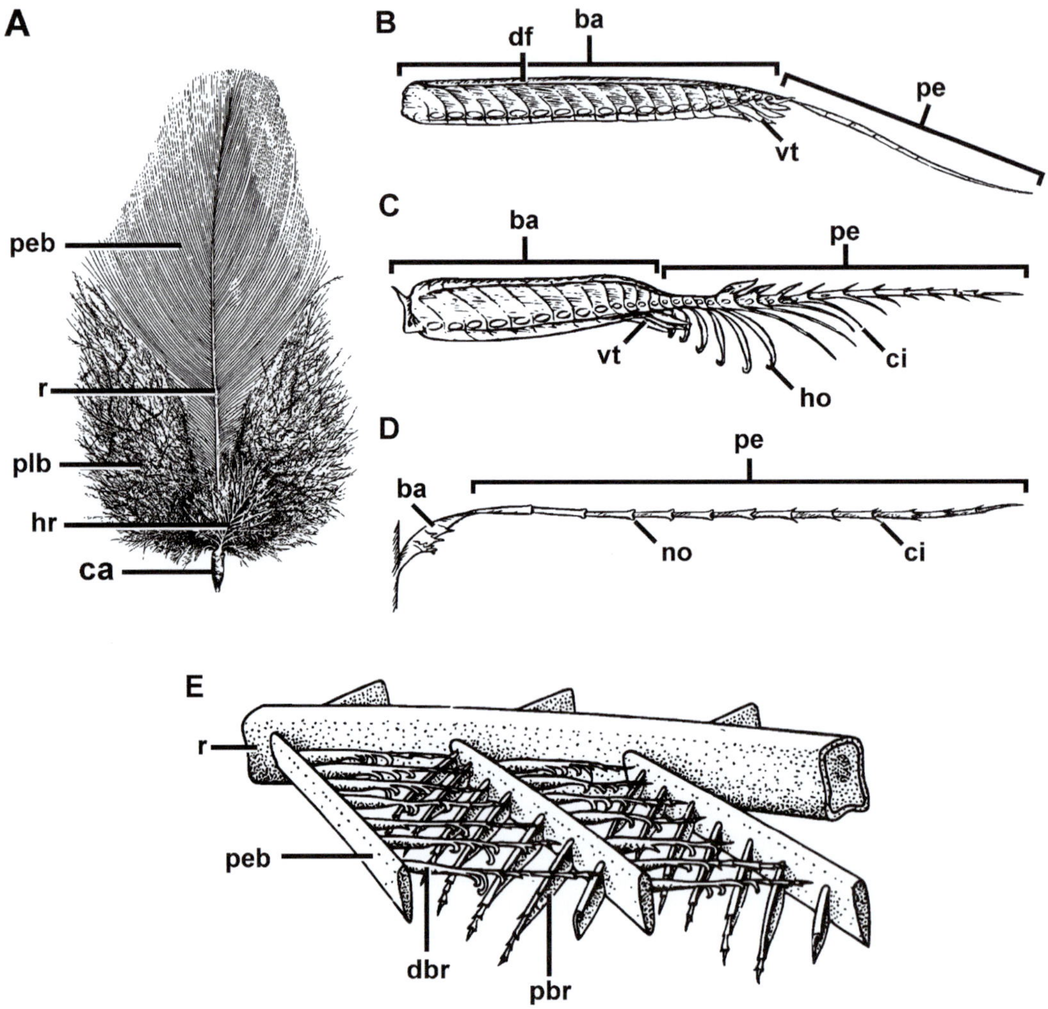

Fig. 1.1 Overview of feather morphology. (**a**) Pennaceous body feather. (**b**) Distal barbule of pennaceous feather. (**c**) Proximal barbule of a pennaceous feather. (**d**) Plumulaceous barbule. (**e**) Detail of a pennaceous feather vane. *ba* basal cells, *ca* calamus, *ci* cilia, *dbr* distal barbules, *df* dorsal flange, *ho* ventral hooks, *hr* hyporhachis, *no* nodes, *pbr* proximal barbules, *pe* pennulum, *peb* pennaceous barbs, *plb* plumulaceous barbs, *r* rhachis, *vt* ventral tooth. (**a**) Modified after Lucas and Stettenheim (1972), (**b**)–(**d**) modified after Chandler (1916), (**e**) modified after Storch and Welsch (1997)

morphogenesis (see below). With the exception of filoplumes and bristle feathers, the calamus of feathers is not associated with any neurons.

The barbs represent the main branching unit of the feathers. Depending on the feather type, they can be stiff or flexible. Barbs are usually serially aligned and proximally fused into a central element, called rhachis, which gives the feather a bilaterally symmetrical organization. The serially aligned barbs form the vane. Depending on the size of the feather, the number of barbs can reach several hundreds. Along the proximal and distal side of the barbs, feathers possess a second, serially aligned branching structure, the barbules (Fig. 1.1b–e). In their main organization, barbules consist of one to multiple basal cells, which are proximally attached to the barb, and multiple distally located pennulum cells. Based on the feather type, the morphology of the barbules can be highly specialized (see below).

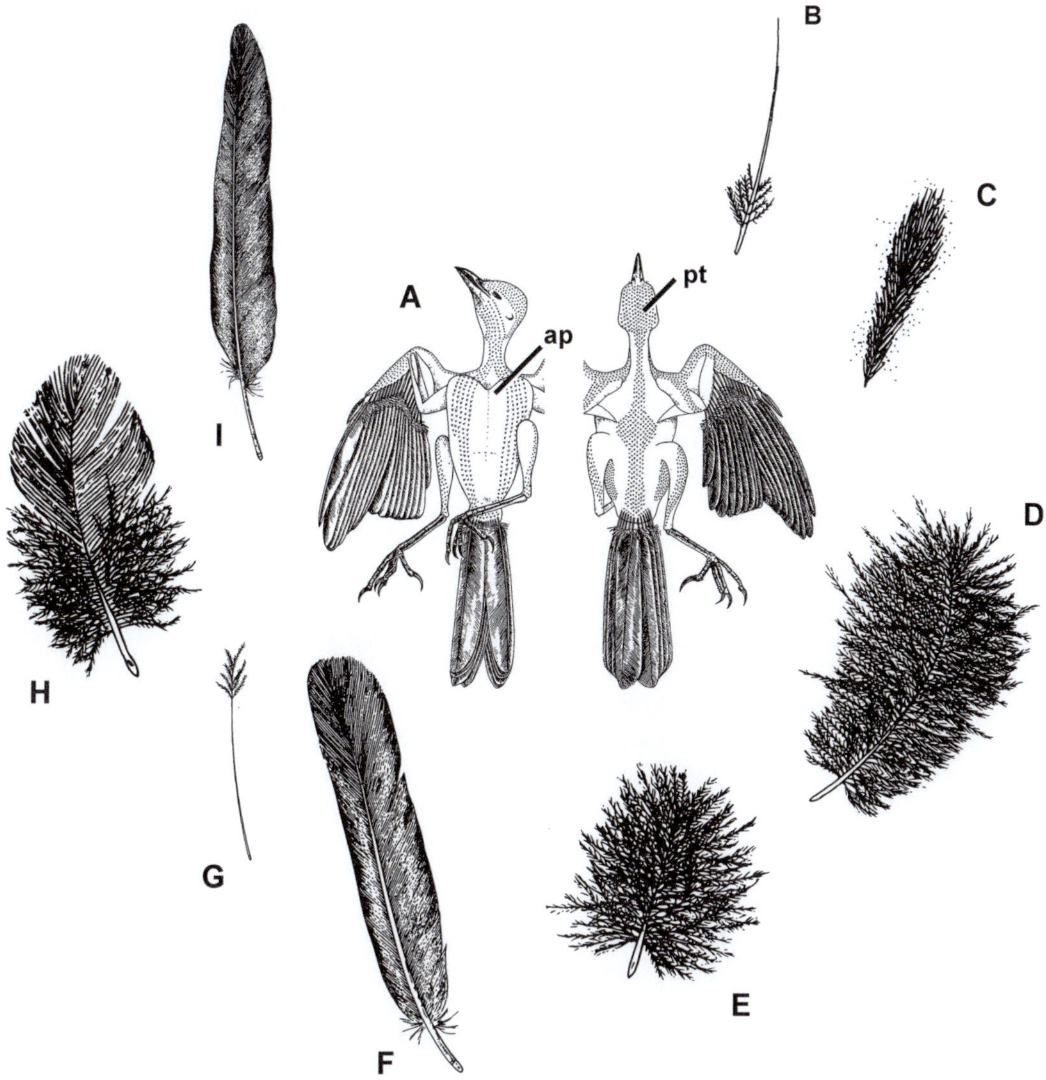

Fig. 1.2 Distribution of feather and main feather types. (**a**) Distribution of pterylae (pt) and apteria (ap) in the common blackbird (*Turdus merula*) (modified after Bergmann 1987). (**b**) Bristle feather. (**c**) Powder down. (**d**) Semiplume. (**e**) Down feather. (**f**) Rectrices. (**g**) Filoplume. (**h**) Pennaceous body feather. (**i**) Primary remiges. (**b**, **c**) modified after Chatterjee (1997)

The rhachis is the central element of a feather. It is anteriorly located, and, depending on the feather type, circular to rectangular in cross section. In contrast to the calamus, the rhachis is not hollow, but filled with a spongy pith consisting of large, polygonal medullary cells. Although the rhachis and the calamus are both tubular in organization, they are not continuous structures, but separated from each other by the superior umbilicus. On the posterior side of the feather, a second rhachis-like structure can be developed, the hyporhachis. If neither rhachis nor hyporhachis are developed, the feather barbs are radial symmetrically organized, and merge into the calamus (Prum and Brush 2002).

Feathers consist of two special forms of the structural protein β-keratin, feather-β-keratin and feather-like-β-keratin, which are significantly shorter in their amino acid sequence length by lacking the characteristic four 13-amino acid repeats present in scale- and claw-β-keratins (Gregg et al. 1984; Greenwold and Sawyer 2011). Feathers furthermore contain α-keratin, which is primarily expressed in the feather sheath (Prin and Dhouailly 2004, see below).

In addition to being morphologically complex, feathers are often extremely colorful and reflective due to the presence of pigmentation and structural coloration. The main pigments are melanins and carotenoids (Lucas and Stettenheim 1972). Melanin is a heterogenous polymer synthesized through oxidation of the amino acid tyrosine. This process occurs in specific cell organelles, the melanosomes, of specialized pigment cells, the melanocytes, which are produced in the epidermal collar during feather morphogenesis (see below). Here, melanosomes are transferred to the epidermal cells of the intermediate layer and embedded in the keratinized cell membrane (Lucas and Stettenheim 1972; Hudon 2005; Prum and Williamson 2002). The two main types of melanins are eumelanins and phaeomelanins. The former have a dark brown to black color and are synthesized in rod-like eumelanosomes, while the latter are yellow to reddish brown and synthesized in small, round phaeomelanosomes (Hudon 2005). By contrast, carotenoids are highly unsaturated hydrocarbons that are dissolved in fats and organic solvents. They cannot be synthesized by birds themselves, but rather are ingested with the diet and slightly modified. Carotenoids are transported to the feather collar via the blood stream and deposited in the intermediate layer during feather morphogenesis in the form of fat droplets (see below). With the beginning of keratinization, the droplets disappear, while the carotenoids are absorbed by the viscous keratin matrix. Carotenoids produce most of the bright red, orange, and yellow colors apparent in birds (Desselberger 1930; Lucas and Stettenheim 1972; Hudon 2005). Further pigments known from recent birds include, among others, psittacofulvins and porphyrins (Völker 1938; Lucas and Stettenheim 1972; Hudon 2005). Computer simulations indicate that the complexity of feather coloration relies on a reaction-diffusion process distributing pigments during feather morphogenesis (Prum and Williamson 2002).

The structural colors produced in feathers are based on absorption, reflection, and refraction of light along special morphological structures, often in interplay with feather pigments. These structures can produce both iridescent and noniridescent colors. Iridescent structural colors are primarily produced in the feather barbules, in which melanosomes are densely packed in multiple, parallelly arranged layers on the upper surface of the barbules (Mason 1923; Rensch 1925). Other melanosome arrangement include squares (Zi et al. 2003) or hexagonal arrays (Eliason and Shawkey 2012; Eliason et al. 2013). For the production of iridescent structural colors, the barbules are broadened and rotated by about 90°, thereby exposing the broad surface of the barbule (Rensch 1925). The iridescence itself results from thin-film interferences, due to the repeated refraction and reflection of light at the surfaces of the single melanosome layers, which in turn is based on contrasting refractive indices and wave impedances of β-keratin, pigments, and air (Lucas and Stettenheim 1972; Shawkey et al. 2006; Doucet et al. 2006). The melanosomes of these specialized barbules are usually pigmented (Rensch 1925), but in at least seven bird clades hollow melanosomes have evolved independently from each other, creating a whitish shine (Eliason et al. 2013; Shawkey et al. 2015). Iridescent structural colors can also be produced in barbs and rhachides by parallel organization of keratin layers or fibers in the cortex (Vigneron et al. 2006; D'Alba et al. 2011).

In contrast, noniridescent structural colors are produced at modified barbs and rhachides. These colors are produced through reflections at keratin-air interfaces between the cortex and the pith (see above). Pigmented melanosomes of the cortex or pith can additionally absorb transmitted light, in contrast to iridescent structural colors, where light is refracted and reflected

(Lucas and Stettenheim 1972; Shawkey and Hill 2006).

1.3 Feather Types and Their Biological Role

The highly variably morphology of feathers is achieved through changes to the absolute growth rate of the feather, the initial number of barb ridges, the location of new barb loci, regions that produce new barb ridges during morphogenesis, within the feather collar, the angle of barb ridges relative to the rhachis ridge, the rate of new barb ridge addition, the barb ridge diameter, and the angle of the barb ramus expansion after emergence from the sheath (Prum and Williamson 2001; Feo and Prum 2014). Nearly all feather shapes are created through modification of these parameters. The main feather types that occur in recent birds are down feathers, semiplumes, pennaceous feathers, bristles feathers, filoplumes, and powder downs. The morphology of these feather types is described below (Fig. 1.2b–i). The different feather types are usually specialized for different biological roles, including body covering, thermoregulation, flight, display and camouflage function, tactile function, and plumage maintenance. Despite these morphotype-specific biological roles, feathers like other integumentary structures also function as toxic storage organs, allowing repetitive excretion due to molting (Reichholf 1996; Metcheva et al. 2006; Dumbacher et al. 2009).

1.3.1 Down Feathers and Semiplumes

Down feathers (or plumes) are primarily used for body heat insulation (thermoregulation) and also function in hatchlings as body covering (Fig. 1.2e). In adult birds, down feathers usually contain a medium number of long and soft barbs forming open vanes, a short calamus, and a short rhachis and hyporhachis, which are circular in cross section. The barbs contain a high number of plumulaceous barbules, which have a short, twisted proximal basis, and a long, flexible distal pennulum (Fig. 1.1d). The pennulum consists of many elongated cells, whose distal end are expanded to nodes or wear one to four cilia (Lucas and Stettenheim 1972; Dove 1997). Because of their long and flexible morphology, barbs and barbules produce electric repulsive forces due to constant friction, which makes these feathers fluffier and increases their insulatory effect (Exner 1895, 1896). In addition to true down feathers, many birds also possess a second type of insulation feathers, semiplumes, which differ in the development of a prolonged rhachis and calamus (Fig. 1.1d). In contrast, the down feathers of hatchlings, so-called neoptile or natal downs, are often characterized by the absence of a rhachis and hyporhachis, leading to radial symmetry. Here, the calamus can be strongly reduced so that the barbs of natal downs are continuous with the distal tips of the barbs of the second feather generation (Schaub 1912; Ewart 1921; Foth 2011).

1.3.2 Pennaceous Feathers

Pennaceous feathers possess a medium to high number of serially aligned barbs, which are proximally attached to an elongated rhachis. The barbs are stiffer than in down feathers. Pennaceous feathers possess two main types of barbules, which are developed at a high density along the barbs. The basis of the barbules is long and stiff and contains many short basal cells. The pennulae of the barbules that are located on the distal side of the barb possess short cells with ventral hooks (Fig. 1.1b), while those located on the proximal side of the barb are often nonspecialized (Fig. 1.1c). Their basis, however, possesses a ventrally recurved flange on their dorsal edge, although a similar flange can also occur in the distal barbules. As a result, the hooks of the distal barbules can interlock with the basis of the proximal barbules (Fig. 1.1e), as in a zipper, forming a close, planar vane (Lucas and Stettenheim 1972). Beyond this ground pattern, the barbules of pennaceous feathers can be very diverse in shape (Chandler 1916; Sick 1937).

Within different body regions, pennaceous feathers vary in their morphology corresponding to varying biological roles. Pennaceous body feathers (Figs. 1.1a and 1.2h) are used for body covering, protection, thermoregulation, and, depending on their coloration, for display or camouflage (Lucas and Stettenheim 1972; Prum and Brush 2002). Their distal portion possesses open vanes with short, nonspecialized barbules along the barbs. In the mid-section the vanes are close, showing the typical pennaceous barbule morphology (see above). Proximally, pennaceous body feathers possess also plumulaceous barbs for thermoregulation (see above). Furthermore, these feathers have a long downy hyporhachis and a short calamus.

The pennaceous feathers of the wing are called remiges (Fig. 1.2i), which can be subdivided into primaries attaching the manus, and secondaries attaching the ulna. The remiges are primarily adapted for flight, but fulfil further biological roles, including display (Darwin 1871) and brooding (Hopp and Orsen 2004). The pennaceous feathers of the tail are called rectrices (Fig. 1.2f) and also play a crucial role in flight and display (Thomas 1997; Aparicio et al. 2003). In contrast to body feathers, remiges and rectrices have elongated rhachis with a rectangular cross section. To fulfil their aerodynamic function, the barbs of remiges and rectrices are stiffer and possess pennaceous barbules (see above) resulting in closed vanes. Proximal downy vanes or a hyporhachis are reduced or absent, while the calamus is elongated and deeply anchored within the skin. Depending on their position within the wing, the remiges vary in symmetry, which relies on the ratio of the width of the inner and outer vane (Busching 2005; Bachmann et al. 2007). In distal primaries and proximal rectrices, the outer vanes are significantly narrower than the inner vane, which is caused by differences in barb length and barb angle (Bachmann et al. 2007; Feo and Prum 2014), increasing the aerodynamic performance of the feathers (Norberg 1985). The degree of asymmetry, however, can also vary within one feather from distal to proximal (Busching 2005). In proximal primaries, secondary remiges, and distal rectrices the vanes are more equal in width.

A special type of pennaceous feathers are display feathers, which can possess complex color patterns (Prum and Williamson 2002; see above) and can be extremely variable in the size and morphology of the rhachis, barb, and barbule, creating aberrant morphologies (Darwin 1871; Brinkmann 1958; Bleiweiss 1987; Bartels 2003; Stavenga et al. 2011). Such feathers are usually developed on the head, breast, wing, and tail region, where they can be easily recognized visually.

The secondary loss of flight can have a significant effect on the morphology of pennaceous feathers, in particular through a decrease of the barb number and a simplification and reduction of barbules, giving the feathers a more down or bristle-like morphology (Lüdicke 1974; McGowan 1989; Livezey 2003).

1.3.3 Bristles Feathers, Filoplumes, and Powder Downs

Only two types of feathers are innervated by nervous cells at the base and function as tactile organs. Bristle feathers (Fig. 1.2b) are usually present in the face around the beak and eyes. The rhachis of these feathers is long and stiff. The barbs are short and stiff, and reduced in number. They are sparsely covered with barbules, which are simplified and stiff. The calamus is short (Lucas and Stettenheim 1972). In contrast, filoplumes (Fig. 1.2g) are short and associated with pennaceous feathers, sensing the position of the latter within the plumage. They have a tiny, flexible rhachis, which possesses a small number of short barbs on its distal end. The barbs themselves have short nonspecialized barbules. A short calamus is developed at the proximal end (Lucas and Stettenheim 1972).

Powder downs are another type of feather (Fig. 1.2c) that play an important role in the maintenance of the plumage. Morphologically, this feather type is very similar to ordinary down feathers and semiplumes, but the barbs are less fluffy. These feathers are coated with a fine powder that is derived from cells that surround the developing barbules during development and is later shed onto neighboring feathers (Lucas and Stettenheim 1972).

1.4 Feather Development

The embryonic development of feathers starts with the formation of feather tracks in particular body regions, the pterylae, which usually develop down-like feathers, so-called neoptile or natal downs, as first feather generation (Lucas and Stettenheim 1972; Foth 2011). In these areas, the skin forms parallel rows of placodes, which are local thickening of dermis and epidermis (Lucas and Stettenheim 1972; Prum and Brush 2002). The placodes do not develop simultaneously, but show regional specifications, depending on the morphotype of later feather generation (e.g., pennaceous feather filoplume, etc.; Burckhardt 1954). After a certain developmental stage, placode formation stops so that their number, and thus also the number of potential feathers, remains constant over the remaining lifetime of the animal (Burckhardt 1954). In the next stage, each placode develops into to a feather bud with a distally located epidermal growth zone (Fig. 1.3a). The dermal core inside the feather bud forms the pulpa (Fig. 1.3b, d), which supplies the feather bud with nutrients via blood vessels, but additionally transfers pigment cells into the epidermis (Lucas and Stettenheim 1972; Yu et al. 2004). The pulpa also expresses signal molecules, which play an important role in the morphogenesis of the epidermis (Yu et al. 2002). During

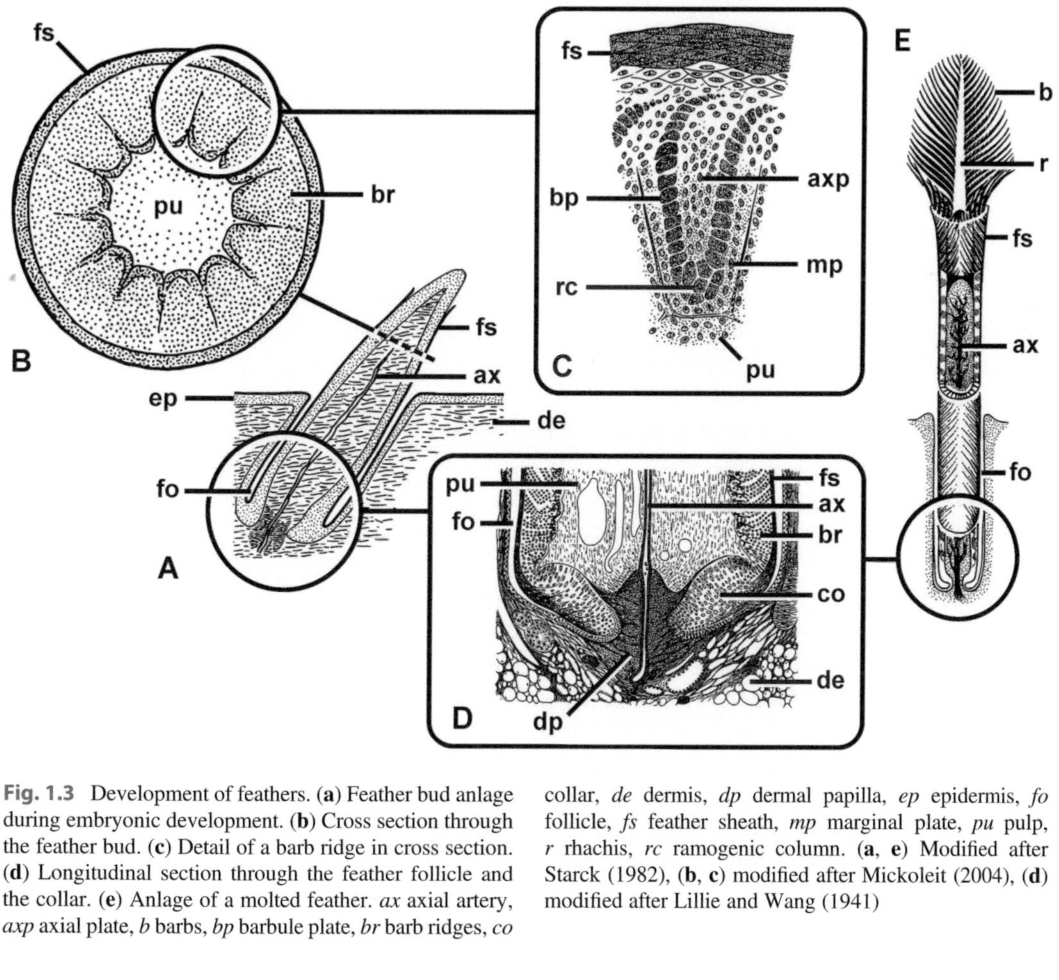

Fig. 1.3 Development of feathers. (**a**) Feather bud anlage during embryonic development. (**b**) Cross section through the feather bud. (**c**) Detail of a barb ridge in cross section. (**d**) Longitudinal section through the feather follicle and the collar. (**e**) Anlage of a molted feather. *ax* axial artery, *axp* axial plate, *b* barbs, *bp* barbule plate, *br* barb ridges, *co* collar, *de* dermis, *dp* dermal papilla, *ep* epidermis, *fo* follicle, *fs* feather sheath, *mp* marginal plate, *pu* pulp, *r* rhachis, *rc* ramogenic column. (**a**, **e**) Modified after Starck (1982), (**b**, **c**) modified after Mickoleit (2004), (**d**) modified after Lillie and Wang (1941)

growth, the dermal pulpa is produced continuously, but reabsorbed periodically, which goes hand in hand with pulp cap formation by the epidermis (Lucas and Stettenheim 1972).

Within the feather bud, the epidermis starts to differentiate into three main layers: the outer layer, the intermediate layer, and the basal layer. The outer layer is homologous with the second periderm of embryonic bird scales and forms the feather sheath, which protects the inside of the feather germ. The feather sheath formation is characterized by strong α-Keratin expression and subsequent apoptosis (Sawyer et al. 2003, 2005). The basal layer forms the pulp caps and the marginal plates (Fig. 1.3c) that separate the barb ridges (Fig. 1.3b–d) from each other and control the morphogenesis of the intermediate layer (Harris et al. 2002; Prum and Dyck 2003) into barbs, barbules, rachis, and calamus. Finally, the intermediate layer is formed between the outer and basal layer due to cell proliferation, forming the barb ridges through a balloon-like expansion into the basal layer (Prum and Dyke 2003).

Within the barb ridge, cells differentiate into a ramogenic column, central axial plate, and two lateral barbule plates so that the axial plate ends up separating the two barbule plates medially (Fig. 1.3c) before disintegrating at the end of this developmental process (Lucas and Stettenheim 1972). The ramogenic columns form the barbs. The barbule plates contain a single row of cells and differentiate into simplified plumulaceous barbules (Lucas and Stettenheim 1972). Here, the innermost cells of the barbule plate become the base and fuse to the ramogenic column, while the more peripheral cells become the elongate distal cells of the pennulum (Lucas and Stettenheim 1972). The process ends with the apoptosis of the cells of the marginal plate and axial plate and the keratinization of the cells of the barb ramus and barbule plate. After keratinization, the remaining cells die as well (Lucas and Stettenheim 1972; Haake et al. 1984; Yu et al. 2002). As written above, barb ridges formation initially starts at the distal end of the feather bud and then moves in a proximal direction (Lucas and Stettenheim 1972). This process goes hand in hand with the delocation of the growth zone to the base of the feather bud, forming the ring-shaped collar. At this point feather embryogenesis can form two different morphologies. The first morphology results from an early stop of barb ridge morphogenesis, resulting in a radially symmetric arrangement. Follicle formation is initiated, while calamus formation is often suppressed and the barbs are held together proximally by the feather sheath. When the second feather generation is formed during the first molting process (see below), the barbs of the first feather generation are continuously connected to the distal barbs of the second generation (Schaub 1912; Foth 2009, 2011). Alternatively, the barb ridges "move" during the proximal delocation of the growth zone in an anterior direction, anteriorly fuse with each other at their proximal end, and form the rhachis ridge. Thus, natal feathers gain a bilaterally symmetric arrangement of the barb ridges. In contrast to later feather generations, the initial number of barb ridges remains and no new barb ridges are formed. As a result, when barb ridge formation is finished the calamus formation is initiated by a stop of differentiation processes in the intermediate layer. As with the barbs, calamus formation, and thereby feather morphogenesis, ends with keratinization and final apoptosis (Lucas and Stettenheim 1972).

During embryogenesis the feather bud grows out, but simultaneously sinks into the skin, forming a follicle (Fig. 1.3a, d, e). The timing of the process is variable between different body regions, but also between species. In *Anas platyrhynchos*, *Anser anser* (both Anseriformes), *Columba livia* (Columbiformes), and *Eudyptes chrysocome* (Sphenisciformes), follicle formation starts after barb ridge formation (Davies 1889; Wohlauer 1901; Hosker 1936). By contrast, in *Struthio camelus* (Struthioformes) the follicle is formed before barb ridge formation (Duerden 1913), while in *Gallus gallus* (Galliformes), follicle formation can happen before, after, or simultaneously with barb ridge formation (Hosker 1936; Lucas and Stettenheim 1972; Chuong and Edelman 1985). Independent from the timing of this process, the collar is finally placed under the skin and divided into two zones (Fig. 1.3d): a proliferation zone and the ramogenic zone

(Lucas and Stettenheim 1972). Due to follicle formation, the outer follicle wall, which surrounds the calamus, comes into contact with the dermal musculature (see above) allowing the movement of the final feather.

Feather embryogenesis can be further varied through placement of the germ under the skin before barb ridge formation is initiated. In this case, all developmental processes rest until hatching. This process can happen regionally or across the entire body so that the chick appears to be partially or fully naked at hatching, as is the case in Coraciiformes, Cuculiformes, Piciformes, and various Passeriformes. Depending on the species, the hatchling develops an ontogenetically delayed neoptile plumage or skips this process entirely forming the second feather generation immediately (Burckhardt 1954).

After the initial development, feather morphogenesis is periodically repeated throughout ontogeny, a process called molting. As part of this cycle, the old feather generation is shed (ecdysis) and the new feather generation is then formed (endysis) (Fig. 1.3e; Watson 1963; Lucas and Stettenheim 1972). In contrast to embryogenesis, the development of later feather generations is initiated in the collar at the base of the follicle, in which the latter can produce different feather types throughout lifetime. However, barb ridge, barbule, rhachis, calamus, and feather sheath formation are basically similar to the embryogenic developmental process described above (except for follicle formation), but can produce very different morphologies of the barbules, barbs, rhachides, or calami by modifying the molecular pathways, which control the developmental processes.

1.5 Summary

The huge morphological variability of recent feathers strongly relies on the modularity of repetitive morphological structures and their underlying developmental processes, in which small changes in various parameters during morphogenesis (due to changes in the molecular pathway) can have a significant impact on the final shape of the feather (Prum and Williamson 2001). The basis of this modularity, however, relies on the ring-shaped collar, which allows the growth of tube-like epidermal structures, which can be transformed into tuft and plane-like structures due to modular cell differentiation and apoptosis processes (Prum and Brush 2002; Prum 2005). Thus, the key innovation related to the origin of feathers was likely the evolution of a follicle with an internal, ring-shaped collar resulting from the secondary invagination of a tubular epidermal outgrowth (Prum 1999; Prum and Brush 2002).

Acknowledgments I thank Walter Joyce (University of Fribourg) for proofreading this chapter. The review was supported by the Swiss National Science Foundation grant PZ00P2_174040.

References

Aparicio JM, Bonal R, Cordero PJ (2003) Evolution of the structure of tail feathers: implications for the theory of sexual selection. Evolution 57:397–405

Bachmann T, Klän S, Baumgartner W, Klaas M, Schröder W, Wagner H (2007) Morphometric characterisation of wing feathers of the barn owl *Tyto alba* pratincola and the pigeon *Columba livia*. Front Zool 4:23

Barrett PM, Evans DC, Campione NE (2015) Evolution of dinosaur epidermal structures. Biol Lett 11:20150229

Bartels T (2003) Variations in the morphology, distribution, and arrangement of feathers in domesticated birds. J Exp Zool (Mol Dev Evol) 298B:91–108

Bergmann H-H (1987) Die Biologie des Vogels. AULA, Wiesbaden

Bleiweiss R (1987) Development and evolution of avian racket plumes: fine structure and serial homology of the wire. J Morphol 194:23–39

Brinkmann A (1958) Die Morphologie der Schmuckfeder von *Aix galericulata* L. Rev Suisse Zool 65:485–608

Burckhardt D (1954) Beitrag zur embryonalen Pterylose einiger Nesthocker. Rev Suisse Zool 61:551–655

Busching W-D (2005) Einführung in die Gefieder- und Rupfungskunde. AULA, Wiebelsheim

Chandler AC (1916) A study of the structure of feathers, with reference to their taxonomic significance. Univ Calif Publ Zool 13:243–446

Chatterjee S (1997) The rise of birds: 225 million years of evolution. The John Hopkins University Press, Baltimore

Chuong C, Edelman GM (1985) Expression of cell-adhesion molecules in embryonic induction. I. Morphogenesis of nestling feathers. J Cell Biol 101:1009–1026

D'Alba L, Saranathan V, Clarke JA, Vinther J, Prum RO, Shawkey MD (2011) Colour-producing ß-keratin nanofibres in blue penguin (*Eudyptula minor*) feathers. Biol Lett 7:543–546

Darwin CR (1871) The descent of man, and selection in relation to sex, vol 2. John Murray, London

Davies HR (1889) Die Entwicklung der Feder und ihre Beziehungen zu anderen Integumentgebilden. Morphol Jahrb 15:560–645

Desselberger H (1930) Ueber das Lipochrom der Vogelfeder. J Ornithol 78:328–376

Doucet SM, Shawkey MD, Hill GE, Montgomerie R (2006) Iridescent plumage in satin bowerbirds: structure, mechanisms and nanostructural predictors of individual variation in colour. J Exp Biol 209:380–390

Dove CL (1997) Quantification of microscopic feather characters used in the identification of the North American plovers. Condor 99:47–57

Duerden JE (1913) Experiments with ostriches. XXII. The development of the feather, showing absence of cruelty in clipping and quilling. Agric J Union of South Africa 6:648–661

Dumbacher JP, Menon GK, Daly JW (2009) Skin as a toxin storage organ in the endemic new Guinean genus *Pitohui*. Auk 126:520–530

Eliason CM, Shawkey MD (2012) A photonic heterostructure produces diverse iridescent colours in duck wing patches. J R Soc Interface 9:2279–2289

Eliason CM, Bitton P-P, Shawkey MD (2013) How hollow melanosomes affect iridescent colour production in birds. Proc R Soc B 280:20131505

Ewart JC (1921) The nestling feathers of the mallard, with observations on the composition, origin, and history of feathers. Proc Zool Soc London 1921:609–642

Exner S (1895) Ueber die elektrischen Eigenschaften der Haare und Federn. Pflugers Arch 61:427–449

Exner S (1896) Ueber die elektrischen Eigenschaften der Haare und Federn. Pflugers Arch 63:305–316

Feo TJ, Prum RO (2014) Theoretical morphology and development of flight feather vane asymmetry with experimental tests in parrots. J Exp Zool (Mol Dev Evol) 322B:240–255

Foth C (2009) Die Morphologie des Erstlingsgefieders ausgewählter Vogeltaxa unter Berücksichtigung der Phylogenie. University of Rostock, Germany

Foth C (2011) The morphology of neoptile feathers: ancestral state reconstruction and its phylogenetic implications. J Morphol 272:387–403

Godefroit P, Sinitsa SM, Dhouailly D, Bolotsky YL, Sizov AV, McNamara ME, Benton MJ, Spagna P (2014) A Jurassic ornithischian dinosaur from Siberia with both feathers and scales. Science 345:451–455

Greenwold MJ, Sawyer RH (2011) Linking the molecular evolution of avian beta (β) keratins to the evolution of feathers. J Exp Zool (Mol Dev Evol) 316:609–616

Gregg K, Wilton SD, Parry DAD, Rogers GE (1984) A comparison of genomic coding sequences for feather and scale keratins: structural and evolutionary implications. EMBO J 3:175–178

Haake AR, König G, Sawyer RH (1984) Avian feather development: relationships between morphogenesis and keratinization. Dev Biol 106:406–413

Harris MP, Fallon JF, Prum RO (2002) Shh-Bmp2 signaling module and the evolutionary origin and diversification of feathers. J Exp Zool (Mol Dev Evol) 294:160–176

Hopp TP, Orsen MJ (2004) Dinosaur brooding behavior and the origin of flight feathers. In: Currie PJ, Koppelhus EB, Shugur MA, Wright JL (eds) Feathered dragons. Indiana University Press, Bloomington, pp 234–250

Hosker A (1936) Studies on the epidermal structures of birds. Philos Trans R Soc Lond Ser B 226:143–188

Hudon J (2005) Considerations in the conservation of feathers and hair, particularly their pigments. In: Brunn M, Burns JA (eds) Fur trade legacy. The preservation of organic materials. Canadian Association for Conservation of Cultural Property, Ottawa, pp 127–147

Lillie FR, Wang H (1941) Physiology of development of the feather. V. Experimental morphogenesis. Physiol Zoöl 14:103–133

Livezey BC (2003) Evolution of flightlessness in rails (Gruiformes: Rallidae): phylogenetic, ecomorphological, and ontogenetic perspectives. Ornithol Monogr 53:1–654

Lucas AM, Stettenheim PR (1972) Avian anatomy. Integument, Part I and II. U.S. Government Printing Office, Washington, DC

Lüdicke M (1974) Radioaktive Markierungsversuche an Federn von *Casuarius casuarius*. J Ornithol 115:348–364

Mason CW (1923) Structural colors in feathers. II. J Phys Chem 27:401–447

McGowan C (1989) Feather structure in flightless birds and its bearing on the question of the origin of feathers. J Zool (Lond) 218:537–547

Metcheva R, Yurukova L, Teodorova S, Nikolova E (2006) The penguin feathers as bioindicator of Antarctica environmental state. Sci Total Environ 362:259–265

Mickoleit G (2004) Phylogenetische Systematik der Wirbeltiere. Verlag Dr. Friedrich Pfeil, München

Norberg RÅ (1985) Function of vane asymmetry and shaft curvature in bird flight feathers; inferences on flight ability of *Archaeopteryx*. In: Hecht MK, Ostrom JH, Viohl G, Wellnhofer P (eds) The beginnings of birds. Freunde des Jura-Museums, Eichstätt, pp 303–318

Norell MA, Xu X (2005) Feathered dinosaurs. Annu Rev Earth Planet Sci 33:277–299

Prin F, Dhouailly D (2004) How and when the regional competence of chick epidermis is established: feathers vs. scutate and reticualte scales, a problem en route to a solution. Int J Dev Biol 48:137–148

Prum RO (1999) Development and evolutionary origin of feathers. J Exp Zool (Mol Dev Evol) 285:291–306
Prum RO (2005) Evolution of the morphological innovations of feathers. J Exp Zool (Mol Dev Evol) 304B:570–579
Prum RO, Brush AH (2002) The evolutionary origin and diversification of feathers. Q Rev Biol 77:261–295
Prum RO, Dyck J (2003) A hierarchical model of plumage: morphology, development, and evolution. J Exp Zool (Mol Dev Evol) 298B:73–90
Prum RO, Williamson S (2001) Theory of the growth and evolution of feather shape. J Exp Zool (Mol Dev Evol) 291:30–57
Prum RO, Williamson S (2002) Reaction-diffusion models of within-feather pigmentation patterning. Proc R Soc Lond B 269:781–792
Rauhut OWM, Foth C, Tischlinger H, Norell MA (2012) Exceptionally preserved juvenile megalosauroid theropod dinosaur with filamentous integument from the late Jurassic of Germany. Proc Natl Acad Sci U S A 109:11746–11751
Reichholf JH (1996) Die Feder, die Mauser und der Ursprung der Vögel. Archaeopteryx 14:27–38
Rensch B (1925) Untersuchungen zur Phylogenese der Schillerstruktur. J Ornithol 73:127–147
Sawyer RH, Salvatore BA, Potylicky T-TF, French JO, Glenn TC, Knapp LW (2003) Origin of feathers: feather Beta (β) keratins are expressed in discrete epidermal cell populations of embryonic scutate scales. J Exp Zool (Mol Dev Evol) 295B:12–24
Sawyer RH, Rogers L, Washington LD, Glenn TC, Knapp LW (2005) Evolutionary origin of the feather epidermis. Dev Dyn 232:256–267
Schaub S (1912) Die Nestdunen der Vögel und ihre Bedeutung für die Phylogenie der Feder. Verh Naturforsch Ges Basel 23:131–182
Shawkey MD, Hill GE (2006) Significance of a basal melanin layer to production of non-iridescent structural plumage color: evidence from an amelanotic Steller's jay (*Cyanocitta stelleri*). J Exp Biol 209:1245–1250
Shawkey MD, Hauber ME, Estep LK, Hill GE (2006) Evolutionary transitions and mechanisms of matte and iridescent plumage coloration in grackles and allies (Icteridae). J R Soc Interface 3:777–786
Shawkey MD, D'Alba L, Xiao M, Schutte M, Buchholz R (2015) Ontogeny of an iridescent nanostructure composed of hollow melanosomes. J Morphol 276:378–384
Sick H (1937) Morphologisch-funktionelle Untersuchung über die Feinstruktur der Vogelfeder. J Ornithol 85:206–372
Starck D (1982) Vergleichende Anatomie der Wirbeltier auf evolutionsbiologischer Grundlage. Bd. 3: Organe des aktiven Bewegungsapparates, der Koordination, der Umweltbeziehung, des Stoffwechsels und der Fortplanzung. Springer, Berlin
Stavenga DG, Leertouwe HL, Marshall NJ, Osorio D (2011) Dramatic colour changes in a bird of paradise caused by uniquely structured breast feather barbules. Proc R Soc B 278:2098–2104
Storch V, Welsch U (1997) Systematische Zoologie. Spektrum Akademischer Verlag, Heidelberg
Thomas ALR (1997) On the tails of birds – what are the aerodynamic functions of birds' tails, with their incredible diversity of form? Bioscience 47:216–225
Vigneron JP, Lousse V, Colomer J-F, Rassart M, Louette M (2006) Complex optical structure in the ribbon-like feathers of the African open-bill stork. Proc SPIE 6320:632014
Völker O (1938) Porphyrin in Vogelfedern. J Ornithol 86:436–456
Watson GE (1963) The mechanism of feather replacement during natural molt. Auk 80:486–495
Wetherbee DK (1957) Natal plumages and downy pteryloses of passerine birds of North America. Bull Am Mus Nat Hist 113:339–436
Wohlauer E (1901) Die Entwicklung des Embryonalgefieders von *Eudyptes chrysocome*. Z Morphol Anthropol 4:149–178
Xu X, Guo Y (2009) The origin and early evolution of feathers: insights from recent paleontological and neontological data. Vertebrata PalAsiatica 47:311–329
Yu M, Wu P, Widelitz RB, Chuong C (2002) The morphogenesis of feathers. Nature 420:308–312
Yu M, Yue Z, Wu P, Wu D, Mayer J-A, Medina M, Widelitz RB, Jiang T, Chuong C (2004) The developmental biology of feather follicles. Int J Dev Biol 48:181–191
Zi J, Yu X, Li Y, Hu X, Xu C, Wang X, Liu X, Fu R (2003) Coloration strategies in peacock feathers. Proc Natl Acad Sci U S A 100:12576–12578

Molecular and Cellular Mechanisms of Feather Development Provide a Basis for the Diverse Evolution of Feather Forms

2

Gee-Way Lin, Ang Li, and Cheng-Ming Chuong

2.1 Diversification of Feather Morphology in Evolution

2.1.1 The Structure of Extant Feathers

Feathers are a skin-associated appendage composed of a matrix of intracellular keratin in the external body parts of birds and feathered dinosaurs. The center of a typical contour feather from modern birds is composed of a hollow shaft (calamus) at the proximal end and a rigid shaft (rachis) in the distal part. The primary branches along the rachis are the barbs. The secondary branches along the main shaft of a barb (ramus) are the barbules. Barbules away from the base of the feather are referred as distal barbules, whereas those toward the base of the feather are referred as proximal barbules. A barbule is a stalk of single cells, beginning with squarish cells at the base and ending with a series of elongated distal cells. The elongated cells of distal barbules differentiate laterally to form hooklets attaching to the groove of adjacent proximal barbules. Thus, the interlocked barbs on each side of the rachis form the vane of feathers (Fig. 2.1a).

2.1.2 Diverse Morphology of Feathers

The diverse morphology of feathers is a consequence of microstructural variations in the rachis, rami, and barbules. From filamentous bristles to sail-shaped remiges, a large spectrum of branching patterns can be found not only among different species of modern Aves, but also at different body regions of one individual bird, which usually serve different functions. For example, the radially or bilaterally symmetric plumulaceous feathers distributed in the breast and belly region mainly help to maintain the body temperature (Fig. 2.1b, b′), whereas the bilaterally symmetric pennaceous feathers covering most parts of the trunk determine the body contour (Fig. 2.1c). The bilaterally asymmetric

G.-W. Lin
Department of Pathology, University of Southern California, Los Angeles, CA, USA

Department of Biochemistry and Molecular Cell Biology, School of Medicine, College of Medicine, Taipei Medical University, Taipei, Taiwan

A. Li
Department of Pathology, University of Southern California, Los Angeles, CA, USA

Department of Kinesiology, University of Texas at Arlington, Arlington, TX, USA

C.-M. Chuong (✉)
Department of Pathology, University of Southern California, Los Angeles, CA, USA

Research Center for Developmental Biology and Regenerative Medicine, National Taiwan University, Taipei, Taiwan
e-mail: cmchuong@med.usc.edu

C. Foth, O. W. M. Rauhut (eds.), *The Evolution of Feathers*, Fascinating Life Sciences,
https://doi.org/10.1007/978-3-030-27223-4_2

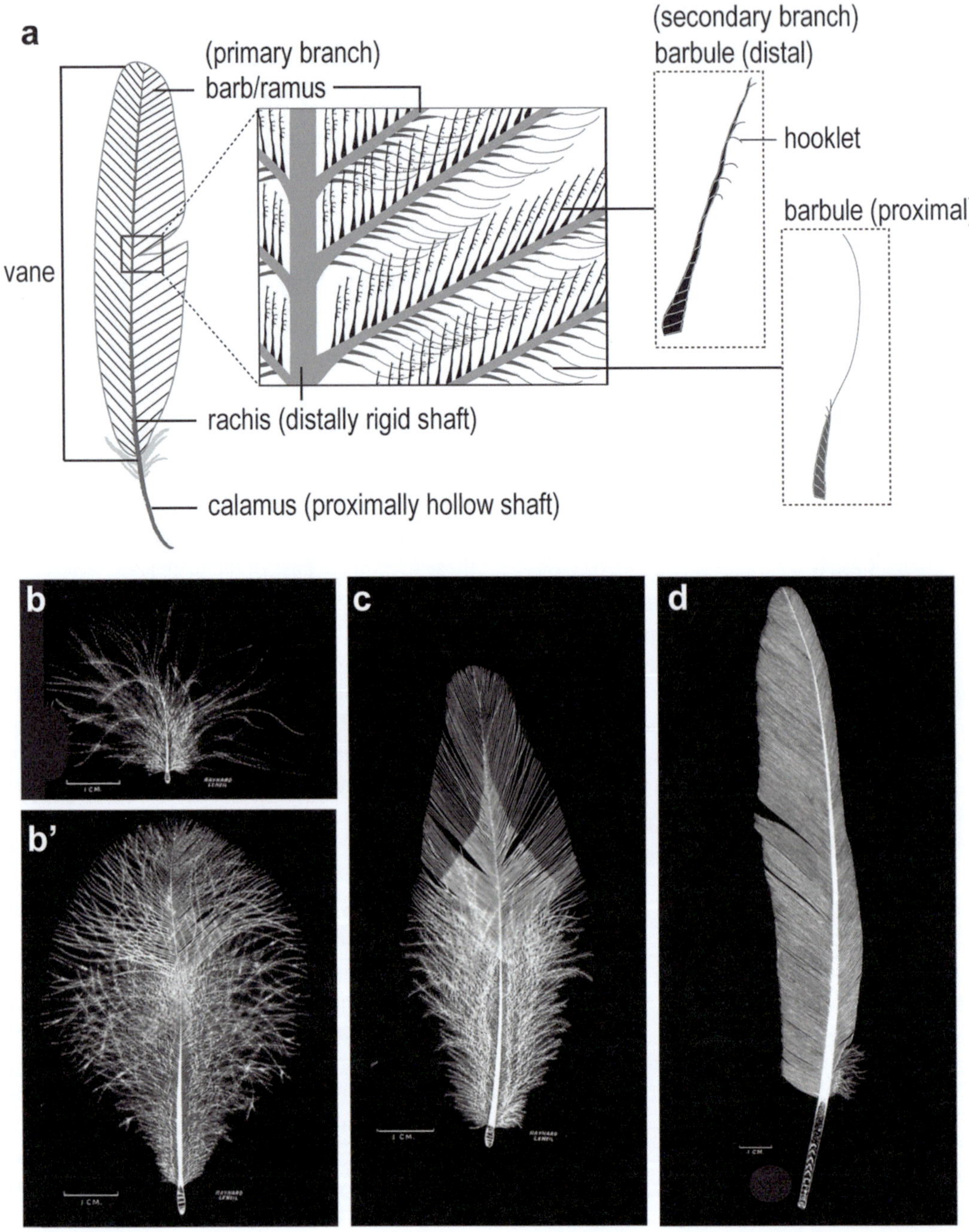

Fig. 2.1 The structure and diverse morphology of feathers: (**a**) hierarchical structure of a feather, (**b**, **b**′) radial symmetric downy feathers from the ventral and lateral sides, (**c**) bilateral symmetric feather from the dorsal trunk, (**d**) bilateral asymmetric feather from the wing (**b–d** are modified from Lucas and Stettenheim 1972)

feathers in the wing allow aerodynamic flight (Fig. 2.1d). Therefore, feathers are considered to be the most complex integumentary structure and are used as a model to study how evolutionary novelties in vertebrate skin appendages emerge (Prum 1999; Chuong et al. 2003).

2.1.3 The Evolution and Development of Feathers

Where feathers came from, how feathers evolved, and how modern feather architectures were achieved have long been debated. The appearance of feathers had defined the appearance of birds before fossil records on feathered dinosaurs were found (Brush 1996). Key events in the evolution of feathers had been reconstructed by linkages of the extant structures of feathers and the developmental process. According to analysis of this kind of linkage, a five-staged feather evolution-development (evo-devo) model was proposed by Prum (1999). In this model, the evolution of a feather follicle is characterized by the presence of a cylindrical, unbranched collar (stage I); the inner layer of the collar then periodically differentiates into barb ridges along a horizontal plane (stage II). Later on, the plane of a barb ridge formation becomes oblique, allowing the emergence of new barbs from one side of the collar (barb generative zone, or BGZ) and the fusion of barbs into rachis at the other side. This results in the helical displacement of barb ridges as they grow (stage III). The next step is the further differentiation of barbule plates into structurally distinct distal and proximal barbules (stage IV), accompanied by lateral displacement or duplication of the BGZ (stage V). These innovations produce the bilaterally asymmetric remiges and after-feathers (Prum 1999; Prum and Williamson 2001; Prum and Brush 2002).

Recent paleontological and neontological studies on feathers and feather-like integumentary structures associated with non-avian and avian theropods have greatly improved our understanding of the origin and early evolution of feathers (Xu 2006; Xu and Guo 2009; Xu et al. 2014). It seems that the five-staged feather evo-devo model is still adopted, even though an intermediate morphotype of the avian feather has been identified from amber (Xing et al. 2016). However, the most developmental criteria for modern feathers are not applicable to fossil records (Xu and Guo 2009). In non-avian dinosaurs, eight morphotypes of feathers have been identified, but they could not be interpreted using extant feather morphogenesis (Xu and Guo 2009). This suggests that the early evolution of feathers would have been more complicated than the morphogenic processes of extant feathers have predicted.

The morphogenic processes of modern feathers start during embryogenesis in conjunction with other skin-associated appendages such as scales and glands (Chang et al. 2009). The inner layer of skin, called the dermis, serves as a center of signaling to induce stratification and further periotic patterning in the outer layer, known as the epidermis (Chuong et al. 2000). The molecular and cellular mechanisms involved in the early stages of developing feathers (before the formation of cylindrical filaments) are conserved in a variety of feathers. In contrast, the emergence of evolutionary novelties—such as branching, stem cell niche formation, and the establishment of anterior-posterior (A-P), proximal-distal (P-D), and medial-lateral (M-L) axes—occur relatively late in development (Fig. 2.2) (Widelitz et al. 2003; Xu et al. 2014). In other words, more morphogenetic processes are gradually added in development to make the feather structure more complex. To date, the evo-devo model proposed by Prum in 1999 is generally consistent with newly discovered molecular and cellular mechanisms of feather morphogenesis. Several key mechanisms are illustrated in Figs. 2.2, 2.3, 2.4, 2.5, 2.6, and 2.7.

2.2 Development of Feathers During Embryogenesis

2.2.1 Tract Field Formation

The morphogenetic processes of feathers start from macro-patterning. Even though the mechanism of the macro-patterning is not well understood, regional specificity, such as different feather tracts and scales in the bird's skin, is quite obvious (Fig. 2.3a) (Gill 1994). On the cellular level, the distinct skin regions are closely associated with the accumulation of dermis underlying the epidermis. For example, a region with relatively dense dermis becomes a competent tract field to form feathers (Wessells 1965). Molecular signals derived from the dorsal neural tube, such as Wnt-1, trigger the formation of feather tracts that are characterized by dense dermis (Olivera-Martinez et al. 2001). After that, the neural cell adhesion molecule (NCAM) and nuclear-enriched β-catenin are observed in the epithelium of the feather tract

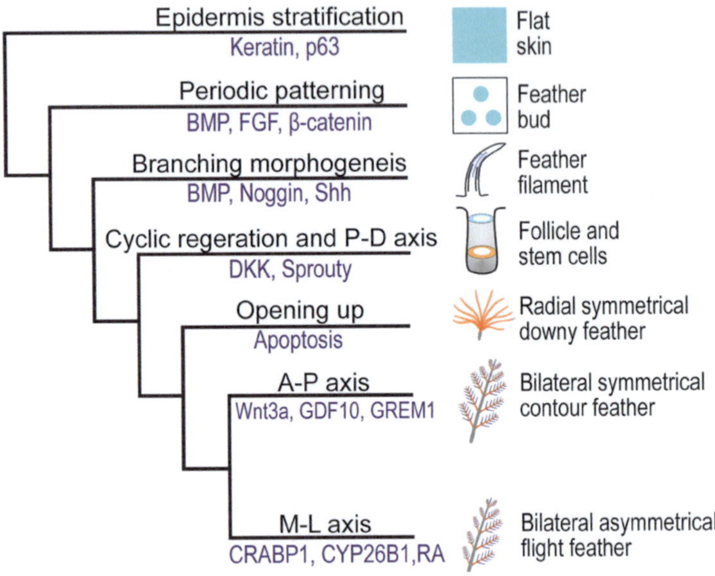

Fig. 2.2 Stepwise evolution of feathers (modified from Xu et al. 2014)

fields (Jiang et al. 1999). Ectopic feather tracts can also be induced by bone morphogenetic protein 2 (BMP2), which up-regulates exogenous expression of cDermo-1 (Twist 2) and then leads to the formation of dense dermis (Fig. 2.3b) (Hornik et al. 2005; Scaal et al. 2002).

2.2.2 Feather Bud Induction

Following the formation of feather tracts, micropatterning takes place within the homogenous tract field to demarcate bud and interbud regions. This periodic patterning can be generated by reaction diffusion (R-D) and interactions of molecules that promote [activators: fibroblast growth factors (FGFs)] and suppress (inhibitors: BMPs) feather bud formation (Fig. 2.3c) (Widelitz et al. 1996; Jung et al. 1998; Noramly and Morgan 1998). During the micro-patterning processing of bud induction, several genes identified as restrictive expression patterns in the bud domain (e.g., Wnt-7a, β-catenin) or interbud domains (e.g., GREM1 and Wnt-11) define the boundaries between neighboring feather buds (Noramly et al. 1999; Ohyama et al. 2001; Chang et al. 2004a).

2.2.3 Establishment of the A-P Axis and Elongation Along the P-D Axis

After the formation of feather bud boundaries, further specification of bud and interbud regions occurs through the de novo activation of certain molecular pathways within specific regions. For example, sonic hedgehog (Shh) is preferentially expressed in the bud region and induces dermal condensation, whereas collagen I is preferentially expressed in the interbud (Ting-Berreth and Chuong 1996; Atit et al. 2003). In summary, these genes expressed de novo are involved in intra-bud morphogenesis, including bud axis specification, growth, and differentiation. At the early stage, feather placodes are radially symmetric. Then, due to the de novo expression of Notch ligand (Delta-1, Serrate-1) and Notch-1 on the posterior and central region of the outgrowing feather bud, an A-P molecular asymmetry emerges (Fig. 2.3d), after which the symmetric short buds will develop into asymmetric elongated buds (Chen et al. 1997). The outgrowth of feather buds proceeds along the posterior direction on the body, accompanied by increased cell

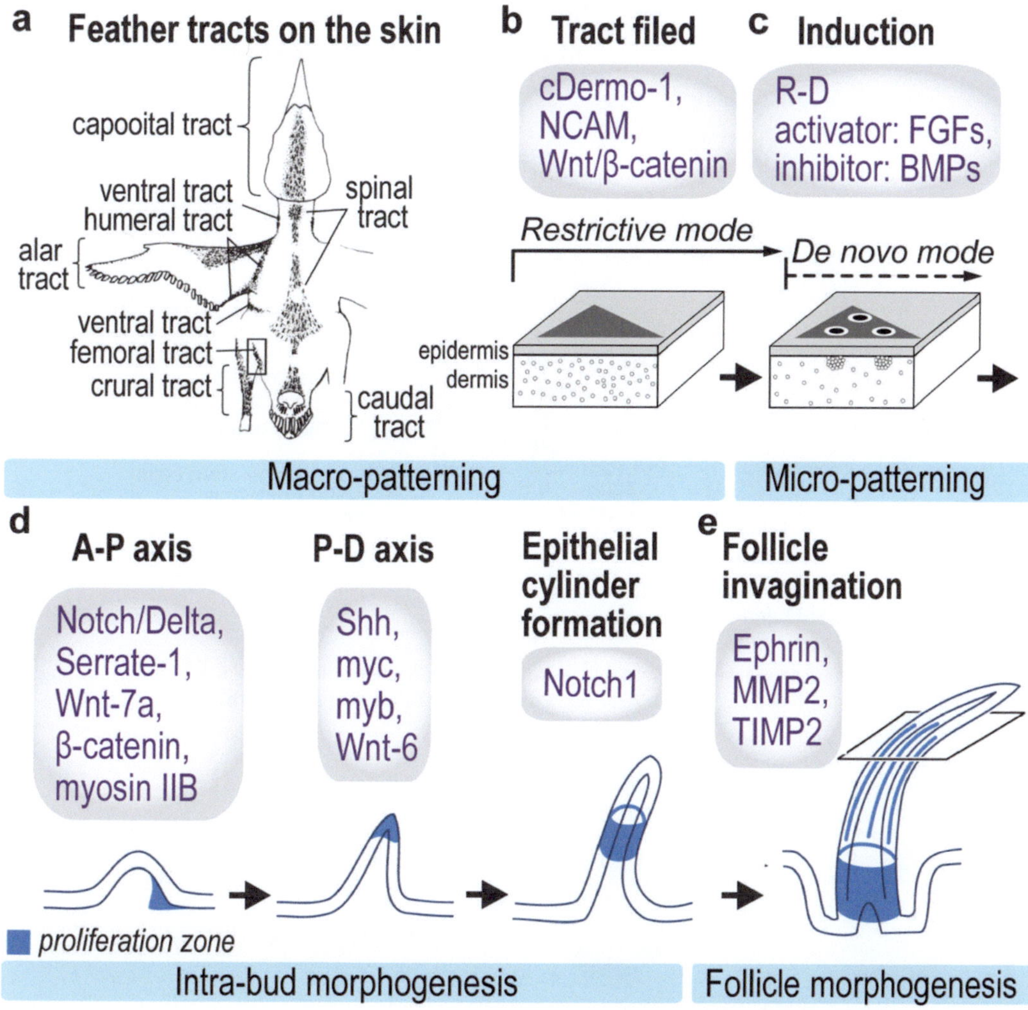

Fig. 2.3 Development of feather follicles during embryogenesis [modified from Gill 1994 (**a**), Chuong et al. 2000 (**b**, **c**), Chen et al. 2015 (**d**, **e**)]

proliferation at the posterior bud epithelium and polarized dermal cell rearrangement (Fig. 2.3d) (Chodankar et al. 2003; Li et al. 2013).

Wnt-7a is initially restrictively expressed at the boundary of feather buds. Later, the asymmetrical expression of Wnt-7a in the posterior-distal bud epidermis becomes an inducer of A-P asymmetry and the elongation of buds along the P-D axis (Widelitz et al. 1999). Wnt-7a induces β-catenin nuclear translocation, which activates non-muscle myosin IIB (NM IIB) and Serrate-1 (Notch ligand) expression. NM IIB enhances cell motility to enable polarized movements, while a positive feedback loop between Wnt and Notch signaling, as well as the lateral inhibition of Serrate-1 and Notch-1, help establish and maintain the spatial configuration of cell rearrangement zones. This ensures the elongation of feathers in a robust manner (Li et al. 2013). In the anterior half of feather buds, Msx-1 and -2 are asymmetrically localized and involved in bud growth and differentiation (Noveen et al. 1995). During feather bud elongation, the localized cell proliferative zone in the epithelium shifts from the posterior to the distal bud end and mediates the expansion of the feather epithelium to adapt to

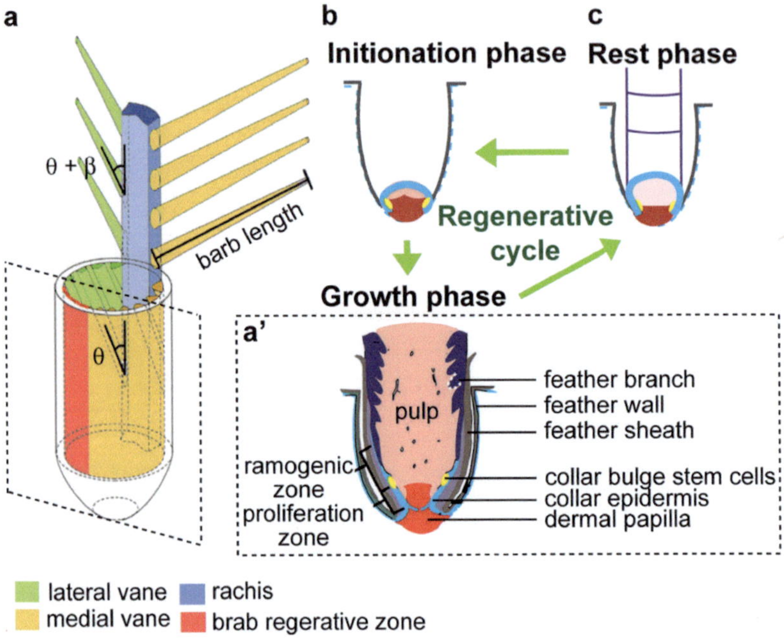

Fig. 2.4 Feather follicle and regenerative cycling: (**a**) structure of a feather follicle, (**a′**) growth phase, (**b**) initiation phase, (**c**) rest phase [modified from Li et al. 2017 (**a**), Yu et al. 2002 (**a′**), Lei et al. 2016 (**b**, **c**)]

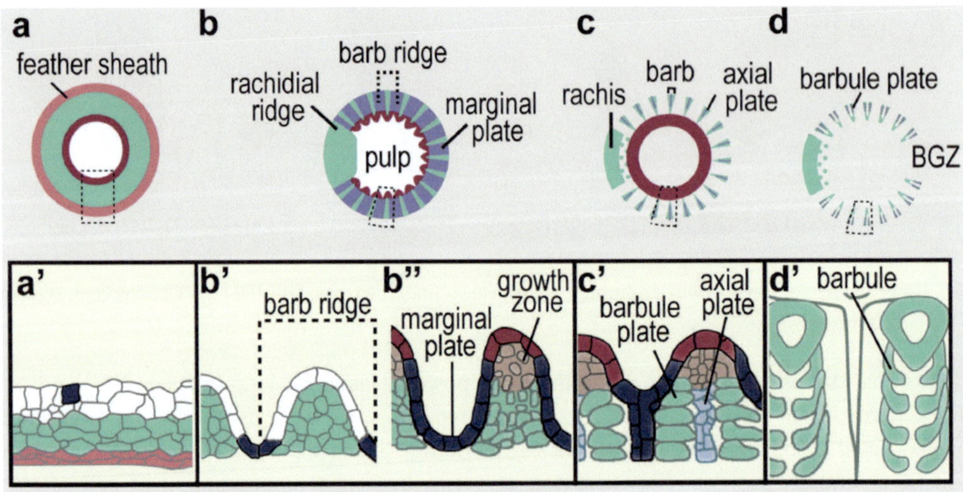

Fig. 2.5 Branching morphogenesis of feathers: (**a–d**) cross sections of feather follicles, (**a′–d′**) magnification of the insets from (**a**) to (**d**), (**a**, **a′**), cross sections at the proliferation zone, (**b–b″**) formation of barb ridges by periodic invagination at ramogenic zone, (**c**, **c′**) barb and rachis formation by differential cell death, (**d**, **d′**) feather follicle open by apoptosis to form a vane (modified from Chuong et al. 2014)

polarized dermal cell movements. Wnt-6, *myb*, and *myc* are concomitantly expressed at the shifting proliferation zone (Fig. 2.3d) (Desbiens et al. 1991; Chodankar et al. 2003). Ectopic

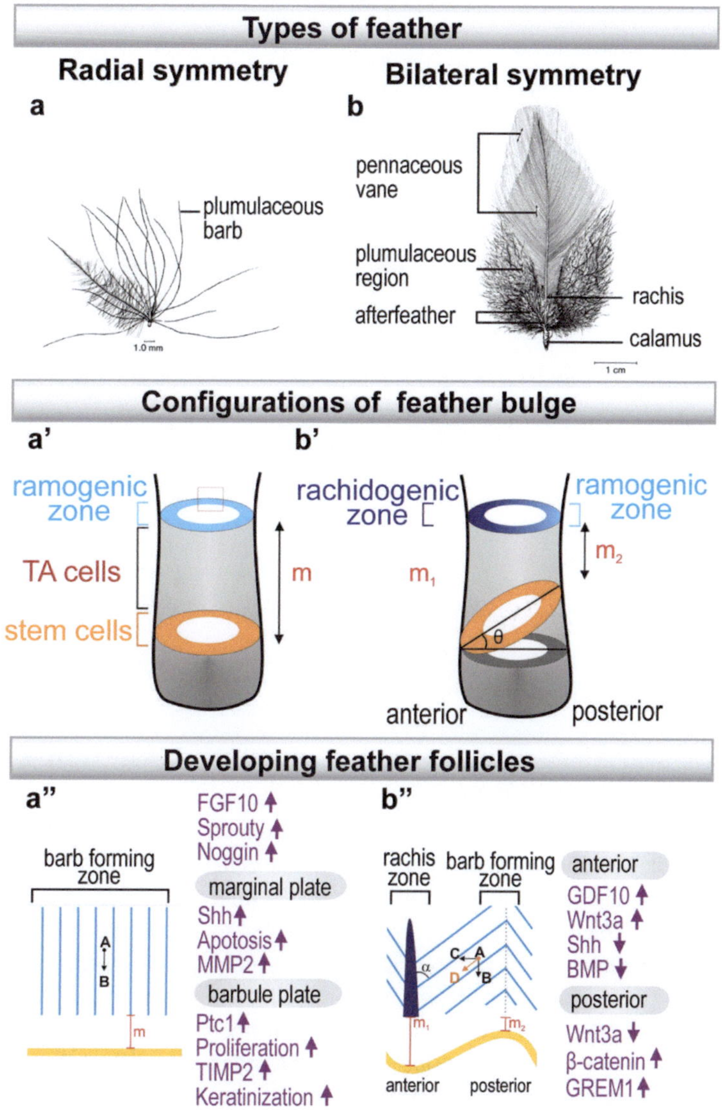

Fig. 2.6 Formation of radial and bilateral symmetric feathers: (**a**–**a**″) radial symmetric feather, (**b**–**b**″) bilateral symmetric feathers, (**a**, **b**) representive feather forms, (**a**′, **b**′) proximal follicle showing the topobiological relationship of stem cells, TA cells and ramogenic zone, (**a**″, **b**″) barb ridge orientation in an open follicle preparation [modified from Lucas and Stettenheim 1972 (**a**, **b**) and Lin et al. 2006 (all others)]

expression of Wnt-6 results in abnormal localized outgrowths (Chodankar et al. 2003).

2.2.4 Invagination of Feather Filaments and Morphogenesis of Feather Follicles

Feather buds elongate into filamentous structures, while the epidermis that surrounds the bud invaginates into the dermis to form the feather follicles (Fig. 2.3e). This process involves the formation of a boundary between the inside and outside of the feather follicle (the future follicle sheath) through Ephrin/Eph signaling in the epidermis at the invagination stage (Suksaweang et al. 2012). A matrix degrading protease (matrix metalloproteinase 2, MMP2) and its inhibitor, tissue inhibitor of metalloproteinase 2 (TIMP2), may also be essential for the invagination of feather buds. Both of them are initially expressed in the feather buds; however, a complementary

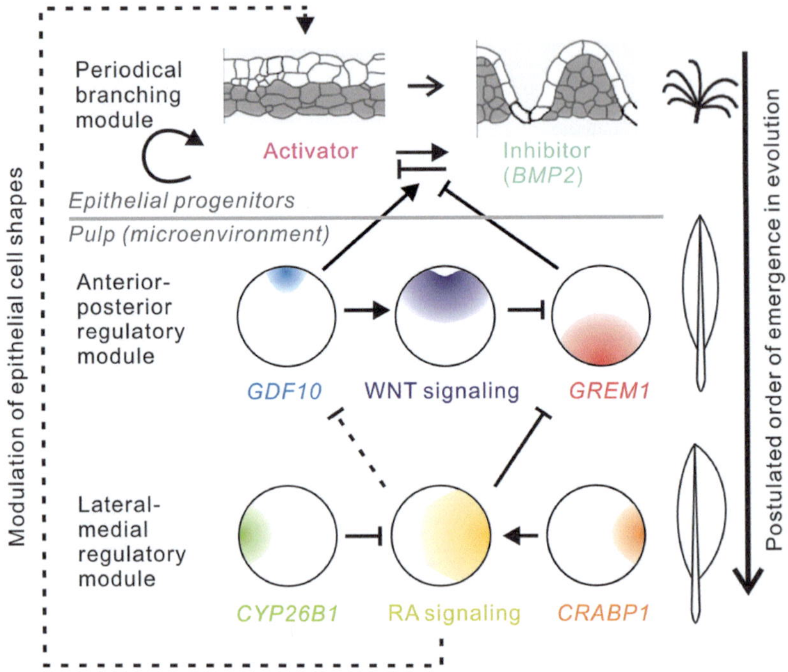

Fig. 2.7 Formation of bilateral asymmetric feather (from Li et al. 2017)

pattern of expression with MMP2 present in marginal plates and TIMP2 in the barb ridges can be detected later (Jiang et al. 2011). At the bottom of the feather follicle, aggregation of dermal cells near the base of the filament becomes the dermal papilla, which serves as the signaling center to trigger feather regeneration (Fig. 2.4a′). In contrast, the sparse population of dermal cells distributed at the more distal part of the feather follicle becomes the pulp, which provides nutrients and signals for the growth and differentiation of the distal epithelium (Fig. 2.4a′). All these processes together shape the feather follicle into a solid foundation so that the feather itself can further elongate along the proximal to distal direction (Fig. 2.4a) (Prum 2005).

2.3 Cyclic Regeneration in Adult Feather Follicles

The natal down feathers usually develop radially symmetric barbs that fuse at the proximal end of the feather (Fig. 2.6a). Contour feathers are bilaterally symmetric with barbs that fuse into the central shaft (rachis), which separates feather vanes into two halves (Fig. 2.6b). These two types of feathers may derive from the same follicle at different regeneration cycles. In nature, the natal down feathers on the newborn chick are replaced with contour feathers after molting. This natural cycle of feather regeneration is constituted by initiation, growth, and rest phases (Fig. 2.4), which occur continuously throughout the life of the bird (Lucas and Stettenheim 1972).

The regeneration of feathers relies on stem cells, which are located within the follicle, in a collar bulge region slightly above the dermal papilla with a ring-shaped distribution (Fig. 2.6a′, b′) (Yue et al. 2005). The initiation phase starts after the shedding of the feather. Feather stem cells will get close to dermal papilla to form a proliferation zone (Fig. 2.4b). During the growth phase, proliferative activity is high at the basal layer of the epithelium and the collar epidermis thickens around the dermal papilla. Transiently amplifying cells (TA cells) are continuously generated towards distal follicles to form a ramogenic zone, where the epidermal cylinder starts to differentiate into feather branches

(Figs. 2.4a′ and 2.6a′, b′). It usually takes a few weeks to accomplish the growth phase and develop a complete feather. After that, the feather follicles stay in the rest phase. During the rest phase, collar epidermis and pulp are degenerated; the stem cells move downward to the region surrounding the dermal papilla again, resulting in the shrinkage of the feather follicle diameter. Only the epithelial pulp caps and the keratinized feather branches are left behind (Fig. 2.4c). Apart from the natural molting cycles, feather regeneration can also be induced by feather plucking.

Several molecules, such as the adhesion molecule Tenascin C, growth factor FGF10, and its inhibitor Sprouty RTK signaling antagonist 4 (Spry4), have been shown to have spatiotemporally dynamic expression patterns during feather cycling. For example, Tenascin C can be detected within the proliferation zone during the initiation phase, but then it shifts to the dermal papilla and proximal pulp during the growth phase (Lin et al. 2013b). In contrast, expression of NCAM is consistent and continuous in the dermal papilla throughout all stages of the feather cycle (Lin et al. 2013a, b). Dermal papilla is essential for feather regeneration. If the dermal papilla is surgically removed, the feather epithelial stem cells alone cannot drive the regeneration (Lillie and Wang 1944).

In recent years, the genome-wide gene expression profile in the dermal papilla of growing feathers has been surveyed (Chu et al. 2014). Many signaling molecules or regulators of major signaling pathways have been identified in the dermal papilla, consistent with its critical role in feather growth and regeneration. For example, the FGF, BMP, and Shh signaling pathways may be involves in various stages of feather growth and regeneration. High levels of Wnt inhibitors, including Dkk2 and Frzb, were also detected in the dermal papilla. These inhibitors interact with the epithelial and pulp Wnt signaling, as well as regulate feather regeneration and axis formation (Chu et al. 2014).

2.4 Branching Morphogenesis of Adult Feathers

Branching morphogenesis begins at the ramogenic zone with a periodic invagination of the multilayered epithelium towards the dermis to form the initial barb ridges (Fig. 2.5b–b″). Both the separation of the barb ridge into individual barbs and the further branching of barbs into barbules are caused by differential cell death. For instance, the invagination of the basal layer of the feather filament flanking each barb ridge becomes a marginal plate that will undergo apoptosis (Fig. 2.5b″). In contrast, the epithelial region located at the summit of the pulp is an actively growing zone, which continuously proliferates during the growth phase (Fig. 2.5c′). Furthermore, the barb ridges at the anterior end of the feather follicle fuse to form the rachis, which is the feather backbone (Fig. 2.5c). Each barb ridge also consists of centrally aligned axial plates and bilaterally positioned barbule plates (Fig. 2.5c′). Axial plate cells will eventually disappear to give space to opening barbules (Fig. 2.5d, d′). Meanwhile, the barbule plates will keratinize. At the follicular level, there is another round of apoptosis that includes the pulp epithelia lying internal to the filament cylinder, the feather sheath enclosing the filament cylinder, and the barb generative zone (BGZ) located opposite to the rachis in bilateral symmetric feathers. Thus, feather barbs open up after the branching pattern is sculpted by programmed cell deaths (Fig. 2.5d) (Chuong and Edelman 1985; Chang et al. 2004b; Chuong et al. 2014).

Apart from the morphological changes, a few molecular pathways have been demonstrated to be involved in the branching morphogenesis. The barb ridge patterning may be regulated by Ephrin B1, which is expressed in the marginal plate epithelium (Suksaweang et al. 2012). The formation of barbs is coordinated by an R-D mechanism in which Shh is an activator and BMP2 is an inhibitor. The striped expression pattern of Shh causes apoptosis in the marginal plates, resulting in periodically arranged barbs (Chang et al. 2004b; Harris et al. 2005). Furthermore, BMP4 and its antagonist Noggin were verified to control the branching of feather epithelia. BMP4 promotes barb fusion and the formation of rachis, whereas Noggin enhances rachis and barb branching (Yu et al. 2002).

2.5 Molecular and Cellular Bases of Diverse Feather Form Formation

2.5.1 Radial and Bilateral Symmetric Feathers

During embryogenesis, the complex shape of feathers develops within the feather germ from the distal to proximal end. In the process of follicle morphogenesis, feather stem cells form a ring-shaped collar located above the dermal papilla. Here, the configuration of the stem cells in the feather follicle is correlated with the morphological type of feather (Fig. 2.6) (Yue et al. 2005). In downy feathers, which have a radially symmetric shape, the stem cells are arranged horizontal to the A-P axis, showing a homogenous expression of Wnt-3a (Fig. 2.6a′). However, a Wnt-3a gradient is responsible for the tilting of the stem cell ring (Yue et al. 2006). This tilting makes the distance between the stem cell ring and the ramogenic zone different at the anterior and posterior side of the feather follicle (marked as m, m1, and m2 in Fig. 2.6). Therefore, the A-P Wnt-3a gradient tilts the barb ridges toward the anterior side, leading to the formation of a rachis on one side but not the other, resulting in the conversion of radially symmetric feathers to bilaterally symmetric ones (Fig. 2.6) (Yue et al. 2006). In a bilaterally symmetric feather, the topologies of rachis and BGZ (setting vane boundaries) are mainly regulated by GDF10 and GREM1, which are diffusible molecules derived from peripheral pulp. Their interactions with the epithelial BMP signaling and the anterior–posterior Wnt-3a gradient modulate the bilateral-symmetric vane configuration, such as the change of feather vane width (Fig. 2.7) (Li et al. 2017).

2.5.2 Bilateral Asymmetric Feathers

The M-L (or bilateral) asymmetric shape of flight feathers (remiges) and tail feathers (rectrices) is a hallmark of flying abilities in modern birds and some feathered non-avian theropod dinosaurs. In the fossil record, the emergence of asymmetric flight feathers precedes that of the specialized bone-musculature for flapping (Feduccia and Tordoff 1979; Dyke et al. 2013), implying a pivotal role for this morphological feature during the early stages of flight evolution. Aerodynamically, the asymmetric flight feathers serve as mini-airfoils that can generate lift. The co-localization of the center of gravity and the center of the lifting force enable more stable flight. These feathers also facilitate the unidirectional pass-through of air during flapping. Additionally, they can separate from each other to minimize wind resistance (Norberg 1985a, b; Paul 2002; Pennycuick 2008; Scott and McFarland 2010).

In the molecular aspect, Li et al. (2017) recently discovered the asymmetric distribution of retinoic acid (RA)–related molecules in the pulp of developing chicken flight feathers. By comparing the gene expression profile of the more asymmetric primary remiges to the less asymmetric secondary remiges, Li et al. revealed localized expression of CYP26B1 (RA degradation enzyme) in the pulp adjacent to the lateral vanes (narrower) of primary remiges. Meanwhile, the RA binding protein CRABP1 has enriched expression in the pulp adjacent to the medial vanes (wider). The nuclear enrichment of CRABP1 protein in these pulp cells indicates that they may facilitate the nuclear transport of RA to activate downstream gene expression. As a consequence, an RA signaling gradient is present in the pulp of primary remiges along the M-L axis. More interestingly, the gradient exhibits position-dependent variation of steepness in remiges along the wing. Remiges closer to the wing tip have higher CYP26B1 expression and less CRABP1, whereas those distant to the wing tip, such as the secondary remiges, barely express any CYP26B1 at all. The bilateral symmetric body plumes have symmetrically distributed CYP26B1 or nuclear CRABP1. Thus, the RA signaling landscape has close correlation with the levels of feather asymmetry.

In functional perturbation experiments, enforced expression of dominant negative forms of RA receptors (to inhibit RA signaling) in plucked remige follicles significantly reduces the vane width of regenerated remiges. The narrower vanes are associated with increased BGZ width, sharper barb-rachis angles, and more elongated feather epithelial cell shapes. Further analysis indicates that RA signaling downregulates GREM1 expression (Fig. 2.7). Because GREM1 is a major determinant of BGZ topology, elevation of RA signaling would inhibit the BGZ and hence expand the vane width during feather growth, whereas the inhibition of RA signaling has the opposite effect. The sharper barb-rachis angle may contribute to the further reduction of vane width. Mathematical modeling revealed a cause-and-effect relationship between elongated epithelial cell shapes and sharper helical growth angles, which is a component of the barb-rachis angle (Prum and Williamson 2001; Feo and Prum 2014). However, the molecular mechanism of how RA signaling modulates epithelial cell shape remains unclear.

The contribution of barb-rachis angles and the asymmetric position of BGZ to asymmetric vane formation has been confirmed in an early morphology study (Feo and Prum 2014). However, the helical growth angle was not found to contribute to vane asymmetry in that study. This difference may be explained by the following factors: (1) the authors did not remove the pulp and flatten the feather cylinder to image under a confocal microscope when they measured the helical growth angles, like we did; (2) the authors only measured the angles at two positions; (3) measurements in their study were taken in rectrices rather than remiges.

Another interesting aspect of the barb-rachis angle is that the narrower lateral vanes of primary remiges in Mesozoic birds, such as *Archaeopteryx* and *Confuciusornis,* have shorter barbs but comparable barb-rachis angles to the wider medial vanes (Feo et al. 2015). Therefore, the utilization of barb-rachis angles to modulate asymmetry levels might be a mechanism that evolved after the Mesozoic period.

2.6 Other Parameters in Feather Complexity

Other parameters can add to the complexity of feather phenotypes. Here, we briefly discuss these topics without a detailed review.

Pigmented stripes were already present in the feathers of feathered dinosaurs (Zhang et al. 2010). With expanded vanes on the protruding feathers, they become the canvas for communication, thus adding one more dimension of complexity to feather phenotypes. Feather coloration can result from the chemical color produced by melanocytes or compounds from the food supply, as well as structural color generated by light refraction through regularly-spaced melanin. The melanocyte-based color patterns are further enriched by the renewal and regulation of feather melanocyte stem cells under sex hormone regulation (Lin et al. 2013a, b). Another parameter is the feather texture, mainly based on the types and organization of feather keratins (Wu et al. 2015). Frizzed chicken feather phenotypes are caused by a mutation in keratin 75 (KRT75) that weakens the rachis medulla structure (Ng et al. 2012).

With more than 30,000 feather primordia on an individual bird, each primordium can have its own pace in development and cycling. Such heterochronic control provides one more mechanism for functional adaptation. The most dramatic example is seen in highly altricial birds, such as the Zebra Finch. Here, the feather development is paused so that the hatchling is naked, allowing a stronger interaction between the chick and mother. The FGF pathway is shown to be involved in this heterochronic control (Chen et al. 2016).

Table 2.1 Definition of true avian feathers in extant birds

No.	Characteristics of a true feather
1.	Feather has a follicle structure. Stem cells and dermal papilla are in the follicle. Hence, it has the ability to go through molting cycles physiologically and to regenerate after plucking.
2.	Feather filament has a mesenchymal core (pulp). When it matures, the two sides of the feather vane face the previous basal and supra-basal layer respectively. The pulp is gone.
3.	There is hierarchical branch formation of rachis, barbs, and barbules. Barbs form by differential cell death. They can be bilaterally or radially symmetric, and bilaterally asymmetric.
4.	Feather has localized zones of proliferating cells positioned proximally. With more mature cells displaced toward the distal end, different feather morphologies can be displayed in different proportions along the proximal-distal axis.

The numbers do not indicate the order of appearance or a prerequisite for subsequent characteristics (Chuong et al. 2003)

2.7 Conclusion

A feather is a good example of hierarchically innovated complexity during evolution and intrigues biologists to investigate the genomic, molecular, and cellular processes underlying its structural complexity. With the progressively increased understanding of feather developmental biology (Chen et al. 2015; Lin et al. 2013a, b), we have revised the definition of true feathers in extant birds (Chuong et al. 2003; Wu et al. 2004). The fossil findings of some extinct feathers may not exhibit all these characteristics, in which case they are protofeathers. Eventually, the useful feather traits were selected and integrated into the feathers of modern Aves. Such is the "Tao" of integuments, which allows diversity to occur in each element and spatio-temporal regulation for complexity to be built collectively (Lai and Chuong 2016) (Table 2.1).

Acknowledgement This work was supported by a Dragon Gate grant from the Taiwan Ministry of Science Technology (MOST) 104-2911-I-002-577 and US NIH Grants AR47364, AR60306, and GM125322. GW. L. was supported by a Taiwan MOST postdoctoral fellowship. A. L. was supported by a CIRM fellowship. We thank Dr. Shuo Wang for reviewing this manuscript.

References

Atit R, Conlon RA, Niswander L (2003) EGF signaling patterns the feather array by promoting the interbud fate. Dev Cell 4(2):231–240

Brush AH (1996) On the origin of feathers. J Evol Biol 9 (2):131–142

Chang CH, Jiang TX, Lin CM, Burrus LW, Chuong CM, Widelitz R (2004a) Distinct Wnt members regulate the hierarchical morphogenesis of skin regions (spinal tract) and individual feathers. Mech Dev 121 (2):157–171. https://doi.org/10.1016/j.mod.2003.12.004

Chang CH, Yu M, Wu P, Jiang TX, Yu HS, Widelitz RB et al (2004b) Sculpting skin appendages out of epidermal layers via temporally and spatially regulated apoptotic events. J Invest Dermatol 122(6):1348–1355. https://doi.org/10.1111/j.0022-202X.2004.22611.x

Chang C, Wu P, Baker RE, Maini PK, Alibardi L, Chuong CM (2009) Reptile scale paradigm: evo-devo, pattern formation and regeneration. Int J Dev Biol 53 (5–6):813–826. https://doi.org/10.1387/ijdb.072556cc

Chen CW, Jung HS, Jiang TX, Chuong CM (1997) Asymmetric expression of notch/Delta/serrate is associated with the anterior-posterior axis of feather buds. Dev Biol 188(1):181–187. https://doi.org/10.1006/dbio.1997.8643

Chen CF, Foley J, Tang PC, Li A, Jiang TX, Wu P et al (2015) Development, regeneration, and evolution of feathers. Annu Rev Anim Biosci 3:169–195. https://doi.org/10.1146/annurev-animal-022513-114127

Chen CK, Ng CS, Wu SM, Chen JJ, Cheng PL, Wu P et al (2016) Regulatory differences in natal down development between altricial zebra finch and precocial chicken. Mol Biol Evol 33(8):2030–2043. https://doi.org/10.1093/molbev/msw085

Chodankar R, Chang CH, Yue Z, Jiang TX, Suksaweang S, Burrus L et al (2003) Shift of localized growth zones contributes to skin appendage morphogenesis: role of the Wnt/beta-catenin pathway. J Invest Dermatol 120(1):20–26. https://doi.org/10.1046/j.1523-1747.2003.12008.x

Chu Q, Cai L, Fu Y, Chen X, Yan Z, Lin X et al (2014) Dkk2/Frzb in the dermal papillae regulates feather regeneration. Dev Biol 387(2):167–178. https://doi.org/10.1016/j.ydbio.2014.01.010

Chuong CM, Edelman GM (1985) Expression of cell-adhesion molecules in embryonic induction.

II. Morphogenesis of adult feathers. J Cell Biol 101 (3):1027–1043
Chuong CM, Chodankar R, Widelitz RB, Jiang TX (2000) Evo-devo of feathers and scales: building complex epithelial appendages. Curr Opin Genet Dev 10 (4):449–456
Chuong CM, Wu P, Zhang FC, Xu X, Yu M, Widelitz RB et al (2003) Adaptation to the sky: defining the feather with integument fossils from mesozoic China and experimental evidence from molecular laboratories. J Exp Zool B Mol Dev Evol 298(1):42–56. https://doi.org/10.1002/jez.b.25
Chuong CM, Bhat R, Widelitz RB, Bissell MJ (2014) SnapShot: branching morphogenesis. Cell 158(5):1212-e1. https://doi.org/10.1016/j.cell.2014.08.019.
Desbiens X, Queva C, Jaffredo T, Stehelin D, Vandenbunder B (1991) The relationship between cell proliferation and the transcription of the nuclear oncogenes *c-myc*, *c-myb* and *c-ets-1* during feather morphogenesis in the chick embryo. Development 111(3):699–713
Dyke G, de Kat R, Palmer C, van der Kindere J, Naish D, Ganapathisubramani B (2013) Aerodynamic performance of the feathered dinosaur microraptor and the evolution of feathered flight. Nat Commun 4:2489. https://doi.org/10.1038/ncomms3489
Feduccia A, Tordoff HB (1979) Feather of *Archaeopteryx*: asymmetric vanes indicate aerodynamic function. Science 203:1021–1022
Feo TJ, Prum RO (2014) Theoretical morphology and development of flight feather vane asymmetry with experimental tests in parrots. J Exp Zool B Mol Dev Evol 322 (4):240–255. https://doi.org/10.1002/jez.b.22573
Feo TJ, Field DJ, Prum RO (2015) Barb geometry of asymmetrical feathers reveals a transitional morphology in the evolution of avian flight. Proc Biol Sci 282 (1803):20142864. https://doi.org/10.1098/rspb.2014.2864
Gill FB (1994) Ornithology, 2nd edn. Freeman, New York
Harris MP, Williamson S, Fallon JF, Meinhardt H, Prum RO (2005) Molecular evidence for an activator-inhibitor mechanism in development of embryonic feather branching. Proc Natl Acad Sci U S A 102 (33):11734–11739. https://doi.org/10.1073/pnas.0500781102
Hornik C, Krishan K, Yusuf F, Scaal M, Brand-Saberi B (2005) cDermo-1 misexpression induces dense dermis, feathers, and scales. Dev Biol 277(1):42–50. https://doi.org/10.1016/j.ydbio.2004.08.050
Jiang TX, Jung HS, Widelitz RB, Chuong CM (1999) Self-organization of periodic patterns by dissociated feather mesenchymal cells and the regulation of size, number and spacing of primordia. Development 126 (22):4997–5009
Jiang TX, Tuan TL, Wu P, Widelitz RB, Chuong CM (2011) From buds to follicles: matrix metalloproteinases in developmental tissue remodeling during feather morphogenesis. Differentiation 81(5):307–314. https://doi.org/10.1016/j.diff.2011.03.004
Jung HS, Francis-West PH, Widelitz RB, Jiang TX, Ting-Berreth S, Tickle C et al (1998) Local inhibitory action of BMPs and their relationships with activators in feather formation: implications for periodic patterning. Dev Biol 196(1):11–23. https://doi.org/10.1006/dbio.1998.8850
Lai YC, Chuong CM (2016) The "tao" of integuments. Science 354(6319):1533–1534. https://doi.org/10.1126/science.aal4572
Lei MI, Inaba M, Chuong CM (2016) Vertebrate embryo: development of the skin and its appendages. https://doi.org/10.1002/9780470015902.a0026601
Li A, Chen M, Jiang TX, Wu P, Nie Q, Widelitz R et al (2013) Shaping organs by a wingless-int/Notch/nonmuscle myosin module which orients feather bud elongation. Proc Natl Acad Sci U S A 110(16): E1452–E1461. https://doi.org/10.1073/pnas.1219813110
Li AFS, Jiang TX, Wu P, Widelitz R, Nie Q, Chuong CM (2017) Diverse feather shape evolution enabled by coupling anisotropic signalling modules with self-organizing branching programme. Nat Commun 8:14139. https://doi.org/10.1038/ncomms14139
Lillie FR, Wang H (1944) Physiology of development of the feather. VII. An experimental study of induction. Physiol Zool 17(1):1–31
Lin CM, Jiang TX, Widelitz RB, Chuong CM (2006) Molecular signaling in feather morphogenesis. Curr Opin Cell Biol 18(6):730–741. https://doi.org/10.1016/j.ceb.2006.10.009
Lin SJ, Foley J, Jiang TX, Yeh CY, Wu P, Foley A et al (2013a) Topology of feather melanocyte progenitor niche allows complex pigment patterns to emerge. Science 340(6139):1442–1445. https://doi.org/10.1126/science.1230374
Lin SJ, Wideliz RB, Yue Z, Li A, Wu X, Jiang TX et al (2013b) Feather regeneration as a model for organogenesis. Develop Growth Differ 55(1):139–148. https://doi.org/10.1111/dgd.12024
Lucas AM, Stettenheim PR (1972) Avian anatomy: integument. Agriculture handbook, vol 362, 1st edn. Washington, Agricultural Research Service
Ng CS, Wu P, Foley J, Foley A, McDonald ML, Juan WT et al (2012) The chicken frizzle feather is due to an α-keratin (KRT75) mutation that causes a defective rachis. PLoS Genet 8(7):e1002748. https://doi.org/10.1371/journal.pgen.1002748
Noramly S, Morgan BA (1998) BMPs mediate lateral inhibition at successive stages in feather tract development. Development 125(19):3775–3787
Noramly S, Freeman A, Morgan BA (1999) Beta-catenin signaling can initiate feather bud development. Development 126(16):3509–3521
Norberg RA (1985a) Function of vane asymmetry and shaft curvanture in bird flight feather: inference on flight ability of *Archaeopteryx*. In: Hecht MK, Ostrom JH, Viohl G, Wellnhofer P (eds) The beginnings of birds. Freunde des Jura-Museums Eichstätt, Willibaldsburg, pp 303–313

Norberg UM (1985b) Evolution of flight in bird: aerodynamic, mechanical and ecological aspects. In: Hecht MK, Ostrom JH, Viohl G, Wellnhofer P (eds) The beginnings of birds. Freunde des Jura-Museums Eichstätt, Willibaldsburg, pp 293–302

Noveen A, Jiang TX, Ting-Berreth SA, Chuong CM (1995) Homeobox genes Msx-1 and Msx-2 are associated with induction and growth of skin appendages. J Invest Dermatol 104(5):711–719

Ohyama A, Saito F, Ohuchi H, Noji S (2001) Differential expression of two BMP antagonists, *gremlin* and *Follistatin*, during development of the chick feather bud. Mech Dev 100(2):331–333

Olivera-Martinez I, Thelu J, Teillet MA, Dhouailly D (2001) Dorsal dermis development depends on a signal from the dorsal neural tube, which can be substituted by Wnt-1. Mech Dev 100(2):233–244

Paul GS (2002) Dinosaurs of the air: the evolution and loss of flight in dinosaurs and birds, 1st edn. Johns Hopkins University Press, Baltimore

Pennycuick CJ (2008) Modelling the flying bird, 1st edn. Academic Press, London

Prum RO (1999) Development and evolutionary origin of feathers. J Exp Zool 285(4):291–306

Prum RO (2005) Evolution of the morphological innovations of feathers. J Exp Zool B Mol Dev Evol 304(6):570–579. https://doi.org/10.1002/jez.b.21073

Prum RO, Brush AH (2002) The evolutionary origin and diversification of feathers. Q Rev Biol 77(3):261–295

Prum RO, Williamson S (2001) Theory of the growth and evolution of feather shape. J Exp Zool B Mol Dev Evol 291(1):30–57. https://doi.org/10.1002/jez.4

Scaal M, Prols F, Fuchtbauer EM, Patel K, Hornik C, Kohler T et al (2002) BMPs induce dermal markers and ectopic feather tracts. Mech Dev 110(1–2):51–60

Scott SD, McFarland C (2010) Bird feathers: a guide to North American species, 1st edn. Stackpole Books, Mechanicsburg

Suksaweang S, Jiang TX, Roybal P, Chuong CM, Widelitz R (2012) Roles of EphB3/ephrin-B1 in feather morphogenesis. Int J Dev Biol 56(9):719–728. https://doi.org/10.1387/ijdb.120021rw

Ting-Berreth SA, Chuong CM (1996) Sonic hedgehog in feather morphogenesis: induction of mesenchymal condensation and association with cell death. Dev Dyn 207(2):157–170. https://doi.org/10.1002/(SICI)1097-0177(199610)207:2<157::AID-AJA4>3.0.CO;2-G

Wessells NK (1965) Morphology and proliferation during early feather development. Dev Biol 12(1):131–153

Widelitz RB, Jiang TX, Noveen A, Chen CW, Chuong CM (1996) FGF induces new feather buds from developing avian skin. J Invest Dermatol 107(6):797–803

Widelitz RB, Jiang TX, Chen CW, Stott NS, Jung HS, Chuong CM (1999) Wnt-7a in feather morphogenesis: involvement of anterior-posterior asymmetry and proximal-distal elongation demonstrated with an in vitro reconstitution model. Development 126(12):2577–2587

Widelitz RB, Jiang TX, Yu M, Shen T, Shen JY, Wu P et al (2003) Molecular biology of feather morphogenesis: a testable model for evo-devo research. J Exp Zool B Mol Dev Evol 298(1):109–122. https://doi.org/10.1002/jez.b.29

Wu P, Hou LH, Plikus M, Hughes M, Scehnet J, Suksaweang S et al (2004) Evo-Devo of amniote integuments and appendages. Int J Dev Biol 48(2–3):249–270. https://doi.org/10.1387/ijdb.15272390

Wu P, Ng CS, Yan J, Lai YC, Chen CK, Lai YT et al (2015) Topographical mapping of α- and β-keratins on developing chicken skin integuments: functional interaction and evolutionary perspectives. Proc Natl Acad Sci U S A 112(49):E6770–E6779. https://doi.org/10.1073/pnas.1520566112

Xing LD, McKellar RC, Xu X, Li G, Bai M, Persons WS et al (2016) A feathered dinosaur tail with primitive plumage trapped in mid-Cretaceous amber. Curr Biol 26(24):3352–3360. https://doi.org/10.1016/j.cub.2016.10.008

Xu X (2006) Feathered dinosaurs from China and the evolution of major avian characters. Integr Zool 1(1):4–11. https://doi.org/10.1111/j.1749-4877.2006.00004.x

Xu X, Guo Y (2009) The origin and early evolution of feathers: insights from recent paleontological and neontological data. Vertebr PalAsiat 47(4):311–329

Xu X, Zhou Z, Dudley R, Mackem S, Chuong CM, Erickson GM et al (2014) An integrative approach to understanding bird origins. Science 346(6215):1253293. https://doi.org/10.1126/science.1253293

Yu M, Wu P, Widelitz RB, Chuong CM (2002) The morphogenesis of feathers. Nature 420(6913):308–312. https://doi.org/10.1038/nature01196

Yue Z, Jiang TX, Widelitz RB, Chuong CM (2005) Mapping stem cell activities in the feather follicle. Nature 438(7070):1026–1029. https://doi.org/10.1038/nature04222

Yue Z, Jiang TX, Widelitz RB, Chuong CM (2006) Wnt3a gradient converts radial to bilateral feather symmetry via topological arrangement of epithelia. Proc Natl Acad Sci U S A 103(4):951–955. https://doi.org/10.1073/pnas.0506894103

Zhang F, Kearns SL, Orr PJ, Benton MJ, Zhou Z, Johnson D et al (2010) Fossilized melanosomes and the colour of cretaceous dinosaurs and birds. Nature 463(7284):1075–1078. https://doi.org/10.1038/nature08740

3 The Origin of Birds: Current Consensus, Controversy, and the Occurrence of Feathers

Oliver W. M. Rauhut and Christian Foth

3.1 Introduction

The transition of one major animal "bauplan" into another and the origin of evolutionary novelty has captured the interest and imagination of scientists and the general public alike, ever since the advent of evolutionary thought in the nineteenth century and its acceptance following the publication of Darwin's epochal book *The Origin of Species* (1859). Birds are arguably the most extremely divergent example of a tetrapod bauplan, as they seem fundamentally different from their living reptilian relatives, crocodiles, turtles and lepidosaurs, in almost any respect, from their anatomy, via neurology and physiology to their behaviour. Although recent research has shown that some of these differences are less marked than originally thought (e.g. important aspects of the avian-type flow-through lung are already present in lepidosaurs and crocodiles; Farmer and Sanders 2010; Schachner et al. 2013; see also Cieri and Farmer 2016), many of these evolutionary novelties of birds require complex and changing functionary scenarios to explain their selective advantages, especially if many must be seen as exaptations to flight, rather than as consequences of this drastic change in locomotor behaviour. Thus, it is not surprising that the question of the origin of birds and the evolutionary history of their novelties has been a "hot topic" in evolutionary biology and palaeontology in the past 150 years. Understanding these evolutionary events requires a good idea of the interrelationships of bird ancestors, the origin of birds, and the phylogenetic relationships between early members of this clade.

The discovery of the first Mesozoic "bird", *Archaeopteryx lithographica*, only 2 years after the publication of Darwin's book (von Meyer 1861a, b) marks a milestone in our quest of understanding bird origins (although, interestingly, both initial descriptions of this animal came from anti-Darwinists, who came to diametrically opposite conclusions: Andreas Wagner (1862) came to the conclusion that this animal clearly represents a somewhat odd lizard, whereas Richard Owen (1863) concluded that the fossil undoubtedly represented a bird). Especially the preservation of feathers in the limestone slab that contained the skeleton of this taxon was taken as a clear indication that this animal was a transitional fossil of importance for the question of the origin

O. W. M. Rauhut (✉)
SNSB, Bayerische Staatssammlung für Paläontologie und Geologie, München, Germany

Department of Earth and Environmental Sciences, Palaeontology and Geobiology, Ludwig-Maximilians-Universität München, München, Germany

GeoBioCenter, Ludwig-Maximilians-Universität München, München, Germany
e-mail: rauhut@snsb.de

C. Foth
Department of Geosciences, Université de Fribourg, Fribourg, Switzerland

Staatliches Museum für Naturkunde Stuttgart, Stuttgart, Germany

C. Foth, O. W. M. Rauhut (eds.), *The Evolution of Feathers*, Fascinating Life Sciences,
https://doi.org/10.1007/978-3-030-27223-4_3

of birds, and soon after the discovery of *Archaeopteryx* (and the discovery of the small theropod dinosaur *Compsognathus* in the same year; Wagner 1861), Darwin's "bulldog" Thomas Henry Huxley published the hypothesis that birds were derived from small theropod dinosaurs (Huxley 1868). However, although this hypothesis fell on fertile ground in the beginning, other possibilities for the ancestry of birds were proposed subsequently, such as the Early Triassic basal archosauromorph *Euparkeria* (Broom 1913).

In his very influential book *The Origin of Birds*, the Danish artist Gerhard Heilmann (1926) summarized the current knowledge on this topic. Although he clearly noticed the many similarities between dinosaurs and birds (especially in *Archaeopteryx*), Heilmann came to the conclusion that birds cannot be derived from dinosaurs, as all dinosaurs then known lacked clavicles, whereas the furcula in birds is generally considered to be derived from a fusion of these bones, which are present in reptiles ancestrally. Heilmann's very detailed and well-illustrated book had a lasting impact on the field and formed the basis for the common consensus for 50 years that birds were derived from some still unknown, probably arboreal, Triassic "Proavis".

In the wake of the "dinosaur renaissance" in the 1960s and 1970s, especially the discovery of the dromaeosaurid *Deinonychus* in North America (Ostrom 1969a, b), and a comparison of this taxon with newly discovered (Ostrom 1970, 1972; Wellnhofer 1974) and already known specimens of *Archaeopteryx* led Ostrom (1973, 1976) to revive the hypothesis of the dinosaur origin of birds. Not surprisingly, the hypothesis was met with scepticism (e.g. Martin et al. 1980; Tarsitano and Hecht 1980; Martin 1983), and a sometimes heated debate ensued over the following two and a half decades (see Witmer 2002; Prum 2002 for a summary).

One important aspect that led to the acceptance of the hypothesis of the theropod origin of birds was the advent of new phylogenetic methods, following the publication of Hennig's book *Phylogenetic Systematics* in 1966. The cladistic methodology outlined in this work first found acceptance in vertebrate palaeontology in the 1980s, and in a very influential paper published in 1986, Jacques Gauthier listed a total of 84 nested synapomorphies that supported the inclusion of birds in the theropod dinosaurs. Gauthier's paper was the first of a long list of phylogenetic analyses that support the inclusion of birds in the Theropoda, and our knowledge of this transition and the successive acquisition of avian characters has considerably increased since (see Chiappe 2009; Brusatte et al. 2015; Cau 2018; Agnolín et al. 2019).

The final push for the theropod hypothesis, however, came from the discovery of abundant feathered dinosaurs in the Cretaceous of China, starting in the late 1990s (Chen et al. 1998; Ji et al. 1998; Xu et al. 1999a, b, 2001), and the subsequent realization that filamentous integumentary structures are widely distributed not only in theropod dinosaurs (Rauhut et al. 2012) but are even found in ornithischians (Zheng et al. 2009; Godefroit et al. 2014, 2020). In some instances, the interpretation of integumentary structures as feathers has been questioned, and the most detailed conflicting analyses interpreted these structures as degraded dermal collagen fibres (e.g. Lingham-Soliar 2003a, b, 2012; Lingham-Soliar et al. 2007) or other tissues (e.g. Lingham-Soliar 2010). However, these studies have been criticised on taphonomic, structural, and methodological grounds (e.g. Mayr 2010; Smith et al. 2015; Smithwick et al. 2017), and thus cannot be sustained. Furthermore, the vast array of taxa in which feathers have now been reported plus the great variety of feather types identified (e.g. Xu and Guo 2009) make these alternative interpretations untenable.

Although the opponents of the theropod origin of birds have questioned the cladistic methodology altogether (e.g. Feduccia 1996, 2013), there is no other hypothesis for avian origins that has been formulated in any comparable detail (see Xu et al. 2014; Brusatte et al. 2015; Cau 2018), and the criticism seems to be rather ideological than scientific (Prum 2003; Smith et al. 2015). Thus, in the absence of contrary evidence, the theropod origin of birds can now be regarded as being firmly established, and it is on this background that we will discuss the current consensus and controversies surrounding the origin of birds. For recent reviews of the overwhelming evidence that birds are theropods see, e.g., Xu et al. (2014), Brusatte et al. (2015), Smith et al. (2015), Mayr (2017), Cau (2018), and Agnolín et al. (2019).

3.2 Current Consensus on the Phylogeny of Theropod Dinosaurs and the Origin of Birds

Since the pioneering work of Gauthier (1986), numerous phylogenetic analyses of the interrelationships of theropod dinosaurs have been published (e.g. Novas 1992; Holtz 1994, 1998; Sereno 1997, 1999; Forster 1999; Rauhut 2003; Smith et al. 2007; Xu et al. 2009; Choiniere et al. 2010; Rauhut et al. 2010; Novas et al. 2015; Wang et al. 2017; Cau 2018), with many more analyses focusing on the different subclades of this clade. Interestingly, several of the main phylogenetic findings of Gauthier (1986) have consistently been confirmed, both in terms of tree topology as well as general taxonomic composition of several major clades, despite widely differing taxon and character sampling. Thus, these aspects of theropod phylogeny can be considered well established and largely uncontroversial.

All phylogenetic analyses including theropod dinosaurs agree that this clade is monophyletic, although there is some controversy as to whether certain basal taxa (e.g., *Eoraptor*, *Herrerasaurus*) are members of Theropoda or not. Nevertheless, at least the monophyly of Neotheropoda (the clade including *Coelophysis* and modern birds [Sereno 1998]; the classical Theropoda before the discovery of a number of basal taxa; see Colbert 1964) has never been questioned. Within Theropoda, a number of mainly Late Triassic and Early Jurassic taxa (sometimes included in a single lineage named Coelophysoidea), but also the clade Ceratosauria, which reached the Cretaceous/Paleogene boundary, are consistently found as basal forms outside a more derived clade which was named Tetanurae by Gauthier (1986). The interrelationships of these basal forms are still debated; although basically all phylogenetic analyses agree in the existence of two monophyletic clades, the Coelophysoidea (Fig. 3.1a) and the Ceratosauria (Fig. 3.1b), the referral of numerous taxa to either one of these clades remains controversial. Furthermore, whereas many early phylogenetic analyses recovered Coelophysoidea and Ceratosauria in a monophyletic clade (for which Gauthier [1986] used the name Ceratosauria; see, e.g., Gauthier 1986; Holtz 1994, 1998; Sereno 1997, 1999; but also Allain et al. 2007 as a more recent example), there is an emerging consensus that Ceratosauria represent the sister-taxon to Tetanurae to the exclusion of Coelophysoidea (e.g. Rauhut, 1998, 2003; Forster 1999; Carrano et al. 2002; Smith et al. 2007; Xu et al. 2009; Novas et al. 2015; Wang et al. 2017; Cau 2018) in a clade that was named Averostra by Paul (2002; see also definition by Ezcurra and Cuny 2007).

Whereas Coelophysoidea seems to represent the first successful radiation of theropod dinosaurs in the Triassic and includes both taxa from the Late Triassic and Early Jurassic, the earliest averostrans are Early Jurassic in age (see Dal Sasso et al. 2018), and there is growing evidence that an explosive radiation of this clade in the latest Early to Middle Jurassic might have been triggered by the Pliensbachian/Toarcian extinction event (Pol and Rauhut 2012; Rauhut and Carrano 2016; Rauhut et al. 2016).

The Tetanurae are the main clade of theropod dinosaurs that include most of the well-known forms and also recent birds. They first occur in the fossil record in the earliest Middle Jurassic, but the clade obviously experienced an explosive radiation soon after its origin, as all major clades, including avialans, are established by the Late Jurassic (Rauhut et al. 2010, 2016; Xu et al. 2010). Basically all recent phylogenetic analyses agree that Tetanurae split into three major lineages early in their evolutionary history, the Megalosauroidea (Spinosauroidea in older literature), Allosauroidea (Fig. 3.1c) and Coelurosauria, although the exact taxonomic composition of the three clades somewhat differs, especially in respect to inclusion or exclusion of basal taxa (e.g. Holtz 1998; Allain 2002; Rauhut

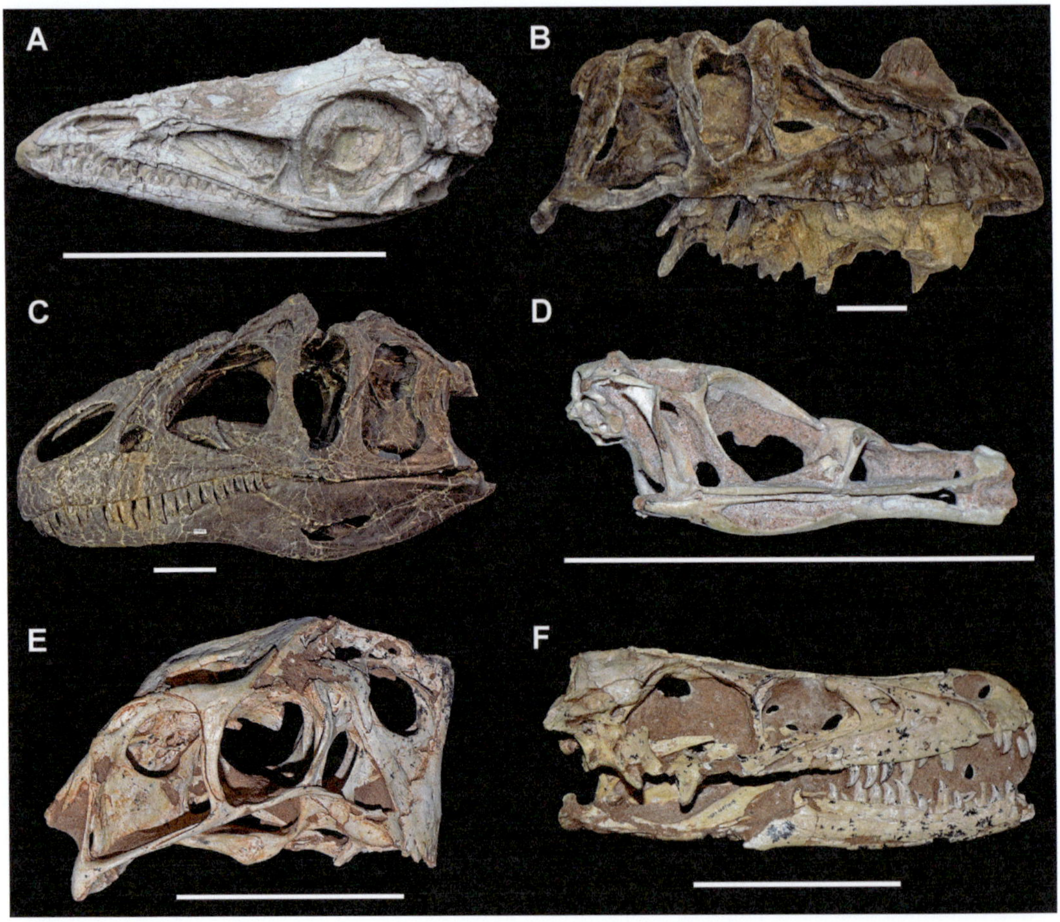

Fig. 3.1 Skulls of representatives of different theropod clades. (**a**) Coelophysid *Coelophysis bauri* (NMNH P-42200; photo courtesy Jörg Schneider) in left dorsolateral view. (**b**) Ceratosaur *Ceratosaurus nasicornis* (USNM 4735) in right lateral view. (**c**) Allosauroid *Allosaurus* sp. (MOR 693; photo courtesy Serjoscha Evers) in left lateral view. (**d**) Alvarezsauroid *Shuvuuia deserti* (IGM 100/1001) in right lateral view. (**e**) Oviraptorosaurid *Citipati osmolskae* (IGM 100/978) in right lateral view. (**f**) Dromaeosaurid *Tsaagan mangas* (IGM 100/1015) in right lateral view. All scale bars are 10 cm

2003; Rauhut and Xu 2005; Smith et al. 2007; Benson 2010; Benson et al. 2010; Choiniere et al. 2010; Carrano et al. 2012; Novas et al. 2015; Wang et al. 2017; Cau 2018).

Megalosauroids include mainly large-bodied and often heavily built megapredators, such as *Megalosaurus* and *Torvosaurus*, that thrived during the Middle and Late Jurassic (Benson 2010; Carrano et al. 2012; Rauhut et al. 2016), but also the highly specialized gigantic spinosaurids of the Cretaceous, which include the largest theropod known, *Spinosaurus*, which probably reached a length of 18 m and up to 10 t in body mass (Stromer 1915; Therrien and Henderson 2007; Hone and Holtz 2017). Likewise, allosauroids were also generally large-bodied and megapredatory theropods that originated in the Middle Jurassic and thrived to at least the early Late Cretaceous, culminating in the gigantic carcharodontosaurids (e.g. Brusatte and Sereno 2008; Benson et al. 2010; Carrano et al. 2012).

In contrast to these two major lineages of tetanurans, the third major clade, the Coelurosauria, includes both large and small forms and saw repeated changes in trophic ecology (Zanno and Makovicky 2011). Coelurosaurs include such iconic animals as *Tyrannosaurus rex* or *Velociraptor mongoliensis*, and more phylogenetic analysis of this clade have probably been published than most other fossil animals (e.g. Makovicky and Sues 1998; Norell et al. 2001, 2006; Xu et al. 2002, 2011, 2015; Makovicky et al. 2003; Senter 2007; Zhang et al. 2008; Choiniere et al. 2010, Choiniere et al. 2014; Turner et al. 2012; Agnolín and Novas 2013; Godefroit et al. 2013a, b; Brusatte et al. 2014; Foth et al. 2014; Cau et al. 2015, 2017). Although there are considerable differences in the placement of many taxa and even whole clades within coelurosaurs (see below), there also exists some consensus about the general topology of the coelurosaur family tree. Thus, basically all analyses of the last 20 years agree that Tyrannosauroidea are one of the most basal clades. Tyrannosauroidea have recently been found to include not only the well-known, gigantic megapredators of the Late Cretaceous, such as *Tyrannosaurus*, but also several other lineages, reaching back to the Middle Jurassic, such as the rather small-bodied, obviously fleet-footed proceratosaurids (Brusatte et al. 2010; Rauhut et al. 2010; Brusatte and Carr 2016).

Tyrannosauroidea and several other basal taxa and clades are outside a derived clade of coelurosaurs that Holtz (1996) named Maniraptoriformes. The most basal group within this clade are the Ornithomimosauria, generally small to medium-sized theropods with small skulls, long necks and elongate hindlimbs. All derived members of this clade are toothless and were probably omnivorous. Although most members of the Ornithomimosauroidea do not exceed 5–6 m in length and weighed less than 600 kg (Benson et al. 2018), the clade also includes the giant *Deinocheirus* that reached more than 11 m in length and more than six tons in weight (Lee et al. 2014).

An important clade within coelurosaurs is the Maniraptora. The clade was originally coined by Gauthier (1986) to include birds (Avialae) and theropods that share characters, especially in the manus that are not present in ornithomimosaurs. The clade has been phylogenetically defined by Holtz (1996) as all coelurosaurs that share a more recent ancestor with birds than with ornithomimids. Ever since the analysis of Gauthier (1986), several clades were consistently found to be maniraptorans, including Oviraptorosauria (Figs. 3.1e and 3.2a), Troodonotidae (Fig. 3.2b), Dromaeosauridae (Fig. 3.1f), and Avialae (including modern birds), together with some taxa that do not seem to be included in a larger clade, such as the genus *Ornitholestes*. A number of clades that have only more recently been recognized (or firmly established as theropodan, in the case of therizinosaurs), including Alvarezsauroidea (Fig. 3.1d), Therizinosauria, Scansoriopterygidae, and Anchiornithidae, are usually also found within Maniraptora, although their detailed relationships differ widely between different analyses (see below).

Within Maniraptora, the Troodontidae, Dromaeosauridae, and Avialae are united in the

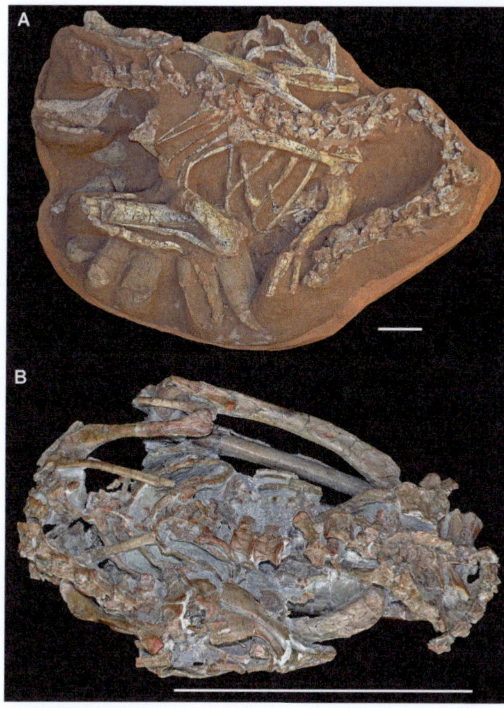

Fig. 3.2 Non-avian theropod skeletons documenting avian-like behaviour. (**a**) Postcranial skeleton of the oviraptorosaurid *Citipati osmolskae* in a brooding position on a nest of eggs (IGM 100/1004). (**b**) Troodontid *Mei long* in an avian-like sleeping position (IVPP V12733). Scale bars are 10 cm

clade Paraves, defined as all maniraptorans that are more closely related to extant birds than to *Oviraptor* (Sereno 1997, 1998). Whereas the recently recognized Anchiornithidae (Xu et al. 2016; Foth & Rauhut 2017) have always also been recovered as Paraves (e.g. Xu and Fucheng 2005; Hu et al. 2009, 2018; Xu et al. 2009, 2011; Godefroit et al. 2013a, b; Foth and Rauhut 2017), the Alvarezsauroidea and Scansoriopterygidae have been found to be Paraves only in some, but not all phylogenetic analyses.

Nested within Paraves is the clade Avialae, which is the most-inclusive clade containing extant birds, but not Dromaeosauridea or Troodontidae (Maryańska et al. 2002). Basel members of this clade are *Archaeopteryx* and *Alcmonavis* from the Late Jurassic of Germany (see Rauhut et al. 2019) and the groups Jeholornithidae, Sapeornithidae, and Confuciusornithidae, which are all known from the Early Cretaceous of China (e.g. Mayr 2017; Wang and Zhou 2017). The clade that embraces Confuciusornithidae and extant birds including all their descendants is the Pygostylia (Chiappe 2002). This clade contains two major groups, the Enantiornithes and Ornithuromorpha (Euornithes), which are summarized as Ornithothoraces (Sereno 1998; Chiappe 2002). The Enantiornithes are small-bodied, toothed Avialae, which represent the most successful group of stem birds during the Cretaceous in terms of species richness as well as temporal and geographic range (Mayr 2017; Wang and Zhou 2017). According to the foot morphology they were primarily arboreal (O'Connor et al. 2011a), but as indicated by differences in the snout shape, tooth morphology, and pedal claw geometry, a certain degree of ecological specializations was present. This includes, for instance, the long-snouted Longipterygidae (O'Connor et al. 2011b) or the raptorial Bohaiornithidae (Li et al. 2014).

In contrast to Enantiornithes, Ornithuromorpha possesses an enormous ecological diversification in terms of habitat and diet preferences, while their species diversity is lower when compared to their sister taxon. The ecological diversity includes semi-to-fully aquatic, but also ground-dwelling, and even secondary flightless taxa. In contrast to other Avialae, they also show a higher degree of tooth reduction and adaptation to piscivory, omnivory, insectivory, and granivory (Mayr 2017; Wang and Zhou 2017). The most successful group of Ornithuromorpha are the Aves (Neornithes), which represent the crown-group of extant birds, and are the only theropod branch that survived the K/T extinction event. The Aves already originated in the Late Cretaceous, showing an initial diversification of the clades Palaeognathae, Galloanseres, and Neovaves (Clarke et al. 2005; Brown et al. 2008; Prum et al. 2015; Mayr 2017). However, the actual radiation of crown group birds happened during the early Cenozoic, after the K/T event (Mayr 2009; Prum et al. 2015).

3.3 Remaining Controversies

Although there is a remarkable consensus in the general hierarchy of theropod interrelationships and the hierarchy levels that most clades belong to, there are numerous controversies about the exact phylogenetic position of numerous taxa and some entire clades. In non-tetanuran theropods, these uncertainties mainly concern the taxonomic composition of the basal clade Coelophysoidea and the question whether Coelophysoidea and Ceratosauria are united in a clade or whether Ceratosauria are closer to Tetanurae. Concerning Coelophysoidea, current hypotheses reach from uniting basically all Late Triassic and the vast majority of Early Jurassic neotheropods in this clade (e.g. Carrano et al. 2005; Allain et al. 2007) to the possibility that a number of Late Triassic and Early Jurassic taxa are more closely related to averostrans (e.g. Rauhut 2003; Smith et al. 2007; Ezcurra and Brusatte 2011; Langer et al. 2014; Ezcurra 2017; Martínez and Apaldetti 2017), including the possibility of another clade of mainly Early Jurassic theropods, the Dilophosauridae (e.g. Smith et al. 2007). In respect to the phylogenetic position of the Ceratosauria, there seems to be a growing consensus that this clade is united with the Tetanurae in a monophyletic Averostra,

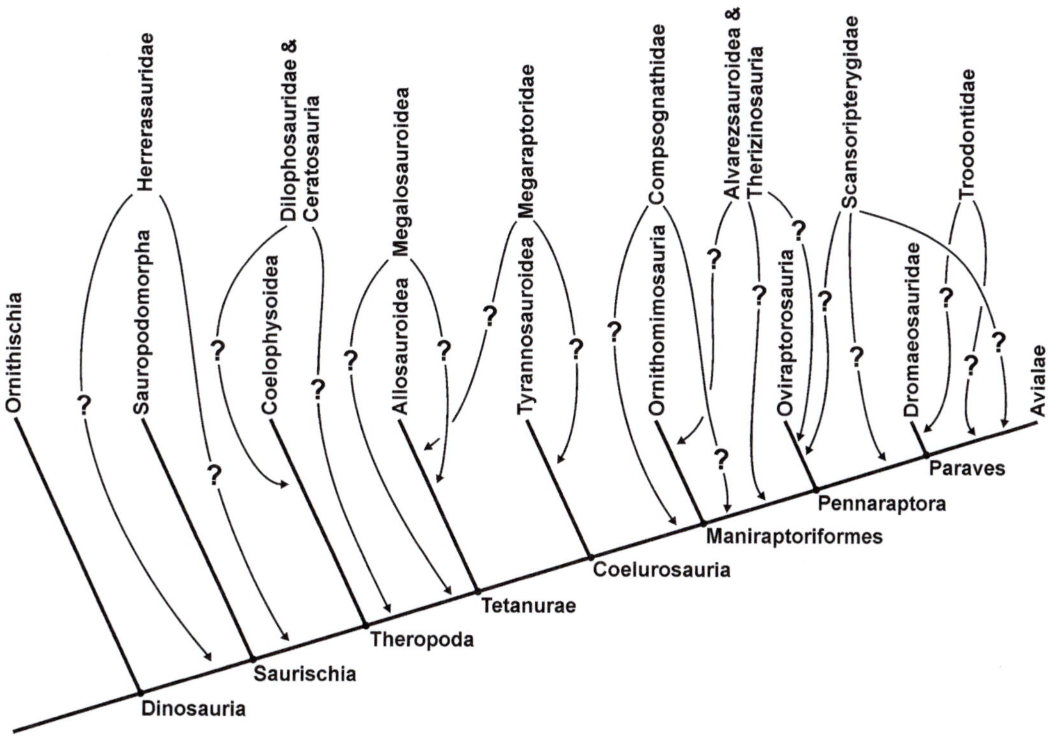

Fig. 3.3 Simplified cladogram of theropod relationships, showing common relationships between theropod dinosaurs and alternative phylogenetic positions for several problematic taxa (modified from Rauhut 2003)

with the last formal analysis finding a Coelophysoidea-Ceratosauria clade being that of Allain et al. (2007), whereas all more recent phylogenies found support for Averostra.

Within basal tetanurans, there is some disagreement on the relationships between the three major lineages, Megalosauroidea, Allosauroidea, and Coelurosauria. Thus, whereas most recent analyses found Allosauroidea and Coelurosauria to be sister taxa in a clade named either Avetheropoda or Neotetanurae (see Carrano et al. 2012), to the exclusion of Megalosauroidea (e.g. Allain 2002; Smith et al. 2007; Benson 2010; Benson et al. 2010; Rauhut et al. 2010, 2016; Carrano et al. 2012; Novas et al. 2015; Wang et al. 2017), some analyses found Allosauroidea and Megalosauroidea to be united in a clade called Carnosauria to the exclusion of Coelurosauria (e.g. Rauhut 2003; Rauhut et al. 2012; Cau 2018).

Apart from uncertainties of the placements of several genera within their respective clades, a further important discrepancy in phylogenetic hypotheses of basal tetanurans concerns the placement of the only recently recognized Megaraptora. The first representatives known of this clade were represented by very fragmentary material (Novas 1998; Azuma and Currie 2000; Calvo et al. 2004), and so their recognition as belonging to a monophyletic clade and an analysis of their phylogenetic relationships only became possible after more complete remains had been found (Sereno et al. 2008; Hocknull et al. 2009). The first work to recognize a monophyletic Megaraptora was Benson et al. (2010), who recovered Megaraptora as part of the Neovenatoridae, the sister taxon to Carcharodontosauridae within the Allosauroidea. This phylogenetic placement was supported by several subsequent analyses (e.g. Carrano et al. 2012; Rauhut et al. 2016), but Novas et al. (2013) argued that Megaraptora were basal coelurosaurs and, more specifically, a mainly Gondwanan radiation of tyrannosauroids. Coelurosaur affinities have been supported by several more recent analyses using new material (Porfiri et al. 2014; Aranciaga Rolando et al. 2019), but the exact position of this interesting clade is still uncertain (e.g. Apesteguía et al. 2016; Coria and Currie 2016; Novas et al. 2016).

Within basal Coelurosauria, an important early clade of uncertain phylogenetic position are the Compsognathidae. This clade might be an early radiation of coelurosaurian theropods that originated in the Late Jurassic at the latest and reached a wide distribution in the Early Cretaceous, although the exact taxonomic composition of the group is also still debated. One problem with the current concept of the Compsognathidae might be that several taxa included in this clade are juveniles (e.g. *Juravenator*: Chiappe and Göhlich 2010; *Scipionyx*: Dal Sasso and Maganuco 2011), and some of the characteristics supporting compsognathid monophyly might be ontogenetically variable (see Rauhut et al. 2012). Thus, *Compsognathus* and its closest relatives are found as the most basal larger clade of coelurosaurs in some analyses (e.g. Rauhut 2003; Holtz et al. 2004; Rauhut et al. 2010; Cau 2018), as sister taxon to Maniraptoriformes (e.g. Senter 2007; Smith et al. 2007; Xu et al. 2009, Xu et al. 2015; Choiniere et al. 2014; Rauhut et al. 2019), or as basal Maniraptora (e.g. Choiniere et al. 2010; Foth et al. 2014).

Another problematic clade within coelurosaurs are the Alvarezsauroidea. Originally thought to be basal birds (e.g. Perle et al. 1994; Novas 1996; Chiappe et al. 1998; Chiappe 2002), most more recent phylogenies have placed these animals as basal maniraptorans (e.g. Clark et al. 2002; Senter 2007; Choiniere et al. 2010, 2014; Foth et al. 2014; Xu et al. 2018), and Sereno (1999) suggested that alvarezsauroids were the sister taxon to Ornithomimosauria. The problem with alvarezsauroids was that most first discoveries of this clade were of highly derived members that have a very aberrant morphology (e.g. Perle et al. 1994; Novas 1996, 1997; Chiappe et al. 1998), making their placement within theropods problematic. However, with the recent discovery of more basal forms (Choiniere et al. 2010, 2014; Xu et al. 2018), our understanding of alvarezsauroid anatomy, phylogeny, and evolution is rapidly improving, and a consensus of this clade being basal maniraptorans seems to be emerging, although the exact phylogenetic position at the base of Maniraptora remains unstable.

A similar problem has affected the Therizinosauria. As with alvarezsauroids, the first discoveries of therizinosaurs were of highly derived forms (e.g. Maleev 1954; Perle 1979, 1982; Barsbold and Perle 1980), and even the placement in one of the principal clades of dinosaurs of these animals was at first unclear (see Paul 1984). Only with the discovery of more basal forms did the theropod affinities of therizinosaurs become firmly established (Russell and Dong 1993). Since then, this clade has repeatedly been found as the sister taxon of the Oviraptorosauria within the Maniraptora (e.g. Makovicky & Sues 1998; Holtz 1998; Clark et al. 2002; Rauhut 2003; Holtz et al. 2004; Xu et al. 2007; Choiniere et al. 2014; Cau 2018), although most recent analyses have favoured a more basal position of therizinosaurs, outside the Pennaraptora (Oviraptorosauria + Paraves; e.g. Senter 2007; Zanno 2010; Xu et al. 2011, 2017; Turner et al. 2012; Agnolín and Novas 2013; Brusatte et al. 2014; Foth et al. 2014; Foth and Rauhut 2017; Hu et al. 2018). This problem remains currently unresolved, as highlighted by the analyses presented by Rauhut et al. (2019): whereas an unweighted analysis found a Therizinosauria-Oviraptorosauria clade, an implicit weight analysis of the same data matrix found the therizinosaurs outside the Pennaraptora.

Another only recently recognized clade of interesting, bird-like theropods are the Scansoriopterygidae. These animals are so far only known from the early Late Jurassic Yanliao Biota of north-eastern China, from where at least four different taxa have been described (Zhang et al. 2002, 2008; Xu et al. 2015; Wang et al. 2019). Scansoriopterygids are small, bird-like theropods that included volant forms with membranous wings (Xu et al. 2015; Wang et al. 2019). The clade was originally regarded as a radiation of basal avialans (Zhang et al. 2008), and this has been supported by some subsequent analyses (e.g. Xu et al. 2011; Foth et al. 2014), whereas a number of more recent analyses regarded scansoriopterygids as basal paravians, outside a Avialae-Deinonychosauria split (Xu et al. 2015; Wang et al. 2019). On the other hand, Agnolín & Novas (2013) recovered scansoriopterygids as basal oviraptorosaurs, which was supported by Brusatte et al. (2014) and Rauhut et al. (2019). Thus, more finds and more detailed studies of the known taxa are necessary to resolve the relationships of these interesting animals.

Another area of conflict concerns basal paravian phylogeny. In most analyses of coelurosaur interrelationships, Troodontidae and Dromaeosauridae are recovered as sister groups, forming the monophyletic Deinonychosauria (e.g. Sereno 1997, 1999; Holtz 1998; Clark et al. 2002; Rauhut 2003; Senter 2007; Turner et al. 2012; Rauhut et al. 2019). In contrast, several recent analyses found Troodontidae and Avialae as sister taxa to the exclusion of Dromaeosauridae (e.g., Godefroit et al. 2013b; Choiniere et al. 2014; Foth et al. 2014; Cau 2018). This uncertainty reflects the great similarity of many of these bird-like dinosaurs and is mirrored by the uncertain phylogenetic position of some other basal paravians, such as the anchiornithids, which are considered to be troodontids in some analyses (e.g. Hu et al. 2009; Turner et al. 2012; Godefroit et al. 2013b; Brusatte et al. 2014), basal deinonychosaurs (e.g. Xu et al. 2011, 2015; Wang et al. 2019), or avialans more basal than *Archaeopteryx* (e.g. Agnolín and Novas 2013; Godefroit et al. 2013a; Foth et al. 2014; Rauhut et al. 2019), apart from other occasional placements within Paraves (e.g. as basal taxon outside the Deinonychosauria-Avialae split; Lefèvre et al. 2017).

These different phylogenetic hypotheses also affected the phylogenetic position of *Archaeopteryx*, which until today represents a yardstick for early bird evolution. Traditionally, *Archaeopteryx* is a basal member of the Avialae (e.g. Sereno 1999; Rauhut 2003, Senter 2007; Turner et al. 2012; Brusatte et al. 2014; see above), while some recent studies placed *Archaeopteryx* together with *Anchiornis* outside Avialae as sister taxon to Deinonychosauria (e.g. Xu et al. 2011, 2015; Xu and Pol 2014; Godefroit et al. 2013a; Hu et al. 2018; Wang et al. 2019). On the other hand, due to a high level of homoplasy in early Paraves, single studies classified *Rahonavis*, *Balaur*, or *Microraptor* to be basal Avialae (e.g. Agnolín and Novas 2011, 2013; Cau et al. 2015, 2017; Foth and Rauhut 2017; Lefèvre et al. 2017), while they are traditionally placed within Dromaeosauridae (see Turner et al. 2012; Brusatte et al. 2013). Further controversies remain regarding the exact relationships between Jeholornithidae, Sapeornithidae, Confuciusornithidae, and more derived Ornithothoraces. Many studies found the long-tailed Jeholornithidae to be the sister taxon of a monophyletic clade Pygostylia containing the short-tailed Sapeornithidae and Confuciusornithidae and more derived Ornithothoraces (e.g. Zhou et al. 2008; O'Connor et al. 2009, 2013; Zhang et al. 2014; Wang et al. 2015). This relationship represents the most parsimonious explanation for the tail evolution in the stem line of birds. However, other analyses found Sapeornithidae to be more basal than Jeholornithidae (e.g. Zhou et al. 2010; Turner et al. 2012; Cau et al. 2017; Foth and Rauhut 2017; Agnolín et al. 2019), which is more parsimonious, explaining the evolution of the pectoral girdle and sternum. In contrast to basal Avialae, the phylogenetic relationship of the main clades within Ornithothoraces are well supported by various phylogenetic analyses (e.g., Clarke et al. 2006; Zhou et al. 2008; Wang et al. 2015; O'Connor et al. 2016).

3.4 The Occurrence of Feathers in the Fossil Record of Theropod Dinosaurs

As feathers are not skeletal tissues, they might only be preserved under exceptional circumstances, such as in Konservat-Lagerstätten. It is therefore not surprising that the fossil record of feathers in general is rather poor, and this is especially also the case for Mesozoic theropods.

The first record of a feather from the Mesozoic was the original isolated feather of *Archaeopteryx*, which von Meyer (1861a) first announced. More importantly, a skeletal specimen including feather impressions was found in the same year (von Meyer 1861b), and it was primarily the feather impressions that led to the identification of this animal as a bird (Owen 1863). The importance of the feather impressions in this iconic fossil was such that their authenticity was questioned as recently as 1985 (Hoyle and Wickramasinghe 1985), although there is no reasonable doubt that these structures are real (see Rietschel 1985; Charig et al. 1986; Wellnhofer 2008).

For many decades, the feathers of *Archaeopteryx* were the only known fossil feathers from the Mesozoic. More importantly, although feathers were later occasionally found as carbonized traces in exceptional lagerstätten or preserved in amber (see Davis and Briggs 1995; Kellner 2002; Prado et al. 2016), these were isolated finds of feathers, which did not allow a taxonomic identification of the animal that they belonged to, and thus provided only limited data on the evolution of these structures. This changed drastically with the discovery of abundant feathered dinosaurs and early birds in the Lower Cretaceous Jehol Biota of China in the 1990s (e.g. Hou et al. 1995; Chen et al. 1998; Ji et al. 1998; Xu et al. 1999a, b). Since then, feathers have been reported from these deposits and the older (late Middle to early Late Jurassic) Yanliao biota for all major clades of maniraptoran theropods (see Xu 2020), and for some more basal coelurosaurian taxa, such as the compsognathid *Sinosauropteryx* (Chen et al. 1998) and the tyrannosauroid *Yutyrannus* (Xu et al. 2012).

Reports of feathers in non-coelurosaurian theropods, in contrast, are still exceedingly rare, mainly because no such taxa have been found in these exceptional lagerstätten. Most reports of the possible presence of feathers in non-coelurosaurian theropods are therefore debatable. Gierlinski (1997) reported feather-like impressions in a theropod resting trace from the Early Jurassic of North America, and this interpretation was more recently supported by Kundrát (2004). As there are no coelurosaurian theropods known from the Early Jurassic, these traces might not only represent the oldest evidence of feathers, but also indirect evidence for non-coelurosaurian feathers. Although Kundrát (2004) made a good case for these imprints to represent feathers, some uncertainty remains, and another problem is, of course, the difficulty in identifying the trackmaker.

Another indirect evidence for feathers in a non-coelurosaurian theropod was presented by Ortega et al. (2010), who reported bumps on the ulna of the carcharodontosaurid *Concavenator*, which they interpreted as feather quill knobs. However, as argued by Foth et al. (2014), these knobs are in a different position than the quill knobs found in some modern volant birds and are irregularly spaced, casting doubt on this interpretation. Although Cuesta et al. (2018), in a study of probable forelimb myology of *Concavenator*, did not find any evidence for these knobs representing attachments of interosseous ligaments, as suggested by Foth et al. (2014), their significance remains controversial.

Probably the best evidence of feathers in a non-coelurosaurian theropod is provided by the exceptionally preserved holotype specimen of *Sciurumimus albersdoerferi* from the Kimmeridgian Torleite Formation in southern Germany (Rauhut et al. 2012). This specimen does have abundant filament impressions above the base of the tail and shows numerous phosphatized filaments in different parts of the body under UV light (Rauhut et al. 2012; see also Foth et al. 2020). In the case of *Sciurumimus*, the question is thus not so much the presence of protofeathers, but there is some uncertainty about its phylogenetic position. Rauhut et al. (2012) recovered this taxon as a basal tetanuran, and probably a megalosauroid, based on an analysis of this taxon in three different phylogenetic matrices. Thus, in this hypothesis, the presence of protofeathers in this taxon extends the record of these structures to at least the base of Tetanurae. However, Godefroit et al. (2013a) recovered *Sciurumimus* as a basal coelurosaur, in which case the origin of protofeathers might well lie within this clade. However, as argued by Rauhut et al. (2012), we strongly suspect that possible coelurosaurian characters in *Sciurumimus* are due to the very early ontogenetic stage of the only known specimen, as heterochrony seems to have played an important role in the evolution of coelurosaurian theropods (e.g. Bhullar et al. 2012; Foth et al. 2016), and thus consider a basal tetanuran placement of this taxon to be more likely.

3.5 Conclusions and Outlook

Although there is still much debate about many details of the phylogenetic relationships of theropod dinosaurs and thus the origin of birds, there is a remarkable consensus on the backbone structure of the family tree of the ancestors of birds and the relative hierarchical placement of almost all major clades that constitute this tree (Fig. 3.3). Thus, disregarding the more problematic (and often smaller) groups, all recent analyses agree that avialans ("birds") are members of the Paraves, together with dromaeosaurids and troodontids; Paraves are a subclade of Maniraptora, together with oviraptorosaurs, therizinosaurs, and alvarezsauroids; Maniraptora is part of Maniraptoriformes, together with ornithomimosaurs; Maniraptoriformes is a subclade of Coelurosauria, together with Tyrannosauroidea; Coelurosauria is part of Tetanurae, together with Megalosauroidea and Allosauroidea; Tetanurae is part of Neotheropoda, together with Ceratosauria and Coelophysoidea (and probably part of Averostra together with Ceratosauria). Thus, this phylogenetic hierarchy forms a solid base for improving our understanding of the evolution of the unique avian body plan (Fig. 3.4), as originally lined out by Gauthier (1986) and elaborated in more detail recently by Brusatte et al. (2014), Xu et al. (2014), and Cau (2018). With the discovery of abundant feathered dinosaurs, mainly from the Late Jurassic and Early Cretaceous of China, but also from other areas, the phylogenetic hierarchy outlined above helps us to extend such evolutionary scenarios to the evolution of feathers (Xu 2006, 2020; Xu and Guo 2009) and novel insights into the possible functional context in which these structures evolved.

New discoveries and more detailed studies of key taxa will certainly help to further improve our understanding of theropod phylogeny. However, there seems to be a trend to increase "birdiness" in several lineages independently, including possible multiple experiments with flight in derived coelurosaurian theropods (e.g. Xu et al. 2003; Foth et al. 2014; Wang et al. 2019). Together with the incomplete preservation of many remains, this marked parallelism—which is also seen in other parts of the theropod family tree (e.g. Rauhut and Pol 2019)—will make detailed reconstructions of the phylogenetic relationships at the origin of birds, in which the relationships of all relevant taxa can be established, difficult, if not impossible. However, such a detailed understanding might not be necessary to improve our understanding of the origin of birds, as the general agreement on the placement of most taxa in the hierarchy leading towards birds provides ample anatomical and functional data for hypothesis formulation and testing.

Likewise, new discoveries of feathered dinosaurs, not only in the now famous Yanliao and Jehol Biota, but also other lagerstätten, such as the limestones of the Solnhofen Archipelago (Chiappe and Göhlich 2010; Rauhut et al. 2012) or in Mesozoic amber (e.g. Xing et al. 2016a, b, 2019), will certainly improve our understanding of feather diversity, evolution, and function. Furthermore, the use of novel techniques, such as laser-stimulated fluorescence (e.g. Kaye et al. 2015, 2019), new microscopic or chemical techniques (Schweitzer et al. 1999, 2008), investigations of the role of melanosomes for both feather colouring and structure (see Smithwick and Vinther 2020, and references therein), and further improvements of methods such as UV photography have great potential to provide new insights into the preservation and structure of feathers in fossil taxa.

Acknowledgements This work was supported by the Volkswagen Foundation under grant I/84 640 (to OR) and the Swiss National Science Foundation under grant PZ00P2_174040 (to CF).

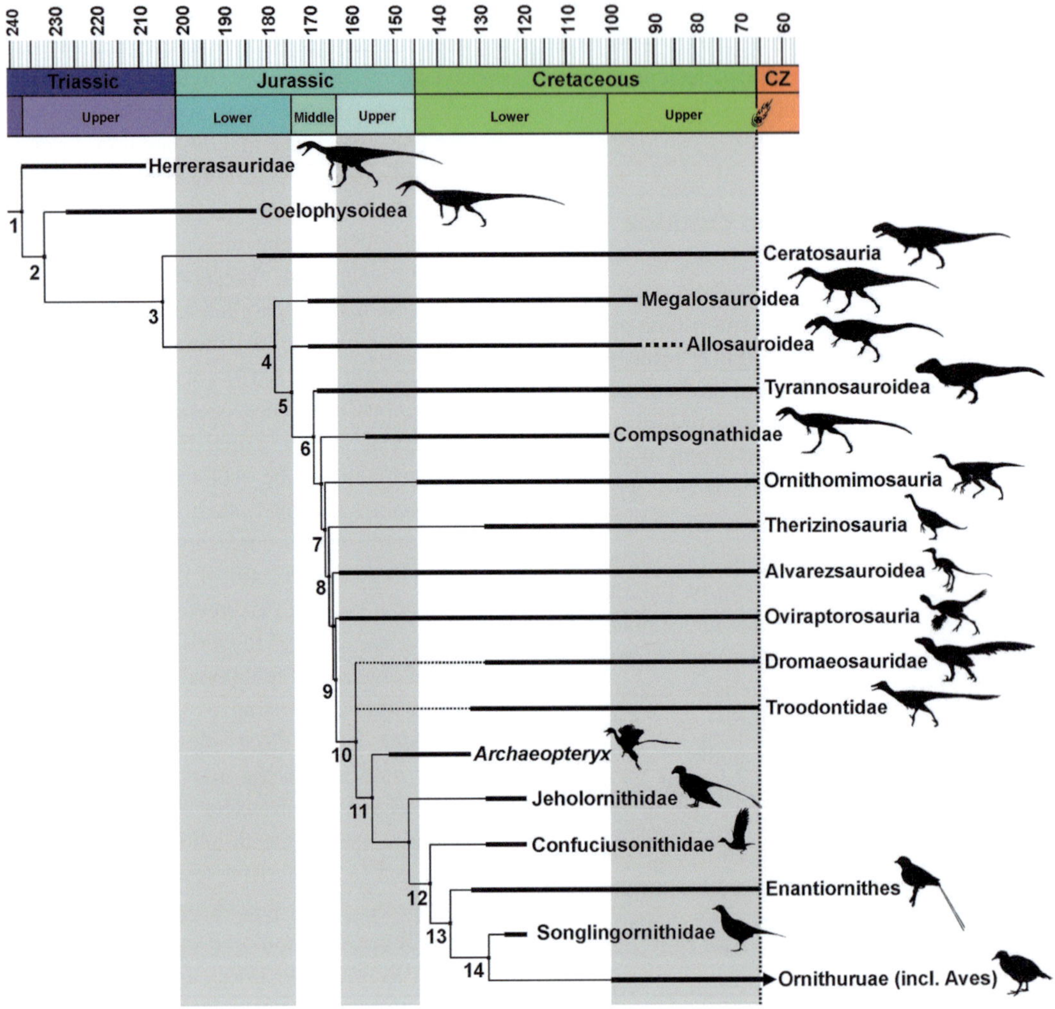

Fig. 3.4 Simplified time-calibrated theropod phylogeny showing the major events of character evolution along the theropod-bird transition. **(1) Theropoda**: bipedal locomotion*, initial vertebral pneumatization and ventilatory air sacs*, increased metabolic rate*, thin bone walls, four-fingered hand (*plesiomorphic characters); **(2) Neotheropoda**: widely arched furcula; (3) Averostra: extended vertebral pneumatization; **(4) Tetanurae/ Orionides**: three-fingered hand with initial lateral folding mechanism; **(5) Avetheropoda**; **(6) Coelurosauria**; **(7) Maniraptoriformes**; **(8) Maniraptora**: semilunate carpal with partial lateral folding mechanism; **(9) Pennaraptora**: cerebral expansion, costosternal ventilator pump, V-shaped furcula, initial forelimb-flapping capabilities, increased manual lateral folding mechanism, two-layered eggshells, brooding behaviour; **(10) Paraves/ Eumaniraptora**: extreme miniaturization, elaborated visual cortex, forelimb elongation and thickening, asymmetric egg shape, egg shells with low porosity and without ornamentation, potential third (external) layer in eggshell; **(11) Avialae**: aerial locomotion, asymmetric pennaceous feathers, lateral facing glenoid, forelimb elongation and thickening with increased flapping capabilities, shortened bony tail; **(12) Pygostylia**: crop, dorsolateral facing glenoid, strut-like coracoid, U-shaped furcula, fused sternum, rod-like pygostyle, posterior pubis orientation, one active ovary and oviduct; **(13) Ornithothoraces**: alula wing feather, sternal keel, synsacrum with 8 or more vertebrae; **(14) Euornithes/Ornithuromorpha**: kinetic skull, full forelimb-flapping capabilities and manual lateral folding mechanism, fused carpometacarpus, fusion of pelvic bones, metatarsal fully fused, increased egg size, ploughshare-shaped pygostyle. All silhouettes taken from (www.phylopic.org)

References

Agnolín FL, Novas FE (2013) Avian ancestors. A review of the phylogenetic relationships of the theropods Unenlagiidae, Microraptoria, Anchiornis and Scansoriopterygidae. Springer, Dordrecht

Agnolín FL, Motta MJ, Egli FB, Lo Coco G, Novas FE (2019) Paravian phylogeny and the dinosaur-bird transition: an overview. Front Earth Sci 6:252

Agnolín FL, Novas FE (2011) Unenlagiid theropods: are they members of the Dromaeosauridae (Theropoda, Maniraptora)? Anais da Academia Brasileira de Ciências 83:117–162

Allain R (2002) Discovery of megalosaur (Dinosauria, Theropoda) in the middle Bathonian of Normandy (France) and its implications for the phylogeny of basal Tetanurae. J Vertebr Paleontol 22:548–563

Allain R, Tykoski RS, Aquesbi N, Jalil N-E, Manbaron M, Russell DA, Taquet P (2007) An Abelisauroid (Dinosauria: Theropoda) from the early Jurassic of the High Atlas Mountains, Morocco, and the radiation of Ceratosaurs. J Vertebr Paleontol 27:610–624

Apesteguía S, Smith ND, Juárez Valieri R, Makovicky PJ (2016) An unusual new theropod with a didactyl manus from the Upper Cretaceous of Patagonia, Argentina. PLoS One 11:e0157793

Aranciaga Rolando AM, Novas FE, Agnolín FL (2019) A reanalysis of *Murusraptor barrosaensis* Coria & Currie (2016) affords new evidence about the phylogenetical relationships of Megaraptora. Cretac Res 99:104–127

Azuma Y, Currie PJ (2000) A new carnosaur (Dinosauria: Theropoda) from the Lower Cretaceous of Japan. Can J Earth Sci 37:1735–1753

Barsbold R, Perle A (1980) Segnosauria, a new infraoder of carnivorous dinosaurs. Acta Palaeontologica Polonica Pol 25:187–195

Benson RBJ (2010) A description of *Megalosaurus bucklandii* (Dinosauria: Theropoda) from the Bathonian of the UK and the relationships of Middle Jurassic theropods. Zool J Linnean Soc 158:882–935

Benson RBJ, Hunt G, Carrano MT, Campione N, Mannion P (2018) Cope's rule and the adaptive landscape of dinosaur body size evolution. Palaeontology 61:13–48

Benson RBJ, Carrano MT, Brusatte SL (2010) A new clade of archaic large-bodied predatory dinosaurs (Theropoda: Allosauroidea) that survived to the latest Mesozoic. Naturwissenschaften 97:71–78

Bhullar B-AS, Marugán-Lobón J, Racimo F, Bever GS, Rowe TB, Norell MA, Abzhanov A (2012) Birds have paedomorphic dinosaur skulls. Nature 487:223–226

Broom R (1913) On the South-African pseudosuchian *Euparkeria* and allied genera. Proc Zool Soc London 83:619–633

Brown JW, Rest JS, Garcia-Moreno J, Sorenson MD, Mindell DP (2008) Strong mitochondrial DNA support for a Cretaceous origin of modern avian lineages. BMC Biol 6:1–18

Brusatte SL, Carr TD (2016) The phylogeny and evolutionary history of tyrannosauroid dinosaurs. Sci Rep 6:20252

Brusatte SL, Sereno PC (2008) Phylogeny of Allosauroidea (Dinosauria: Theropoda): comparative analysis and resolution. J Syst Palaeontol 6:155–182

Brusatte SL, Norell MA, Carr TD, Erickson GM, Hutchinson JR, Balanoff AM, Bever GS, Choiniere JN, Makovicky PJ, Xu X (2010) Tyrannosaur paleobiology: new research on ancient exemplar organisms. Science 329:1481–1485

Brusatte SL, Vremir M, Csiki-Sava Z, Turner AH, Watanabe A, Erickson GM, Norell MA (2013) The osteology of *Balaur bondoc*, an island-dwelling dromaeosaurid (Dinosauria: Theropoda) from the Late Cretaceous of Romania. Bull Am Museum Nat Hist 374:1–100

Brusatte SL, Lloyd GT, Wang SC, Norell MA (2014) Gradual assembly of avian body plan culminated in rapid rates of evolution across the dinosaur-bird transition. Curr Biol 24:2386–2392

Brusatte SL, O'Connor JK, Jarvis ED (2015) The origin and diversification of birds. Curr Biol 25:R888–R898

Calvo JO, Porfiri JD, Veralli C, Novas FE, Poblete F (2004) Phylogenetic status of *Megaraptor namunhuaiquii* Novas based on a new specimen from Neuquén, Patagonia, Argentina. Ameghiniana 41:565–575

Carrano MT, Sampson SD, Forster CA (2002) The osteology of *Masiakasaurus knopfleri*, a small Abelisauroid (Dinosauria, Theropoda) from the Late Cretaceous of Madagascar. J Vertebr Paleontol 22:510–534

Carrano MT, Hutchinson JR, Sampson SD (2005) New information on *Segisaurus halli*, a small theropod dinosaur from the Early Jurassic of Arizona. J Vertebr Paleontol 25:835–849

Carrano MT, Benson RBJ, Sampson SD (2012) The phylogeny of Tetanurae (Dinosauria: Theropoda). J Syst Palaeontol 10:211–300

Cau A (2018) The assembly of the avian body plan: a 160-million-year long process. Bollettino della Società Paleontologica Italiana 57:1–25

Cau A, Brougham T, Naish D (2015) The phylogenetic affinities of the bizarre Late Cretaceous Romanian theropod *Balaur bondoc* (Dinosauria, Maniraptora): dromaeosaurid or flightless bird? PeerJ 3:e1032

Cau A, Beyrand V, Voeten FAE, Fernandez V, Tafforeau P, Stein K, Barsbold R, Tsogtbaatar K, Currie PJ, Godefroit P (2017) Synchrotron scanning reveals amphibious ecomorphology in a new clade of bird-like dinosaurs. Nature 552:395–399

Charig AJ, Greenaway F, Milner AC, Walker CA, Whybrow PJ (1986) *Archaeopteryx* is not a forgery. Science 232:622–626

Chen P, Dong Z, Zhen S (1998) An exceptionally well-preserved theropod dinosaur from the Yixian Formation of China. Nature 391:147–152

Chiappe LM (2002) Basal bird phylogeny: problems and solutions. In: Chiappe LM, Witmer LM (eds)

Mesozoic birds: Above the heads of dinosaurs. University of California Press, Berkeley, pp 448–472
Chiappe LM (2009) Downsized dinosaurs: the evolutionary transition to modern birds. Evol Educ Outreach 2:248–256
Chiappe LM, Göhlich UB (2010) Anatomy of *Juravenator starki* (Theropoda: Coelurosauria) from the Late Jurassic of Germany. Neues Jahrbuch für Geologie und Paläontologie, Abhandlungen 258:257–296
Chiappe LM, Norell MA, Clark JM (1998) The skull of a relative of the stem-group bird *Mononykus*. Nature 392:275–278
Choiniere JN, Xu X, Clark JM, Forster CA, Guo Y, Han F (2010) A basal alvarezsauroid theropod from the Early Jurassic of Xinjiang, China. Science 327:571–574
Choiniere JN, Clark JM, Forster CA, Norell MA, Eberth DA, Erickson GM, Chu H, Xu X (2014) A juvenile specimen of a new coelurosaur (Dinosauria: Theropoda) from the Middle–Late Jurassic Shishugou Formation of Xinjiang, People's Republic of China. J Syst Palaeontol 12:177–215
Cieri RL, Farmer CG (2016) Unidirectional pulmonary airflow in vertebrates: a review of structure, function, and evolution. J Comp Psychol B 186:541–552
Clark JM, Norell MA, Makovicky PJ (2002) Cladistic approaches to the relationships of birds to other theropod dinosaurs. In: Chiappe LM, Witmer LM (eds) Mesozoic birds: above the heads of dinosaurs. University of California Press, Berkeley, pp 31–61
Clarke JA, Tambussi CP, Noriega JI, Erickson GM, Ketcham RA (2005) Definitive fossil evidence for the extant avian radiation in the Cretaceous. Nature 433:305–308
Clarke JA, Zhou Z, Zhang F (2006) Insight into the evolution of avian flight from a new clade of Early Cretaceous ornithurines from China and the morphology of *Yixianornis grabaui*. J Anat 208:287–308
Colbert EH (1964) Relationships of the saurischian dinosaurs. Am Museum Novitates 2181:1–24
Coria RA, Currie PJ (2016) A new megaraptoran dinosaur (Dinosauria, Theropoda, Megaraptoridae) from the Late Cretaceous of Patagonia. PLoS One 11:e015797
Cuesta E, Ortega F, Sanz JL (2018) Appendicular osteology of *Concavenator corcovatus* (Theropoda; Carcharodontosauridae; Early Cretaceous; Spain). J Vertebr Paleontol 38:e1485153
Dal Sasso C, Maganuco S (2011) *Scipionyx samniticus* (Theropoda: Compsognathidae) from the Lower Cretaceous of Italy. Memorie della Società Italiana di Scienze Naturali e del Museo Civico di Storia Naturale di Milano 37:1–281
Dal Sasso C, Maganuco S, Cau A (2018) The oldest ceratosaurian (Dinosauria: Theropoda), from the Lower Jurassic of Italy, sheds light on the evolution of the three-fingered hand of birds. PeerJ 6:e5976
Darwin CR (1859) On the origin of species by means of natural selection. John Murray, London
Davis PG, Briggs DEG (1995) Fossilization of feathers. Geology 23:783–786
Ezcurra MD (2017) A new early coelophysoid neotheropod from the Late Triassic of northwestern Argentina. Ameghiniana 54:506–538
Ezcurra MD, Brusatte SL (2011) Taxonomic and phylogenetic reassessment of the early neotheropod dinosaur C *amposaurus arizonensis* from the Late Triassic of North America. Palaeontology 54:763–772
Ezcurra MD, Cuny G (2007) The coelophysoid *Lophostropheus airelensis*, gen. nov.: a review of the systematics of '*Liliensternus*' *airelensis* from the Triassic-Jurassic outcrops of Normandy (France). J Vertebr Paleontol 27:73–86
Farmer CG, Sanders K (2010) Unidirectional airflow in the lungs of alligators. Science 327:338–340
Feduccia A (1996) The origin and evolution of birds. Yale University Press, New Haven
Feduccia A (2013) Bird origins anew. Auk 130:1–12
Forster CA (1999) Gondwanan dinosaur evolution and biogeographic analysis. J Afr Earth Sci 28:169–185
Foth C, Rauhut OWM (2017) Re-evaluation of the Haarlem *Archaeopteryx* and the radiation of maniraptoran theropod dinosaurs. BMC Evol Biol 17:236
Foth C, Tischlinger H, Rauhut OWM (2014) New specimen of *Archaeopteryx* provides insights into the evolution of pennaceous feathers. Nature 511:79–82
Foth C, Hedrick BP, Ezcurra MD (2016) Cranial ontogenetic variation in early saurischians and the role of heterochrony in the diversification of predatory dinosaurs. PeerJ 4:e1589
Foth C, Haug C, Haug JT, Tischlinger H, Rauhut OWM (2020) Two of a feather: a comparison of the preserved integument in the juvenile theropod dinosaurs *Sciurumimus* and *Juravenator* from the Kimmeridgian Torleite Formation of southern Germany. In: Foth C, Rauhut OWM (eds) The evolution of feathers: from their origin to the present. Springer, Cham
Gauthier JA (1986) Saurischian monophyly and the origin of birds. Memoirs Calif Acad Sci 8:1–55
Gierlinski G (1997) What type of feather could nonavian dinosaur have, according to an Early Jurassic ichnological evidence from Massachusetts? Przeglad Geologiczny 54:419–422
Godefroit P, Demuynck H, Dyke GJ, Hu D, Escuillié F, Claeys P (2013a) Reduced plumage and flight ability of a new Jurassic paravian theropod from China. Nat Commun 4:1394
Godefroit P, Cau A, Hu D, Escuillié F, Wu W, Dyke GJ (2013b) A Jurassic avialan dinosaur from China resolves the early phylogenetic history of birds. Nature 498:359–362
Godefroit P, Sinitsa SM, Dhouailly D, Bolotsky YL, Sizov AV, McNamara ME, Benton MJ, Spagna P (2014) A Jurassic ornithischian dinosaur from Siberia with both feathers and scales. Science 345:451–455
Godefroit P, Sinitsa SM, Cincotta A, McNamara ME, Reshetova SA, Dhouailly D (2020) Integumentary structures in *Kulindadromeus zabaikalicus*, a Basal Neornithischian Dinosaur from the Jurassic of Siberia.

In: Foth C, Rauhut OWM (eds) The evolution of feathers: from their origin to the present. Springer, Cham

Heilmann G (1926) The origin of birds. Witherby, London

Hennig W (1966) Phylogenetic systematics. University of Illinois Press, Urbana

Hocknull SA, White MA, Tischler TR, Cook AC, Calleja ND, Sloan T, Elliot DA (2009) New Mid-Cretaceous (Latest Albian) Dinosaurs from Winton, Queensland, Australia. PLoS One 4:e6190

Holtz TRJ (1994) The phylogenetic position of the Tyrannosauridae: implications for theropod systematics. J Paleontol 68:1100–1117

Holtz TR (1996) Phylogenetic taxonomy of the Coelurosauria (Dinosauria: Theropoda). J Paleontol 70:536–538

Holtz TRJ (1998) A new phylogeny of the carnivorous dinosaurs. Gaia 15:5–61

Holtz TRJ, Molnar RE, Currie PJ (2004) Basal Tetanurae. In: Weishampel DB, Dodson P, Osmólska H (eds) The Dinosauria. University of California Press, Berkeley, pp 71–110

Hone DWE, Holtz TRJ (2017) A century of spinosaurs—a review and revision of the Spinosauridae with comments on their ecology. Acta Geologica Sinica 91:1120–1132

Hou L, Zhou Z, Martin LD, Feduccia A (1995) A beaked bird from the Jurassic of China. Nature 377:616–618

Hoyle F, Wickramasinghe C (1985) *Archaeopteryx*, the primordial bird: a case of fossil forgery. Christopher Davies, Swansea

Hu D, Hou L, Zhang L, Xu X (2009) A pre- *Archaeopteryx* troodontid theropod from China with long feathers on the metatarsus. Nature 461:640–643

Hu D, Clarke JA, Eliason CM, Qiu R, Li Q, Shawkey MD, Zhao C, D'Alba L, Jiang J, Xu X (2018) A bony-crested Jurassic dinosaur with evidence of iridescent plumage highlights complexity in early paravian evolution. Nat Commun 9:217

Huxley TH (1868) On the animals which are most nearly intermediate between birds and reptiles. Ann Mag Nat Hist 4:66–75

Ji Q, Currie PJ, Norell MA, Ji S (1998) Two feathered dinosaurs from northeastern China. Nature 393:753–761

Kaye TG, Falk AR, Pittman M, Sereno PC, Martin LD, Burnham DA, Gong E, Xu X, Wang Y (2015) Laser-stimulated fluorescence in paleontology. PLoS One 10: e0125923

Kaye TG, Pittman M, Marugán-Lobón J, Martín-Abad H, Sanz JL, Buscalioni AD (2019) Fully fledged enantiornithine hatchling revealed by laser-stimulated fluorescence supports precocial nesting behavior. Sci Rep 9:5006

Kellner AWA (2002) A review of avian Mesozoic fossil feathers. In: Chiappe LM, Witmer LM (eds) Mesozoic birds: above the heads of Dinosaurs. University of California Press, Berkeley, pp 389–404

Kundrát M (2004) When did theropods become feathered? Evidence for pre-*Archaeopteryx* feathery appendages. J Exp Zool (MOL DEV EVOL) 302B:355–364

Langer MC, Rincón AD, Ramezani J, Solórzano A, Rauhut OWM (2014) New dinosaur (Theropoda, stem-Averostra) from the earliest Jurassic of the La Quinta Formation, Venezuelan Andes. R Soc Open Sci 1:140184

Lee Y-N, Barsbold R, Currie PJ, Kobayashi Y, Lee H-J, Godefroit P, Escuillié F, Tsogtbaatar C (2014) Resolving the long-standing enigmas of a giant ornithomimosaur *Deinocheirus mirificus*. Nature 515:257–260

Lefèvre U, Cau A, Cincotta A, Hu D, Chinsamy A, Escuillié F, Godefroit P (2017) A new Jurassic theropod from China documents a transitional step in the macrostructure of feathers. Sci Nat 104:74

Li Z, Zhou Z, Wang M, Clarke JA (2014) A new specimen of large-bodied basal Enantiornithine *Bohaiornis* from the early Cretaceous of China and the inference of feeding ecology in Mesozoic birds. J Paleontol 88:99–108

Lingham-Soliar T (2003a) The dinosaurian origin of feathers: perspectives from dolphin (Cetacea) collagen fibers. Naturwissenschaften 90:563–567

Lingham-Soliar T (2003b) Evolution of birds: ichthyosaur integumental fibers conform to dromaeosaur protofeathers. Naturwissenschaften 90:428–432

Lingham-Soliar T (2010) Dinosaur protofeathers: pushing back the origin of feathers into the Middle Triassic. J Ornithol 151:193–200

Lingham-Soliar T (2012) The evolution of the feather: *Sinosauropteryx*, life, death and preservation of an alleged feathered dinosaur. J Ornithol 153:699–711

Lingham-Soliar T, Feduccia A, Wang X (2007) A new Chinese specimen indicates that 'protofeathers' in the Early Cretaceous theropod dinosaur *Sinosauropteryx* are degraded collagen fibres. Proc R Soc Lond B 274:1823–1829

Makovicky PJ, Sues H-D (1998) Anatomy and phylogenetic relationships of the theropod dinosaur Microvenator celer from the Lower Cretaceous of Montana. Am Museum Novitates 3240:1–27

Makovicky PJ, Norell MA, Clark JM, Rowe TB (2003) Osteology and relationships of *Byronosaurus jaffei* (Theropoda: Troodontidae). Am Museum Novitates 3402:1–32

Maleev EA (1954) Noviy chyeryepoobrazniy yashschyer Mongolii [A new turtle-like reptile from Mongolia]. Priroda 1954:106–108

Martin LD (1983) The origin and the early radiation of birds. In: Brush AH, Clark GA (eds) Perspectives in ornithology. Cambridge University Press, Cambridge, pp 291–338

Martínez RN, Apaldetti C (2017) A late Norian-Rhaetian coelophysid neotheropod (Dinosauria, Saurischia) from the Quebrada Del Barro Formation, northwestern Argentina. Ameghiniana 54:488–505

Martin LD, Stewart JD, Whetstone KN (1980) The origin of birds: structure of the tarsus and teeth. Auk 97:86–93

Maryańska T, Osmólska H, Wolsan M (2002) Avialian status for Oviraptorosauria. Acta Palaeontologica Polonica 47:97–116

Mayr G (2009) Paleogene fossil birds. Springer, Berlin

Mayr G (2010) Response to Lingham-Soliar: Dinosaur protofeathers: pushing back to origin of feathers into the Middle Triassic. J Ornithol 151:523–524

Mayr G (2017) Avian evolution. Wiley, Chichester

Norell MA, Clark JM, Makovicky PJ (2001) Phylogenetic relationships among coelurosaurian theropods. In: Gauthier JA (ed) New perspectives on the origin and early evolution of birds: Proceedings of the international symposium in Honor of John H. Ostrom. Yale University, New Haven, pp 49–67

Norell MA, Clark JM, Turner AH, Makovicky PJ, Barsbold R, Rowe TB (2006) A new droameosaurid theropod from Ukhaa Tolgod (Ömnögov, Mongolia). Am Museum Novitates 3545:1–51

Novas FE (1992) Phylogenetic relationships of the basal dinosaurs, the Herrerasauridae. Palaeontology 35:51–62

Novas FE (1996) Alvarezsauridae, Cretaceous basal birds from Patagonia and Mongolia. Mem Queensland Mus 39:675–702

Novas FE (1997) Anatomy of *Patagonykus puertai* (Theropoda, Avialae, Alvarezsauridae), from the Late Cretaceous of Patagonia. J Vertebr Paleontol 17:137–166

Novas FE (1998) *Megaraptor namunhuaiquii*, gen. et so. nov., a large-clawed, Late Cretaceous theropod from Patagonia. J Vertebr Paleontol 18:4–9

Novas FE, Agnolín FL, Ezcurra MD, Porfiri JD, Canale JI (2013) Evolution of the carnivorous dinosaurs during the Cretaceous: the evidence from Patagonia. Cretac Res 45:174–215

Novas FE, Salgado L, Suárez M, Agnolín FL, Ezcurra MD, Chimento NR, de al Cruz R, Isasi MP, Vargas AO, Rubular-Rogers D (2015) An enigmatic plant-eating theropod from the Late Jurassic period of Chile. Nature 522:331–334

Novas FE, Aranciaga Rolando AM, Agnolín FL (2016) Phylogenetic relationships of the Cretaceous Gondwanan theropods *Megaraptor* and *Australovenator*: the evidence afforded by their manual anatomy. Mem of Mus Vic 74:49–61

O'Connor JK, Wang X, Chiappe LM, Gao C, Meng Q, Cheng X, Liu J (2009) Phylogenetic support for a specialized clade of Cretaceous enantiornithine birds with information from a new species. J Vertebr Paleontol 29:188–204

O'Connor JK, Chiappe LM, Bell A (2011a) Pre-modern birds: avian divergences in the Mesozoic. In: Dyke GJ, Kaiser GW (eds) Living dinosaurs: the evolutionary history of modern birds. John Wiley, Chichester, pp 39–119

O'Connor JK, Chiappe LM, Gao C, Zhao B (2011b) Anatomy of the Early Cretaceous enantiornithine bird *Rapaxavis pani*. Acta Palaeontologica Polonica 56:463–475

O'Connor JK, Zhang Y, Chiappe LM, Meng Q, Quanguo L, Di L (2013) A new enantiornithine from the Yixian Formation with the first recognized avian enamel specialization. J Vertebr Paleontol 33:1–12

O'Connor JK, Wang M, Hu H (2016) A new ornithuromorph (Aves) with an elongate rostrum from the Jehol Biota, and the early evolution of rostralization in birds. J Syst Palaeontol 14:939–948

Ortega F, Escaso F, Sanz JL (2010) A bizarre, humped Carcharodontosauria (Theropoda) from the Lower Cretaceous of Spain. Nature 467:203–206

Ostrom JH (1969a) A new theropod dinosaur from the Lower Cretaceous of Montana. Postilla 128:1–17

Ostrom JH (1969b) Osteology of *Deinonychus antirrhopus*, an unusual theropod from the Lower Cretaceous of Montana. Peabody Museum Nat Hist Bull 30:1–165

Ostrom JH (1970) *Archaeopteryx*: notice of a 'new' specimen. Science 170:537–538

Ostrom JH (1972) Description of the *Archaeopteryx* specimen in the Teyler Museum, Haarlem. Proceedings Koninklijk Nederlandse Akademie van Wetenschappen, B 75:289–305

Ostrom JH (1973) The ancestry of birds. Nature 242:136

Ostrom JH (1976) *Archaeopteryx* and the origin of birds. Biol J Linnean Soc 8:91–182

Owen R (1863) On the *Archaeopteryx* of von Meyer, with a description of the fossil remains of a long-tailed species from the lithographic stone of Solenhofen. Philos Trans R Soc Lond 153:33–47

Paul GS (1984) The segnosaurian dinosaurs: relics of the prosauropod-ornithischian transition. J Vertebr Paleontol 4:507–515

Paul GS (2002) Dinosaurs of the air: the evolution and loss of flight in dinosaurs and birds. The John Hopkins University Press, Baltimore

Perle A (1979) Segnosauridae - a new family of theropods from the Late Cretaceous of Mongolia. Trudy Sovmestnay Sovetsko-Mongolskay Paleontologiyeskay Ekspeditsiy 8:45–55

Perle A (1982) A hind limb of *Therizinosaurus* from the Upper Cretaceous of Mongolia. Probl in Mongolian Geol 5:94–98

Perle A, Chiappe LM, Barsbold R, Clark JM, Norell MA (1994) Skeletal morphology of *Mononykus olecranus* (Theropoda: Avialae) from the Late Cretaceous of Mongolia. Am Mus Novit 3105:1–19

Pol D, Rauhut OWM (2012) A Middle Jurassic abelisaurid from Patagonia and the early diversification of theropod dinosaurs. Proc R Soc B 279:3170–3175

Porfiri JD, Novas FE, Calvo JO, Agnolín FL, Ezcurra MD, Cerda IA (2014) Juvenile specimen of *Megaraptor* (Dinosauria, Theropoda) sheds light about tyrannosauroid radiation. Cretac Res 51:35–55

Prado GMEM, Anelli LE, Petri S, Romero GR (2016) New occurrences of fossilized feathers: systematics and taphonomy of the Santana Formation of the Araripe Basin (Cretaceous), NE, Brazil. PeerJ 4:e1916

Prum RO (2002) Perspectives in ornithology. Why ornithologist should care about the theropod origin of birds. Auk 119:1–17

Prum RO (2003) Are current critiques of the theropod origin of birds science? Rebuttal to Feduccia (2002). Auk 120:550–561

Prum RO, Berv JS, Dornburg A, Field DJ, Townsend JP, Moriarty Lemmon E, Lemmon AR (2015) A comprehensive phylogeny of birds (Aves) using targeted next-generation DNA sequencing. Nature 526:569–573

Rauhut OWM (1998) *Elaphrosaurus bambergi* and the early evolution of theropod dinosaurs. J Vertebr Paleontol 18:71A

Rauhut OWM (2003) The interrelationships and evolution of basal theropod dinosaurs. Spec Pap Palaeontol 69:1–213

Rauhut OWM, Pol D (2019) Probable basal allosauroid from the early Middle Jurassic Cañadón Asfalto Formation of Argentina highlights phylogenetic uncertainty in tetanuran theropod dinosaurs. Sci Rep 9:18826

Rauhut OWM, Carrano MT (2016) The theropod dinosaur *Elaphrosaurus bambergi* Janensch, 1920, from the Late Jurassic of Tendaguru, Tanzania. Zool J Linnean Soc 178:546–610

Rauhut OWM, Xu X (2005) The small theropod dinosaurs *Tugulusaurus* and *Phaedrolosaurus* from the Early Cretaceous of Xinjiang, China. J Vertebr Paleontol 25:107–118

Rauhut OWM, Milner AC, Moore-Fay S (2010) Cranial osteology and phylogenetic position of the theropod dinosaur *Proceratosaurus bradleyi* (Woodward, 1910) from the Middle Jurassic of England. Zool J Linnean Soc 158:155–195

Rauhut OWM, Foth C, Tischlinger H, Norell MA (2012) Exceptionally preserved juvenile megalosauroid theropod dinosaur with filamentous integument from the Late Jurassic of Germany. Proc Natl Acad Sci USA 109:11746–11751

Rauhut OWM, Hübner TR, Lanser K-P (2016) A new megalosaurid theropod dinosaur from the late Middle Jurassic (Callovian) of north-western Germany: implications for theropod evolution and faunal turnover in the Jurassic. Palaeontologia Electronica 19:26A

Rauhut OWM, Tischlinger H, Foth C (2019) A non-archaeopterygid avialan theropod from the Late Jurassic of southern Germany. eLife 8:e43789

Rietschel S (1985) Feathers and wings of *Archaeopteryx* and the question of her fligth ability. In: Hecht MK, Ostrom JH, Viohl G, Wellnhofer P (eds) The beginnings of birds. Freunde des Jura-Museums, Eichstätt, pp 251–260

Russell DA, Dong Z (1993) The affinities of a new theropod from the Alxa Desert, Inner Mongolia, People's Republic of China. Can J Earth Sci 30:2107–2127

Schachner ER, Hutchinson JR, Farmer CG (2013) Pulmonary anatomy in the Nile crocodile and the evolution of unidirectional airflow in Archosauria. PeerJ 1:e60

Schweitzer MH, Watt JA, Avci R, Knapp LW, Chiappe LM, Norell MA, Marshall M (1999) Beta-keratin specific immunological reactivity in feather-like structures of the Cretaceous alvarezsaurid, *Shuvuuia deserti*. J Exp Zool (MOL DEV EVOL) 285:146–157

Schweitzer MH, Avci R, Collier T, Goodwin MB (2008) Microscopic, chemical and molecular methods for examining fossil perservation. C R Palevol 7:159–184

Senter P (2007) A new look at the phylogeny of Coelurosauria (Dinosauria: Theropoda). J Syst Palaeontol 5:429–463

Sereno PC (1997) The origin and evolution of dinosaurs. Annu Rev Earth Planetary Sci 25:435–489

Sereno PC (1998) A rationale for phylogenetic definitions with application to the higher-level taxonomy of Dinosauria. Neues Jahrbuch für Geologie und Paläontologie, Abhandlungen 210:41–83

Sereno PC (1999) The evolution of dinosaurs. Science 284:2137–2147

Sereno PC, Martinez RN, Wilson JA, Varricchio DJ, Alcober OA, Larsson HCE (2008) Evidence for avian intrathoracic air sacs in a new predatory dinosaur from Argentina. PLoS One 3:e3303

Smith ND, Makovicky PJ, Hammer WR, Currie PJ (2007) Osteology of *Cryolophosaurus ellioti* (Dinosauria: Theropoda) from the Early Jurassic of Antarctica and implications for early theropod evolution. Zool J Linnean Soc 151:377–421

Smith NA, Chiappe LM, Clarke JA, Edwards SV, Nesbitt SJ, Norell MA, Stidham TA, Turner AH, van Tuinen M, Vinther J, Xu X (2015) Rhetoric vs. reality: a commentary on "Bird origins anew" by A. Feduccia. Auk 132:467–480

Smithwick F, Vinther J (2020) Review on color patterns of fossil feathers. In: Foth C, Rauhut OWM (eds) The evolution of feathers: from their origin to the present. Springer, Cham

Smithwick FM, Mayr G, Saitta ET, Benton MJ, Vinther J (2017) On the purported presence of fossilized collagen fibres in an ichthyosaur and a theropod dinosaur. Palaeontology 60:409–422

Stromer E (1915) Ergebnisse der Forschungsreisen Prof. Stromers in den Wüsten Ägyptens. II Wirbeltier-Reste der Baharîje-Stufe (unterstes Cenoman). 3. Das Original des Theropoden *Spinosaurus aegyptiacus* nov. gen., nov. spec. Abhandlungen der Königlich Bayerischen Akademie der Wissenschaften. Mathematisch-physikalische Klasse 28:1–32

Tarsitano S, Hecht MK (1980) A reconsideration of the reptilian relationships of *Archaeopteryx*. Zool J Linnean Soc 69:149–182

Therrien F, Henderson DM (2007) My theropod is bigger than yours...or not: estimating body size from skull length in theropods. J Vertebr Paleontol 27:108–115

Turner AH, Makovicky PJ, Norell MA (2012) A review of dromaeosaurid systematics and paravian phylogeny. Bull Am Museum Nat Hist 371:1–206
von Meyer H (1861a) Vogel-Federn und *Palpipes priscus* von Solnhofen. Neues Jahrbuch für Mineralogie, Geognosie, Geologie und Petrefakten-Kunde 1861:561
von Meyer H (1861b) *Archaeopteryx lithographica* und *Pterodactylus*. Neues Jahrbuch für Mineralogie, Geognosie, Geologie und Petrefakten-Kunde 1861:678–679
Wagner A (1861) Beiträge zur Kenntnis der urweltlichen Fauna des lithographischen Schiefers. 2. Abt. Schildkröten und Saurier. Abhandlungen der Königlich Bayerischen Akademie der Wissenschaften. Mathematisch-physikalische Klasse 9:67–124
Wagner A (1862) On a new fossil reptile supposed to be furnished with feathers. Ann Mag Nat Hist 9:261–267
Wang M, Zhou Z (2017) The evolution of birds with implications from new fossil evidences. In: Maina JN (ed) The biology of the avian respiratory system. Springer, Cham, pp 1–26
Wang M, Zheng X, O'Connor JK, Lloyd GT, Wang X, Wang Y, Zhang X, Zhou Z (2015) The oldest record of Ornithuromorpha from the early cretaceous of China. Nat Commun 6:6987
Wang S, Stiegler J, Amiot R, Wang X, Du G, Clark JM, Xu X (2017) Extreme ontogenetic changes in a ceratosaurian theropod. Curr Biol 27:1–5
Wang M, O'Connor JK, Xu X, Zhou Z (2019) A new Jurassic scansoriopterygid and the loss of membranous wings in theropod dinosaurs. Nature 569:256–259
Wellnhofer P (1974) Das fünfte Skelettexemplar von *Archaeopteryx*. Palaeontographica Abt. A 147:169–216
Wellnhofer P (2008) *Archaeopteryx*: Der Urvogel von Solnhofen. Verlag Dr Friedrich Pfeil, München
Witmer LM (2002) The debate on avian ancestry: phylogeny, function, and fossils. In: Chiappe LM, Witmer LM (eds) Mesozoic birds: above the heads of dinosaurs. University of California Press, Berkeley, pp 3–30
Xing L, McKellar RC, Wang M, Bai M, O'Connor JK, Benton MJ, Zhang J, Wang Y, Tseng K, Lockley MG, Li G, Zhang W, Xu X (2016a) Mummified precocial bird wings in mid-Cretaceous Burmese amber. Nat Commun 7:12089
Xing L, McKellar RC, Xu X, Li G, Bai M, Persons WS IV, Miyashita T, Benton MJ, Zhang J, Wolfe AP, Yi Q, Tseng K, Ran H, Currie PJ (2016b) A feathered dinosaur tail with primitive plumage trapped in mid-Cretaceous amber. Curr Biol 26:1–9
Xing L, McKellar RC, O'Connor JK, Niu K, Mai H (2019) A mid-Cretaceous enantiornithine foot and tail feather preserved in Burmese amber. Sci Rep 9:15513
Xu X (2006) Feathered dinosaurs from China and the evolution of major avian characters. Integr Zool 1:4–11
Xu X (2020) Filamentous integuments in nonavialan theropods and their kin: advances and future perspectives for understanding the evolution of feathers. In: Foth C, OWM R (eds) The evolution of feathers: from their origin to the present. Springer, Cham
Xu X, Fucheng F (2005) A new maniraptoran dinosaur from China with long feathers on the metatarsus. Naturwissenschaften 92:173–177
Xu X, Guo Y (2009) The origin and early evolution of feathers: insights from recent paleontological and neontological data. Vertebrata PalAsiatica 47:311–329
Xu X, Pol D (2014) *Archaeopteryx*, paravian phylogenetic analyses, and the use of probability-based methods for palaeontological datasets. J Syst Palaeontol 12:323–334
Xu X, Wang X, Wu X (1999a) A dromaeosaurid dinosaur with filamentous integument from the Yixian Formation of China. Nature 401:262–266
Xu X, Tang Z, Wang X (1999b) A therizinosauroid dinosaur with integumentary structures from China. Nature 399:350–354
Xu X, Zhou Z, Prum RO (2001) Branched integumental structures in *Sinornithosaurus* and the origin of feathers. Nature 410:200–204
Xu X, Norell MA, Wang X, Makovicky PJ, Wu X (2002) A basal troodontid from the Early Cretaceous of China. Nature 415:780–784
Xu X, Zhou Z, Wang X, Kuang X, Zhang F, Du X (2003) Four-winged dinosaurs from China. Nature 421:335–340
Xu X, Tan Q, Wang J, Zhao X, Tan L (2007) A gigantic bird-like dinosaur from the Late Cretaceous of China. Nature 447:844–847
Xu X, Clark JM, Mo J, Choiniere JN, Forster CA, Erickson GM, Hone DWE, Sullivan C, Eberth DA, Nesbitt SJ, Zhao Q, Hernandez R, Jia C, Han F, Guo Y (2009) A Jurassic ceratosaur from China helps clarify avian digital homologies. Nature 459:940–944
Xu X, Ma Q, Hu D (2010) Pre-*Archaeopteryx* coelurosaurian dinosaurs and their implications for understanding avian origins. Chin Sci Bull 55:3971–3977
Xu X, You H, Du K, Han F (2011) An *Archaeopteryx*-like theropod from China and the origin of Avialae. Nature 475:465–470
Xu X, Wang K, Zhang K, Ma Q, Xing L, Sullivan C, Hu D, Cheng S, Wang S (2012) A gigantic feathered dinosaur from the Lower Cretaceous of China. Nature 484:92–95
Xu X, Zhou Z, Dudley R, Mackem S, Chuong C, Erickson GM, Varricchio DJ (2014) An integrative approach to understanding bird origins. Science 346:1253293
Xu X, Zheng X, Sullivan C, Wang W, Xing L, Wang Y, Zhang X, O'Connor JK, Zhang F, Pan Y (2015) A bizarre Jurassic maniraptoran theropod with preserved evidence of membranous wings. Nature 521:70–73
Xu X, Zhou Z, Sullivan C, Wang Y, Ren D (2016) An updated review of the Middle-Late Jurassic Yanliao Biota: chronology, taphonomy, paleontology and paleoecology. Acta Geol Sin 90:2229–2243

Xu X, Choiniere JN, Tan Q, Benson RBJ, Clark JM, Sullivan C, Zhao Q, Han F, Ma Q, He Y, Wang S, Xing H, Tan L (2018) Two Early Cretaceous fossils document transitional stages in alvarezsaurian dinosaur evolution. Curr Biol 28:1–8

Zanno LE (2010) A taxonomic and phylogenetic re-evaluation of Therizinosauria (Dinosauria: Maniraptora). J Syst Palaeontol 8:503–543

Zanno LE, Makovicky PJ (2011) Herbivorous ecomorphology and specialization patterns in theropod dinosaur evolution. Proc Natl Acad Sci USA 108:232–237

Zhang F, Zhou Z, Xu X, Wang X (2002) A juvenil coelurosaurian theropod from China indicates arboreal habits. Naturwissenschaften 89:394–398

Zhang F, Zhou Z, Xu X, Wang X, Sullivan C (2008) A bizarre Jurassic maniraptoran from China with elongate ribbon-like feathers. Nature 455:1105–1108

Zhang Y, O'Connor JK, Di L, Meng Q, Sigurdsen T, Chiappe LM (2014) New information on the anatomy of the Chinese Early Cretaceous Bohaiornithidae (Aves: Enantiornithes) from a subadult specimen of *Zhouornis hani*. PeerJ 2:e407

Zheng X, You H, Xu X, Dong Z (2009) An Early Cretaceous heterodontosaurid dinosaur with filamentous integumentary structures. Nature 458:333–336

Zhou Z, Clarke JA, Zhang F (2008) Insight into diversity, body size and morphological evolution from the largest Early Cretaceous enantiornithine bird. J Anat 212:565–577

Zhou Z, Zhang Z, Li Z (2010) A new Lower Cretaceous bird from China and tooth reduction in early avian evolution. Proc R Soc B 277:219–227

Integumentary Structures in *Kulindadromeus zabaikalicus*, a Basal Neornithischian Dinosaur from the Jurassic of Siberia

4

Pascal Godefroit, Sofia M. Sinitsa, Aude Cincotta, Maria E. McNamara, Svetlana A. Reshetova, and Danielle Dhouailly

4.1 Introduction

Pre-1998, feathers were thought to be an exclusively avian characteristic, shared by all birds and extending back to the earliest representative of the clade, *Archaeopteryx*, in the Late Jurassic, ca. 150 million years ago (Witmer 2009). Since the description of *Sinosauropteryx* (Ji and Ji 1996; Chen et al. 1998), Middle-Late Jurassic and Early Cretaceous deposits from northeastern China have yielded numerous theropod dinosaurs bearing simple filamentous "protofeathers." Further, true pennaceous feathers featuring a rachis (shaft) and vanes were reported in nonavian pennaraptoran theropods that are closely related to birds (e.g., Xu et al. 2001, 2003; Hu et al. 2009; Godefroit et al. 2013). These discoveries of feather-like structures in theropods are consistent with current understanding of the evolutionary origin of birds, which are now almost unanimously regarded as representatives of the theropod and maniraptoran clades.

Subsequent reports of projecting, possibly hollow, bristle-like structures on the tail of the basal ceratopsian *Psittacosaurus*, also from the Early Cretaceous of Liaoning Province in northeastern China (Mayr et al. 2002, 2016), raised the possibility that filamentous epidermal structures may have been present in a broader clade that includes ornithischian dinosaurs. Patches of long filaments, reminiscent of structures present in theropods and thought to be the evolutionary precursors of feathers, were subsequently described in the heterodontosaurid ornithischian *Tianyulong*, from the Middle-Late Jurassic Yinliao Biota of Liaoning Province (Zheng et al. 2009; Sereno 2012). Whether these filaments in ornithischian dinosaurs are epidermal in origin, or represent remains of dermal collagen fibers, however, has been disputed (Lingham-Soliar 2010a, b; Mayr 2010). Even if these filaments can be confirmed as representing

P. Godefroit (✉)
Directorate 'Earth and History of Life', Royal Belgian Institute of Natural Sciences, Brussels, Belgium
e-mail: Pascal.Godefroit@naturalsciences.be

S. M. Sinitsa · S. A. Reshetova
Institute of Natural Resources, Ecology, and Cryology, Siberian Branch of the Russian Academy of Sciences, Chita, Russia

A. Cincotta
Directorate 'Earth and History of Life', Royal Belgian Institute of Natural Sciences, Brussels, Belgium

Department of Geology, University of Namur, Namur, Belgium

School of Biological, Earth and Environmental Sciences, University College Cork, Cork, Ireland

M. E. McNamara
School of Biological, Earth and Environmental Sciences, University College Cork, Cork, Ireland
e-mail: maria.mcnamara@ucc.ie

D. Dhouailly
Department of Biology, Université Grenoble-Alpes, Grenoble, France

C. Foth, O. W. M. Rauhut (eds.), *The Evolution of Feathers*, Fascinating Life Sciences,
https://doi.org/10.1007/978-3-030-27223-4_4

epidermal structures, it is still unclear whether these simple monofilaments are part of the evolutionary lineage of feathers, or represent independent evolution of projecting epidermal appendages (Witmer 2009; Barrett et al. 2015).

The discovery of *Kulindadromeus zabaikalicus* from the base of the Ukureyskaya Formation (Middle Jurassic; Cincotta et al., 2019) of Kulinda (Cherynyshevsky District of Chita Region, southeastern Siberia, Russia) sheds light on the origin and early diversification of integumentary appendages in dinosaurs. Various epidermal structures are well preserved adjacent to, and sometimes directly associated with, the skeletal elements. These include small scales along the distal tibia and on the foot, larger imbricated scales on the tail, long unbranched filaments on the head and thorax, and compound "protofeather"-like structures on the humerus, femur, and the proximal part of the tibia (Godefroit et al. 2014a). Given the position of *Kulindadromeus* near the evolutionary base of ornithischian dinosaurs, the presence of not only monofilaments, but also of branched epidermal structures, suggests that compound feather-like structures were potentially widespread among the whole dinosaur clade, or at least within the Ornithoscelida (Theropoda + Ornithischia) clade according to Baron et al.'s (2017) recent phylogenetic classification of dinosaurs. Here we describe in detail the integumentary structures in *Kulindadromeus zabaikalicus* and discuss their potential function and importance for dinosaur evolution.

4.2 Geological Setting and Taphonomy of the Kulinda Locality

The Kulinda locality is in the Chernyshevsky District of the Chita Region (Zabaikalsky Krai), about 220 km to the east of Chita city (Fig. 4.1).

Fig. 4.1 Location of Kulinda dinosaur locality (Chitinskaya Oblast, Russia). Inset map: Zabaikalsky Krai (in yellow)

The site was discovered by Sofia M. Sinitsa and her team from the Institute of Natural Resources, Ecology, and Cryology, Siberian Branch of the Russian Academy of Sciences, while they were conducting a geological survey in the Olov Depression along the Kulinda River, close to Chernyshevsk village (Sinitsa 2011). Four trenches in the lower part of the Ukureyskaya Formation revealed a succession of massive and alternating sandstones, siltstones, tuffaceous sandstones, tuffaceous siltstones, and tuffites. Based on comparisons of the paleoentomological and the microfaunal contents with the Glushkovo Formation in the Unda-Daya Depression, the Ukureyskaya Formation has been dated as Late Jurassic–Early Cretaceous (Sinitsa and Starukhina 1986; Sinitsa 2011). However, new U-Pb ages, together with palynological data, provide evidence of a Bathonian age—between 168.3 ± 1.3 Ma and 166.1 ± 1.2 Ma—for *Kulindadromeus* (Cincotta et al. 2019). Three fossiliferous horizons rich in ornithischian skeletal remains have been excavated to date at Kulinda, one in each of trenches 3 and 4, and an additional bonebed between those trenches (Fig. 4.2).

Bonebed 3, in trench 3, is 100–200 mm thick and consists of well-preserved disarticulated bones within a gray, silty matrix. Articulated elements and integumentary structures are rare. The sediments in trench 4 are probably slightly older than those in trench 3. Bonebed 4 comprises finely laminated, organic-rich claystone. Some of the bones in this horizon are articulated, and delicate integumentary structures are preserved as a thin film of carbonaceous material. This horizon was deposited in a quiescent environment far from clastic sources. The matrix of this bonebed is highly indurated, and laminae are occasionally deformed; some skeletal elements are preserved as external molds. This contrasts with the lithology and style of preservation of the material from bonebed 3 and suggests localized chemical environments during diagenesis.

Trench 3-3, between trenches 3 and 4, has been excavated since 2012. The section mostly contains fine-grained deposits (siltstones) along with poorly sorted sandstones and breccia. Seven fossiliferous layers with bone remains have been identified so far. The bone material is dominated by vertebrae, pelvic, and limb bones. Soft tissues are rather rare although faint traces of protofeathers are present, along with well-preserved scales.

Most fossils discovered in bonebeds 3, 3-3, and 4 belong to small individuals, most likely juveniles or sub-adults; larger individuals are rare. The overrepresentation of younger individuals in the bonebeds could suggest an attritional accumulation of carcasses leading to the formation of the bonebeds and not a single catastrophic event (Lyman 1994). Confirmation of this hypothesis requires a detailed age-frequency distribution of the long bones and further taphonomic investigation.

4.3 Biodiversity of the Kulinda Dinosaur Fauna

Most of the skeletal elements recovered from the three bonebeds at Kulinda are isolated or only partially articulated. The integumentary elements typically occur as isolated patches but can be directly associated with the skeletal elements. These features hamper attempts to reconstruct the skeleton and the external aspect of *Kulindadromeus*, and to estimate the taxonomic diversity of the bonebed material.

Except for a single shed tooth from a medium-sized theropod (which was found in bonebed 3), Godefroit et al. (2014a) hypothesized that the bonebeds at Kulinda are monospecific: detailed analysis of the skeletal elements (in particular, the partially articulated remains) preserved within and between the bonebeds reveals no evidence for multiple taxa of basal ornithischians in the Ukureyskaya Formation of the Kulinda locality. Each individual skeletal element is represented by a single morphotype; all the observed minor differences can easily be explained by ontogenetic and normal intraspecific variations.

Kulindadromeus zabaikalicus was a small, 1.5 m long, bipedal herbivore with a short skull, teeth adapted for herbivory, short forelimbs and elongate hindlimbs and tail (Fig. 4.3). Phylogenetic analysis (Fig. 4.4) recovers *Kulindadromeus* as a

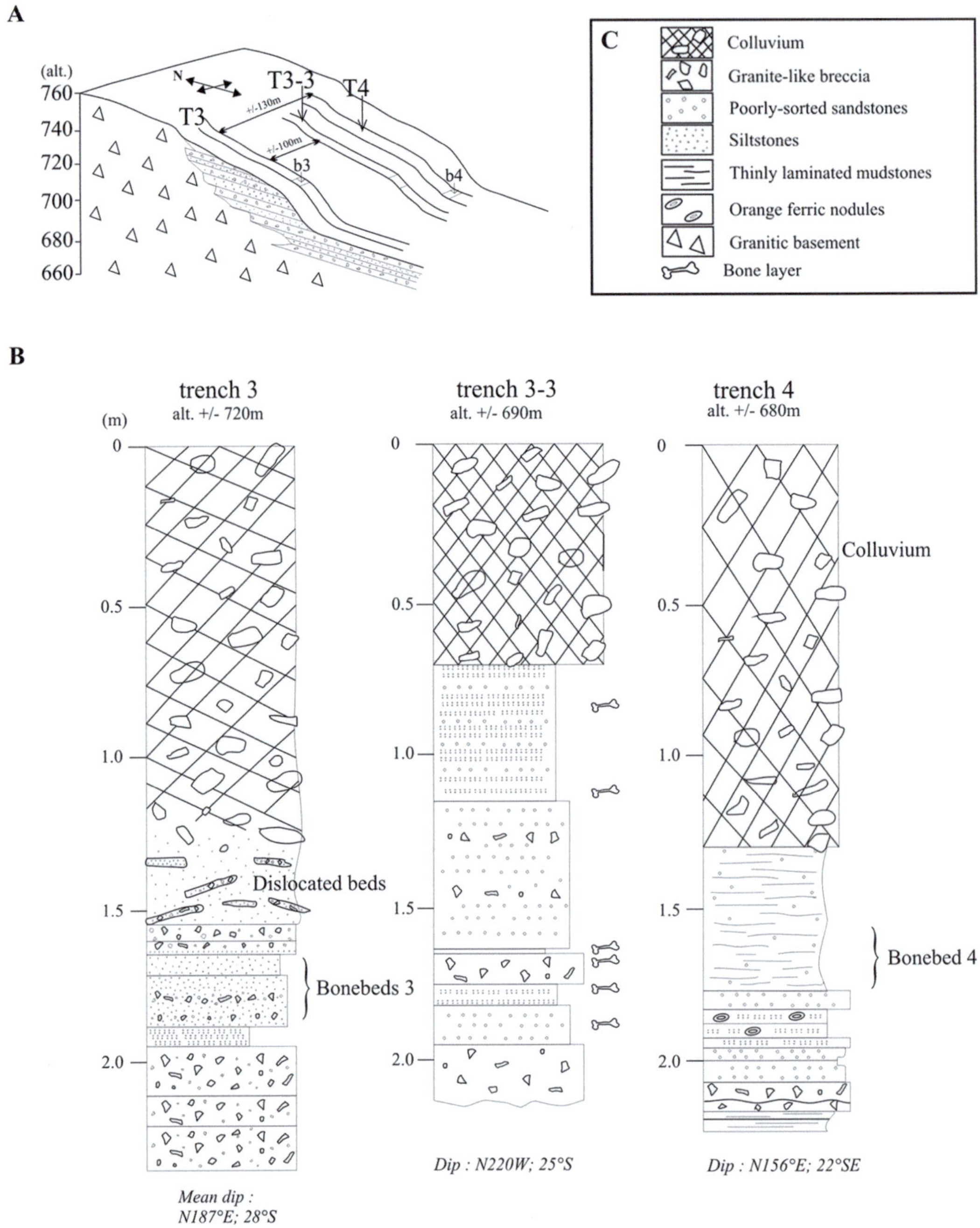

Fig. 4.2 (**a**) Schematic positioning of the trenches (T3, T3-3, and T4) and of the bonebeds (b3 and b4) at the Kulinda dinosaur locality, Ukurey Fm (Middle to Late Jurassic); (**b**) lithological logs of the sediments in trenches 3, 3-3, and 4, with the positions of the bonebeds. Bonebed 4 was only identified in the south front of trench 4; (**c**) lithological legend of the figures

basal member of Neornithischia (all genasaurians more closely related to *Parasaurolophus walkeri* than to *Ankylosaurus magniventris* or *Stegosaurus stenops* [Butler et al. 2008]) and the sister taxon for Cerapoda (*Parasaurolophus walkeri*, *Triceratops horridus*, their most recent common ancestor and

Fig. 4.3 Osteological reconstruction of *Kulindadromeus zabaikalicus*. Model by Jonica dos Remedios Esteves

all descendants [Butler et al. 2008]) (Godefroit et al. 2014a).

Alifanov and Saveliev (2014, 2015) proposed an alternative interpretation for the dinosaur fauna at Kulinda and named three new taxa from this locality: the 'hypsilophodontian' ornithopods *Kulindapteryx ukureica* and *Daurosaurus olovus* (Alifanov and Saveliev 2014), and the 'nqwebasaurid' ornithomimosaur *Lepidocheirosaurus natalis* (Alifanov and Saveliev 2015). Detailed description of the osteology of the dinosaurs from Kulinda is beyond the scope of this chapter, but some brief comments are made below.

Alifanov and Saveliev (2014, 2015) do not apply modern taxonomic standards for elaborating their classification schemes: they do not use cladistic methods for inferring the phylogenetic relationships between taxa. *Kulindapteryx* and *Daurosaurus* only differ in the structure of their ischia, but those differences can easily be explained by differences in the preservation and orientation of the bones, falling within the intraspecific variation of the *Kulidadromeus zabaikalicus* hypodigm.

Per Alifanov and Saveliev (2015), the caudal vertebrae and associated scales referred by Godefroit et al. (2014a) to *Kulindadromeus zabaikalicus* belong to a new ornithomimosaur, *Lepidocheirosaurus natalis*. This interpretation is based on analysis of one partially articulated manus and caudal vertebrae associated by caudal scales. The caudal vertebrae show a spool-shaped centrum, well-developed postzygapophyses and weakly developed neural spines; these features are characteristic of theropods and contrast with the vertebrae of bipedal Ornithischia, which possess neural spines, a cylindrical centrum and weakly developed postzygapophyses. However, Alifanov and Saveliev's (2015) interpretation is apparently based on direct comparisons with the ornithopod *Hypsilophodon foxii* and lacks a broader phylogenetic context. For example, caudal vertebrae of more basal ornithischians, e.g., the heterodontosaurid *Tianyulong confuciusi* (see Sereno, 2012, fig. 25), closely resemble those discovered at Kulinda: from about the tenth vertebra, the centrum is elongate and spool shaped in lateral view, the neural spines are reduced to a ridge, and both the pre- and postzygapophyses are long, extending beyond the level of the articular surfaces of the centra. Except for the absence of evidence for ossified tendons, the caudal structure in *Tianyulong* is remarkably similar to that in dromaeosaurid theropods (Sereno 2012) Furthermore, the hand of *Lepidocheirosaurus natalis*

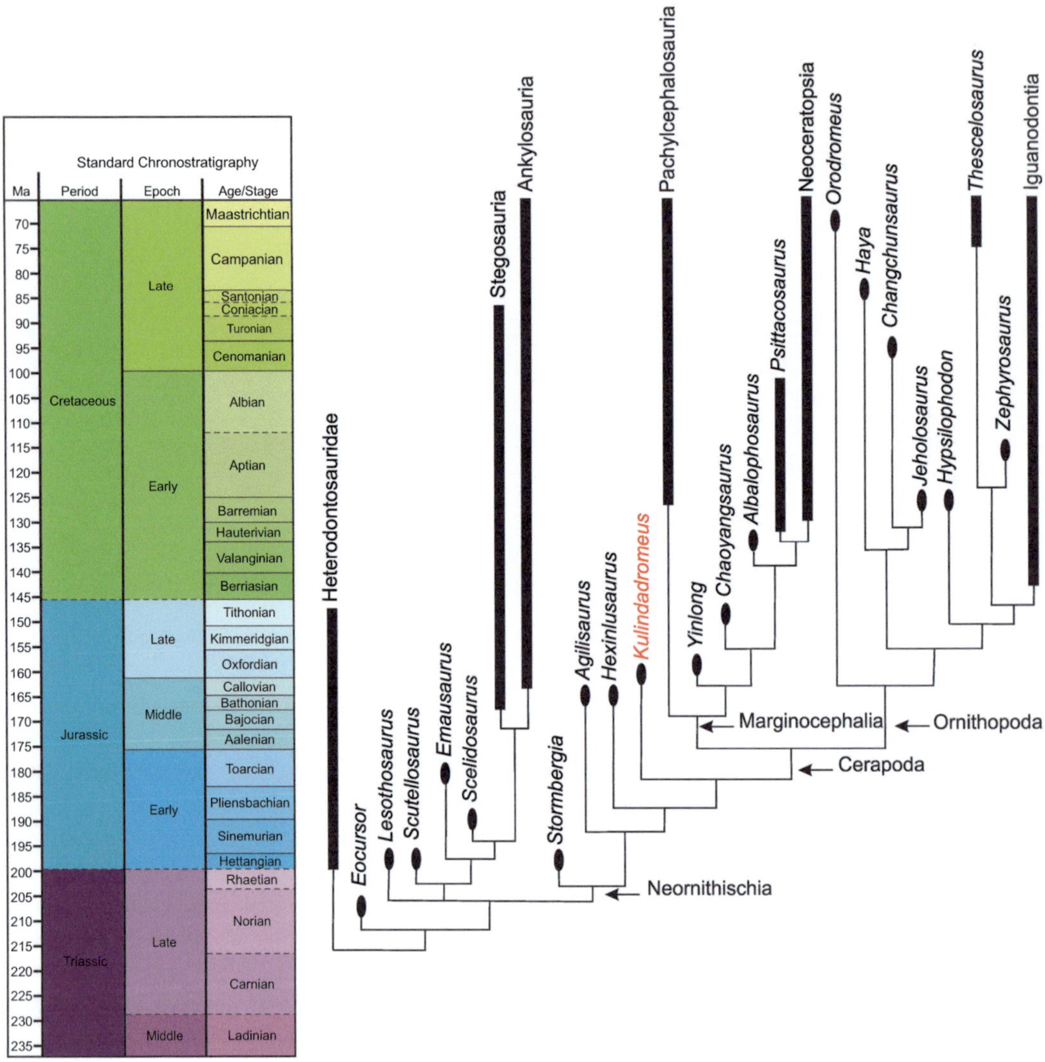

Fig. 4.4 Phylogenetic relationships of *Kulindadromeus zabaikalicus* among ornithischian dinosaurs (after Godefroit et al. 2014a). Time-calibrated strict consensus tree of the four most parsimonious trees (tree length = 571; consistency index excluding uninformative characters = 0.42; retention index = 0.7). In this hypothesis, *Kulindadromeus* is the sister-taxon of Cerapoda

from Kulinda closely resembles that of *Tianyulong* (see fig. 27 in Sereno 2012). Combined, these observations strongly indicate a lack of support for the hypothesis that basal ornithomimosaurs were present at Kulinda and that the caudal and manus material described by Alifanov and Saveliev (2015) can be confidently attributed to basal ornithischians, such as *Kulindadromeus zabaikalicus*.

The most parsiminous interpretation of the Kulinda bonebeds is thus that they represent accumulation of a monospecific dinosaur assemblage, as hypothesized by Godefroit et al. (2014a).

4.4 Diversity of Epidermal Structures in *Kulindadromeus*

The various epidermal structures preserved in *Kulindadromeus* are discussed below.

Pedal and Manual Scales Small (3.5–5 mm long) imbricated and hexagonal scales that resemble the scutella in modern birds (Lucas and Stettenheim 1972) are associated with the distal part of the tibia and usually show high relief (Fig. 4.5d). Smaller (<1 mm long), rounded, and nonoverlapping scales occur around the tarsus (Fig. 4.5d), metatarsus, and pes (Fig. 4.5e); these resemble the reticula along the plantar face of the pes in modern birds (Lucas and Stettenheim 1972). In INREC K4-6a-18/35 (part and counterpart of the same specimen), thick hemispherical pads, covered by tiny rounded reticula, occur on the ventral side of the manual phalanges, with one pad corresponding to each phalanx. A ventral pad is also present below the proximal part of the unguals of digits I and II (Fig. 4.5a–c). Similar pads are also present in *Concavenator* (Cuesta et al. 2015) and *Psittacosaurus* (Vinther et al. 2016).

Alifanov and Saveliev (2015) described similar reticulate scales around the manus of PIN 5395/1 ('*Lepidocheirosaurus natalis*'), herein tentatively referred to *Kulindadromeus zabaikalicus*. These authors also described a series of eight wide crescentic scales, approximately 1.5–2 times as wide as the corresponding phalanges, on phalanges I-1, III-2, and 3 and on metacarpal I. These scales abut but do not overlap. Similar scales occur on the scuta on the dorsal side of the metacarpus and pes in modern birds (Lucas and Stettenheim 1972). The crescentic scales in the fossils therefore likely covered the dorsal side of the phalanges, whereas the smaller reticulate scales likely covered the ventral side of the manus (*contra* Alifanov and Saveliev 2015).

Pad-like scaled structures also occur in bonebeds 3/3 and 4 but are not associated with skeletal material. The pads preserve tiny (<0.5 mm) reticulate scales that are organized into thin parallel sinuous lines, e.g., in specimen INREC K3/3-4-628 (Fig. 4.5f). The taphonomic conditions responsible for the formation of the Kulinda fossil locality were clearly conducive to the exceptional preservation of exquisite soft tissue structures (*contra* Lingham-Soliar 2014).

Scutate Caudal Scales The tail of *Kulindadromeus* was surrounded by a dozen longitudinal rows of scales (Figs. 4.6 and 4.7). It is, however, unclear whether the ventral side of the tail was completely covered by this scaly armor. The largest scales (ca. 20 mm long and 10 mm wide) occur along the proximal part of the tail. The paired dorsal series comprises scales that are significantly larger, and with a more complex morphology, than the lateral series. The caudal scales of *Kulindadromeus* are thin (<100 μm thick) and unornamented, and the dorsal series are slightly imbricated. Those caudal scales are clearly different from the thicker, usually sculptured, and nonoverlapping osteoderms in thyreophoran ornithischians and from the proportionally thicker and smaller scales in iguanodontian ornithopods (Bell 2012). Instead, the caudal scales of *Kulindadromeus* more closely resemble the epidermal scutae that cover the dorsal side of the metatarsus and pes in modern birds (Lucas and Stettenheim 1972). The preservation of the scales as carbonaceous remains does not support their interpretation as osteoderms because the bones (which also comprise calcium phosphate *in vivo*) clearly differ in their mode of preservation.

Scales from the paired dorsal series are arched and rectangular, whereby each scale covers a small part of the adjacent distal one (Fig. 4.6c, e, f). In the distal part of the tail, three scales from each series cover one vertebra (Fig. 4.6b). The proximal dorsal scales are much wider mediolaterally than long proximodistally. Some paired proximal dorsal scales are apparently fused along the median axis of the tail (Fig. 4.6c). Distal dorsal scales are proportionally narrower mediolaterally throughout the series. Notably, each scale from the paired dorsal series forms a triangular proximal spur that covers part of the preceding scale, forming a clip-like system that connects adjacent elements (Fig. 4.6e). Proximally, near the base of the tail (Fig. 4.6c), the dorsal scales are progressively smaller, more rounded, and do not overlap.

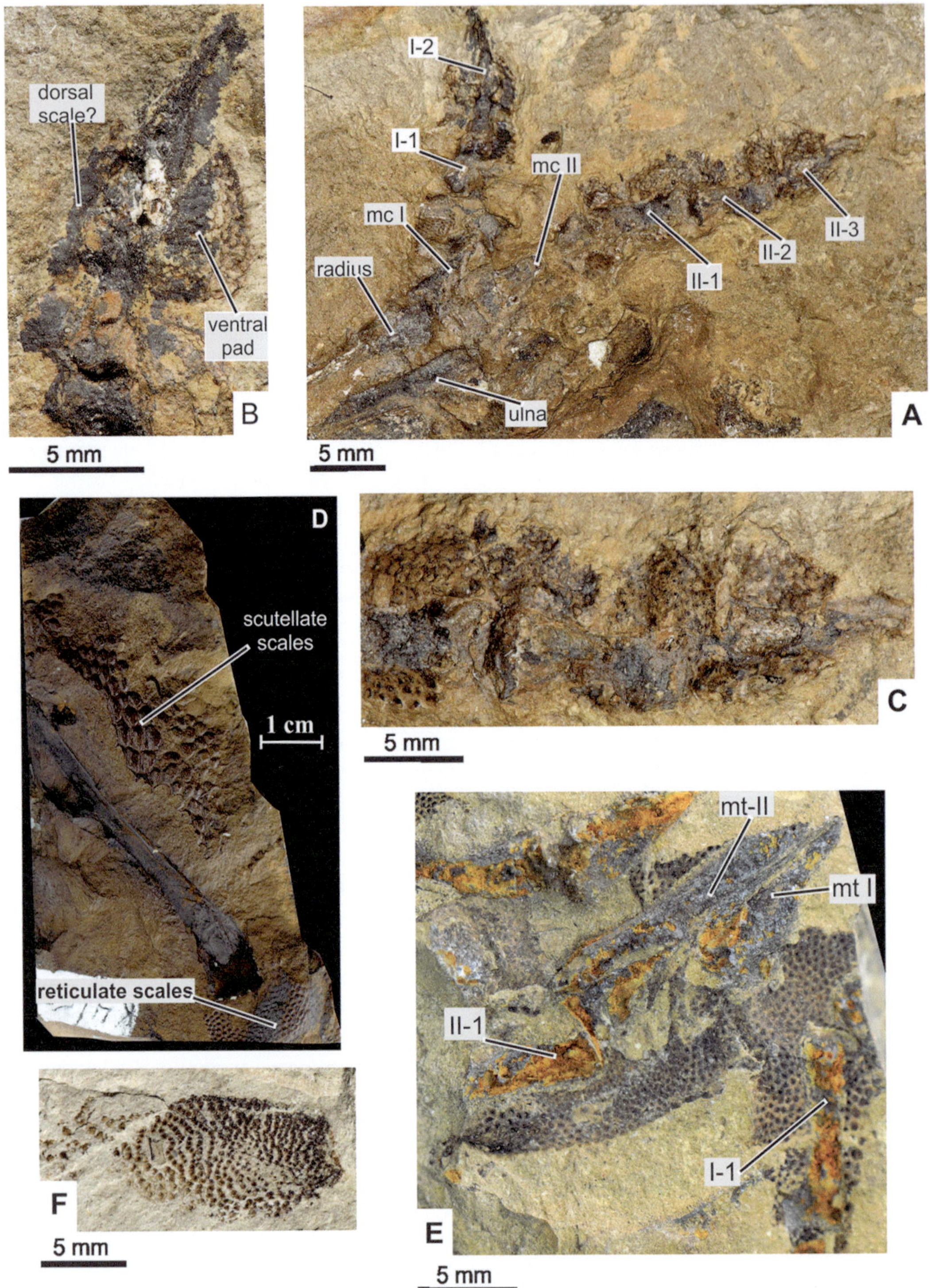

Fig. 4.5 Scales on the appendicular skeleton of *Kulindadromeus zabaikalicus*. (**a**) Scales around the right manus (INREC K4/6a-18); (**b**) close-up of the distal part of right digit I (INREC K4/6a-18); (**c**) close-up of the distal part of the right digit II (INREC INREC K4/6a-18); (**d**) scales around the distal tibia and around the tarsus (INREC K4/57); (**e**) scales around the right metatarsus and pes (INREC K4/118); (**f**) isolated pad-like reticulate structure (INREC K3/3-4-628). *mc* metacarpal, *mt* metatarsal

Fig. 4.6 Scales around the tail of *Kulindadromeus zabaikalicus*. (**a**) Scales around distal portion of the tail in left lateral view (INREC K4/5-6-127); (**b**) distal caudal vertebrae (left lateral view) and displaced dorsal scales (ventral view) (INREC K4/159); (**c**) double row of dorsal scales above the proximal part of the tail (INREC K4/9-41) in dorsal view; (**d**) close-up of ventral bundles of bristle-like structures (INREC K4/9-41); (**e**) close-up of the left dorsal row of caudal scales (INREC K4/117) in dorsal view; (**f**) partial reconstruction of the caudal scales in laterodorsal view

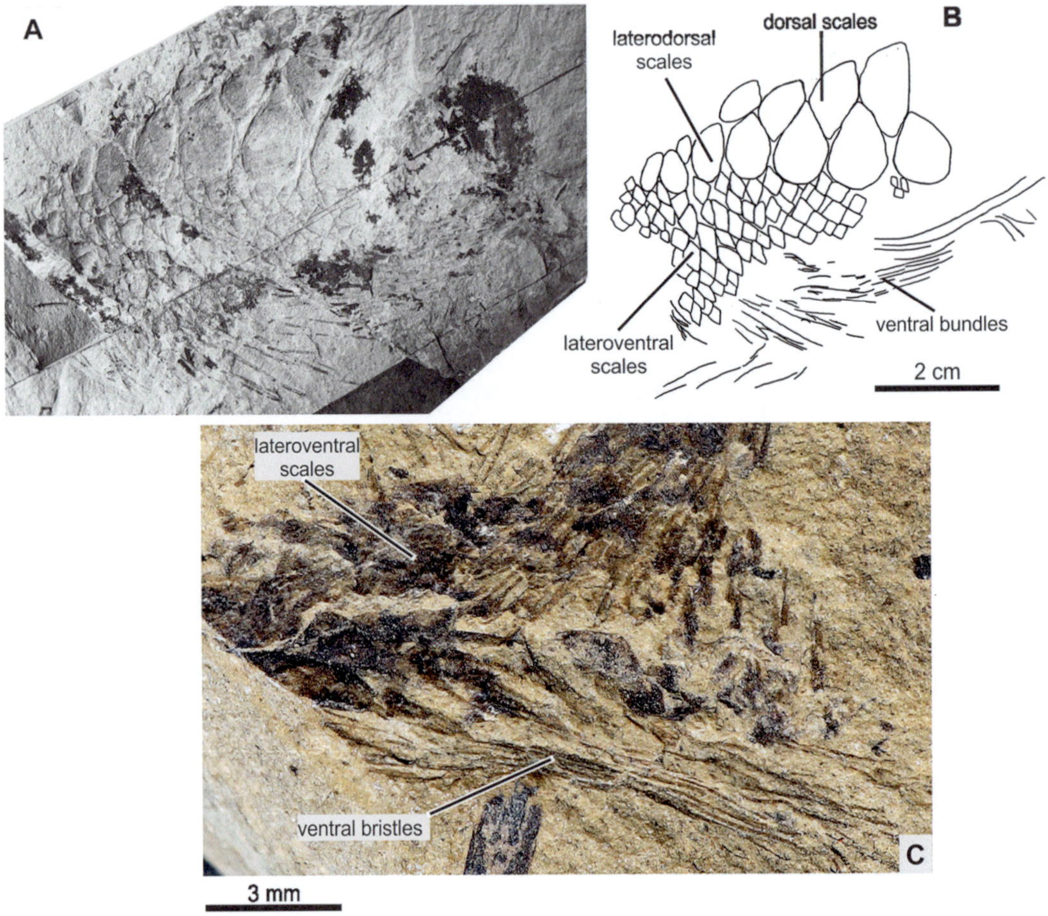

Fig. 4.7 Scales around the tail of *Kulindadromeus zabaikalicus*. Photograph (**a**) and interpretative drawing (**b**) of the right lateral side of a partial tail armor (INREC K3-3/95); (**c**) close-up of lateroventral scales (INREC K4/6d-1), showing the ventral bristle-like structures

The scales from the dorsolateral series are significantly narrower dorsoventrally (or mediolaterally) than the corresponding scales of the dorsal series (Fig. 4.7a, b). The ventral border of the dorsolateral scales is rounded, whereas their dorsal border is triangular, so each lateral scale imbricates between adjacent dorsal scales. The number of dorsolateral scales therefore corresponds to the number of dorsal scales. Unlike the dorsal scales, the dorsolateral scales do not overlap each other and there is no anterior spur.

The ventral scales are much smaller (<5 mm) than those of the other caudal series and are lozenge-shaped and imbricate, whereby each scale slightly overlaps the adjacent distal one (Fig. 4.7a, b). The scales progressively decrease in size distally and towards the ventral side of the tail, and it is unclear whether the ventral part of the tail was completely covered by scales. The ventralmost caudal scales are highly modified: their proximoventral edge extends as extremely elongated, bristle-like structures (up to four per scales, e.g., INREC K-4-4/5-28 and K-4-6d-34; Fig. 4.7c) that form dense and anteriorly oriented bundles (see INREC K4/9-41). In the latter specimen, the bristles are particularly thick (up to 1.5 mm wide) and long (>140 mm) (Fig. 4.6d). In smaller specimens and in the distal part of the tail, the bristles are much thinner (0.3 mm wide; Fig. 4.7c). Regardless of their size, these bristles typically exhibit multiple fractures along their length; this suggests that they were rigid (Fig. 4.6d). The unusual ventral position and anteroventral orientation of these bristle-like structures would have limited their role in protection.

Monofilaments Monofilaments are widely distributed around the head (Fig. 4.8a, b), thorax (Fig. 4.8c–e) and back. Their diameter is constant throughout their length. They do not show a hollow structure as they appear as cords, similar to hairs and individual barbs of downy feathers, and also to stage 1 in Xu et al.'s (2010) model for feather evolution. Whether the internal void space in hollow tubes, such as filament-like feathers in extant birds, can be retained during fossilization is, however, unknown. The monofilaments above the head are thin (ca. 0.15 mm wide), short (10–15 mm long), and curved, with no preferred orientation (Fig. 4.8b). The thoracic and abdominal filaments are wider (0.2–0.3 mm wide) and longer (20–30 mm). In INREC K4-6a-22, most of the monofilaments associated with the ribcage are consistently oriented craniocaudally within a 10°–15° interval (Fig. 4.8c–e). The typically curved geometry and continuous nature of these monofilaments indicates that they were likely flexible. The monofilaments in *Kulindadromeus* are thus shorter and thinner than those on the proximal part of the tail in *Psittacosaurus* (Mayr et al. 2002) and thinner than the filamentous structures in *Tianyulong* (Zheng et al. 2009).

Lingham-Soliar (2014) interpreted the monofilaments in *Kulindadromeus* as dermal collagen fibers or fiber bundles and not as epidermal structures, as it was also hypothesized for the thinner (<0.05 mm) monofilaments in the basal theropod *Sinosauropteryx* (Lingham-Soliar et al. 2007). Interpretations of the monofilaments in *Kulindadromeus* were based on comparative analysis with the thick bundles of collagen fibers present in the integument of marine tetrapods. The validity of a direct comparison between integumentary collagen in marine tetrapods and in a terrestrial dinosaur is, however, uncertain given the marked differences in mechanical stresses acting upon the skin in these animals. Indeed, Lingham-Soliar and Plodowski (2007) and Lingham-Soliar and Wesley-Smith (2008) showed previously that the organization of collagen fibers into thick bundles is linked to high tensile stiffness and efficiency of the locomotory organs in high-speed marine tetrapods.

Further, the morphology and arrangement of the fossilized filaments in *Kulindadromeus* is not consistent with that of *in vivo* or degraded collagen. *In vivo* integumentary collagen fibers typically occur in layered arrays of parallel, densely packed fibers where fibers in successive layers are oblique to one another. The filamentous structures in *Kulindadromeus* are straight to slightly curved, show constant width along their length, show a constant orientation, remain well separated from each other, and lack evidence for arrangement of fibers into successive vertical layers (Fig. 4.8). It is also difficult to envisage how the preserved arrangement of monofilaments could be generated during decay of collagen. Our ongoing experiments simulating the degradation of avian collagen reveal that the collagen fiber bundles of the dermal stratum compactum progressively dissociate and ultimately degrade into a dense tangled mass of highly anastomosing fibrils (Godefroit et al. 2014b).

Filaments Associated with a Basal Plate Importantly, *Kulindadromeus* shows compound, nonshafted integumentary structures, which can be directly associated with the proximal parts of the humerus (Fig. 4.9a–c) and femur (Fig. 4.9g–i). These occur as groups of six or seven filaments, each 10–40 mm long, that converge proximally and arise from the central, or even the basal region of a basal plate. This contrasts with the bristle-like structures that emerge from the edge of the lateroventral caudal scales (*contra* Saveliev and Alifanov 2014). Filaments on the humerus are thick (0.2–0.4 mm wide), straight and are extensively fractured (and therefore probably relatively rigid; Fig. 4.9b, c), whereas those on the femur are thinner (0.1–0.2 mm wide; Fig. 4.9d–f, h, i). The basal plates are larger on the humerus (3–4 mm wide; Fig. 4.9b, c) than on the femur (2–3 mm wide; Fig. 4.9e, f, h, i). On both the humerus and the femur, they show a hexagonal arrangement but

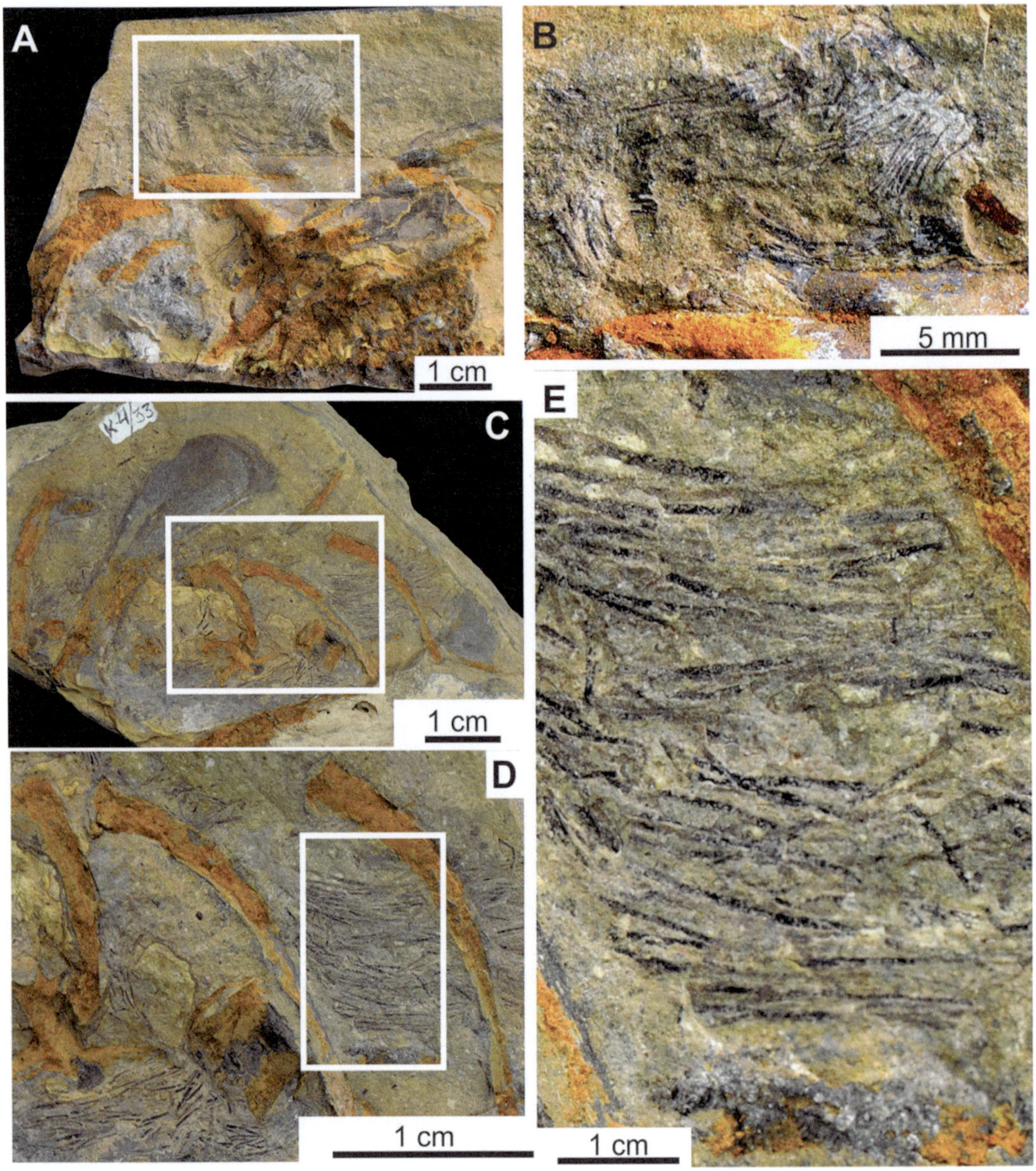

Fig. 4.8 Monofilaments in *Kulindadromeus zabaikalicus*. (**a**) partial skull (INREC K4/22) in right lateral view, with **b**, detail of areas indicated in (**a**); (**c**) left part of ribcage (INREC K4/33), with **d–e**, details of areas indicated in (**c**) and (**d**) showing filamentous structures

are mutually separated. This contrasts with the contiguous distribution of scales on the distal forelimb, hindlimb, and tail in *Kulindadromeus*, and also with that of the feathered scales that cover the tarsometatarsus of living birds (Lucas and Stettenheim 1972). Godefroit et al. (2014a) hypothesized that these groups of filaments are similar to the stage 2 feathers of Prum's (1999) and Xu's (2006b) model for feather evolution, in which the feather comprises a tuft of unbranched barbs joined proximally. This type of "protofeather" has been described in several feathered nonavian theropods or basalmost birds from northeastern China, including *Sinornithosaurus millenii* (Xu et al. 2001, fig. 3), *Microraptor gui* (Xu et al. 2003), *Anchiornis*

Fig. 4.9 Filaments associated with a basal plate in *Kulindadromeus zabaikalicus*. (**a**) Right humerus and proximal part of right radius and ulna (INREC K4/115), with detail (**b**) and interpretative drawing (**c**) of compound structures around the right humerus; (**d**) (INREC K4/88) and (**e**) (INREC K/117), compound structures, presumably around femur (part and counterpart of INREC K4/116) with, (**f**) interpretative drawing of (**e**); (**g**) distal part of right femur; (**h**) detail of the counterpart of INREC K4/116 showing compound structures; (**i**) interpretative drawing of (**h**). *bpl* basal plate, *fdc* femoral distal condyles, *icd* intercondylar groove

huxleyi (Hu et al. 2009, fig. 3h), *Eosinopteryx brevipenna* (Godefroit et al. 2013, fig. 2i), and *Similicaudipteryx yixianensis* (Xu et al., 2010), and was also likely present, according to Xu and Guo, in *Beipiaosaurus inexpectus*. Filamentous parallel barbs arising from the edge of a membranous structure also occur in *Epidexipteryx hui* (Zhang et al. 2008, fig. 2d, e).

According to Prum (1999), stage 2 plumulaceous feathers are radially symmetrical. Because the stage 2-like structures in *Kulindadromeus* and feathered nonavian theropods from Liaoning are preserved as discrete, two-dimensional outlines, whether those structures were already planar *in vivo*, or were originally circular and subsequently compressed during fossilization remains unclear. The basal plates in *Kulindadromeus* unlikely correspond to feathered scales, as observed in domestic pigeons, chickens, and barn owls, because they are not contiguous or overlapping. In *Kulindadromeus*, the distal filaments of those compound structures are usually slightly divergent and did not overlap each other during fossilization; this is consistent with the arrangement of similar groups of filaments in nonavian feathered theropods, but not with diagenetic compression of filaments arranged in a circular fashion, which would have resulted in overlapping filaments. The recurrent geometry of the filaments in the fossil compound structures suggests that the structures originally were arranged in a planar fashion *in vivo*.

An alternative interpretation of the fossil compound structures is offered by Lingham-Soliar (2014), in which the compound structures around the humerus and femur are interpreted as collagen support fibers associated with badly degraded scales. Godefroit et al. (2014b) considered this hypothesis highly unlikely due to the taphonomy and morphology of the preserved structures. An analysis of more than 1000 isolated bones and partial skeletons has revealed that different integumentary structures are systematically associated with specific anatomical regions: small nonoverlapping scales occur in the distal hindlimb (distal tibia + pes) and the manus; larger imbricated scales occur above the tail; long bristle-like structures emerge from the ventrodistal edge of small overlapping scales along the ventrolateral side of the tail; monofilaments occur in the head and the thorax, grouped filaments, and basal plate; in the proximal parts of the limbs (humerus and femur); and clusters of ribbon-shaped elements occur around the proximal tibia (see below). If the basal plates represent badly degraded scales (Lingham-Soliar (2014), it is difficult to envisage a scenario in which scales were selectively degraded and/or disarticulated around the humerus and femur, but never around the distal tibia, pes, and tail. The different anatomy and mode of preservation of the structures preserved in the tail and distal parts of the limbs, and those preserved in the rest of the body, support our interpretation that the two sets of features represent fundamentally different structures—the fossilized remains of epidermal scales and "protofeather"-like structures, respectively. The suggestion that collagen fibers or fiber bundles would be better preserved than scales (Lingham-Soliar 2014), is inconsistent with previous studies demonstrating that keratin has a higher preservation potential than collagen (Bjelland et al. 1988; Davis and Briggs 1995; Brown and Brown 2011). Our own decay experiments using feathers reveal that these are much more robust than the collagenous dermis and retain gross morphological characteristics for a longer period during decay (Godefroit et al. 2014b). The excellent state of preservation of the filaments in *Kulindadromeus* thus supports our interpretation that they represent keratinous structures rather than collagen fibers or fiber bundles.

Ribbon-Shaped Structures Ribbon-shaped structures associated proximally into clusters of six to seven elements occur along the proximal part of the tibia (Fig. 4.10). Each individual element is 15–20 mm long and 1.5–3 mm wide with a dark axial lineation (Fig. 4.10b–d). A thin superficial carbonaceous sheet envelops ca. 10 thin internal (50–100 μm) parallel filaments (Fig. 4.10d).

Similar broad (2–3 mm) monofilaments have been reported in the basal therizinosaur *Beipiosaurus inexpectus*, from the Early Cretaceous Yixian Formation of Liaoning Province, China (Xu et al. 2009), but are much longer (about 100–150 mm) than those in *Kulindadromeus* and not bundled proximally. Tail feathers with a ribbon-like proximal portion and a pennaceous distal tip have been described in a juvenile specimen of the oviraptorosaur *Similicaudipteryx yixianensis* (Xu et al. 2010), and also in the basal maniraptoran *Epidexipteryx*

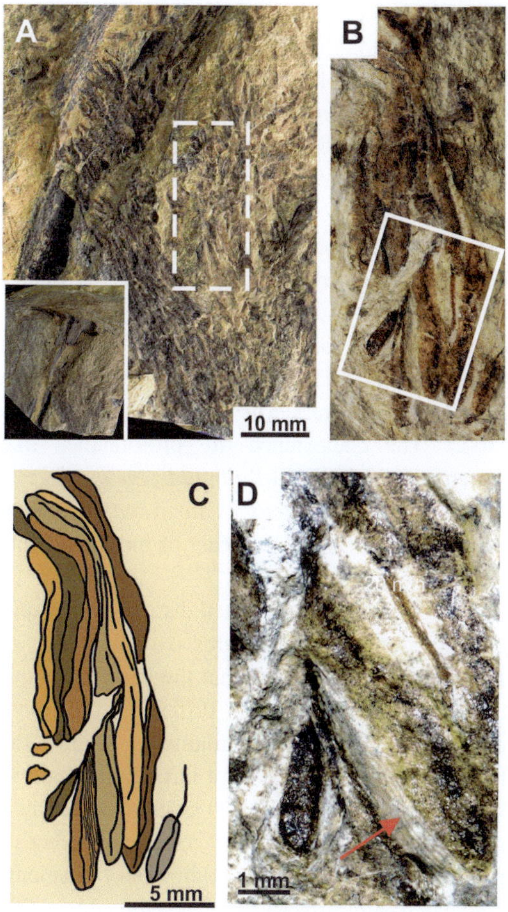

Fig. 4.10 (**a**) Ribbon-like structures around proximal part of tibia (INREC K4/44); inset shows slab in lower magnification; (**b**) detail of area indicated in **a** with interpretative drawing (**c**) and further details of ribbon-like structures (**d**); the superficial carbonaceous sheet has been removed during preparation, revealing an internal structure of thin parallel filaments (arrow)

hui (Zhang et al. 2008; Xu et al. 2010). Prum (2010) and Foth (2012) observed that juvenile tail feathers of *Similicaudipteryx* closely resemble the morphology of molting feathers of living birds and that those proximally ribbon-like pennaceous feathers could represent "pin feathers" or developing feather germs. The tubular feather germ is surrounded by a keratinized sheath that is lost during development to expose the mature feather. Typically, the sheath begins to dissociate from the mature distal tip of the feather before the ensheathed tubular base of the feather is fully developed. For a pennaceous feather, this intermediate stage of growth appears as a limited distal vane emerging from a smooth tubular base, corresponding to the morphotype observed in the juvenile *Similicaudipteryx* (Prum 2010).

Lingham-Soliar (2014) stated that ribbon-like structures, similar to those described in *Kulindadromeus*, occur as structural collagen in blood vessels, the *linea alba* (which separates the central band of abdominal muscles), and the rectus sheath. Godefroit et al. (2014b) noted that the arrangement, size, and morphology of the ribbon-like structures is strongly dissimilar to that of a network of circulatory vessels, which typically shows frequent branching and a reduction in diameter distally. Secondly, thick bands of connective tissue in the *linea alba* and rectus sheath occur in the human body where large muscles connect either to each other or to bone. It is highly unlikely that *Kulindadromeus* possessed similar, particularly large muscles in the region of the tibia and thus it is unlikely that the ribbon-like structures represent structural collagen (Godefroit et al. 2014b).

4.5 Discussion

The presence of both simple and compound filamentous structures in *Kulindadromeus* supports previous reports of simple "protofeather"-like structures in the ornithischian dinosaurs *Psittacosaurus* (Mayr et al. 2002, 2016) and *Tianyulong* (Zheng et al. 2009; Sereno 2012). Critically, this discovery confirms that monofilaments (i.e., stage 1 of Xu et al. 2010) and grouped filaments arising from a basal plate (similar to stage 2 of Prum 1999, and stage 3 of Xu et al. 2010) were not a synapomorphy of coelurosaurian theropods but were also present, and could form extensive body coverings, in ornithischians. Given the phylogenetic position of these taxa near the evolutionary base of ornithischians, these findings imply that the common ancestor of theropods and ornithischians also possessed simple feather structures. Regardless of the adopted phylogenetic hypothesis (Baron 2017), this common ancestor lived during the Upper Triassic. Scaly, nonfeathered, skin has been described in several groups on both the ornithischian and theropods branches of the dinosaur family tree (Witmer 2009; Barrett et al. 2015), which suggests that protofeathers were probably lost several times in dinosaur evolution. Furthermore, the ability to form monofilaments and more complex compound epidermal structures is potentially nested within the archosauromorph clade, as exemplified by *Longisquama* (Buchwitz and Voigt 2012) and pterosaurs (Bakhurina and Unwin 1995; Wang et al. 2002; Yang et al. 2019). The ultimate question of whether these monofilaments and compound filaments of primitive ornithischians represent part of the evolutionary lineage of feathers in paravian dinosaurs or, as hypothesized by Barrett et al. (2015), independent evolution of projecting epidermal appendages, remains unclear. Monofilaments, as the first step towards the development of "true" feathers, are similar to hair shafts, the other great category of filamentous epidermal structures in tetrapods. Contrasting with feathers, the geometry and anatomy of hairs has not changed during mammal evolution, as it was already present more than 160 million years ago, in Middle-Late Jurassic basal mammaliaformes (e.g., Ji et al. 2006). The important point is that the flight feather vane in living birds is the most complex epidermal appendage known in modern and ancient animals. One of the main reasons why hairs have never evolved branched structures might be the shape of their collar: pointed in hairs and ring shaped in feathers (Prum and Brush 2002). Current data is insufficient to dismiss the hypothesis that other complex integumentary structures, different to the feather vane, might have independently evolved in nonavian archosauromorphs.

As in living birds, *Kulindadromeus* confirms other evidence that scales and "protofeather"-like structures are not mutually exclusive in a single dinosaur taxon. Small, tubercle-like scales are known from the limb and tail of a psittacosaur that bears bristle-like structures above its tail (Mayr et al. 2002, 2016; Vinther et al. 2016). Patches of scaly structures have been reported from the tail and tibiae in the compsognathid theropod *Juravenator starki*, from the Late Jurassic of Bavaria, Germany (Göhlich and Chiappe 2006). Given that the "scaly" *Juravenator* is closely related to the "fuzzy" *Sinosauropteryx*, Göhlich and Chiappe (2006) hypothesized that the early evolution of feather structures may have been flexible and experimental (see also Xu 2006a, b). Per Xu (2006a), the presence of protofeathers in *Sinosauropteryx* and their absence in *Juravenator* might also be explained by slight differences in the phylogenetic position of those coelurosaurians, *Juravenator* being slightly more primitive. However, the latter hypothesis was subsequently invalidated by the discovery of protofeathers in the more basal megalosauroid *Sciurumimus albersdoeferi*, also from the Late Jurassic of Bavaria (Rauhut et al. 2012). The coexistence of "protofeather"-like structures and scales in *Kulindadromeus* suggests the potential for the body of *Juravenator* to have had a similar "fuzzy" covering of monofilaments that are not preserved.

The diversity and localized distribution of specific integumentary structures over the body of *Kulindadromeus* indicate that the various integumentary structures had different functional adaptations. The simple filaments around the head, thorax, and back may have functioned in insulation, but it is difficult to assess whether their density was sufficient to form an effective insulating layer. The compound filaments associated with a basal plate and the ribbon-like structures occur in the proximal parts of the limbs, and therefore probably did not function primarily in insulation. The regular organization of the basal plates and the clustered arrangement of the ribbon-like structures suggests that they may have functioned in visual display. It is also possible that these integumentary structures had ancillary functions in balance or insulation of eggs. The scales around the tail and the bristle-like structures on the lateroventral aspect of the tail were obviously too thin for efficient defensive functions but rather likely stiffened the tail dorsoventrally, the latter acting as a pendulum.

4.6 Conclusions

The body of *Kulindadromeus zabaikalicus* was covered by diverse integumentary structures, including different types of scales and filaments

Fig. 4.11 Tentative reconstruction of *Kulindadromeus zabaikalicus*, showing the distribution of the various integumentary structures. Model by Mostfa Mohammed

(Fig. 4.11). The different structures are systematically associated with specific anatomical regions: (1) small nonoverlapping scales are localized to the distal hindlimb (distal tibia + pes) and the manus, (2) larger imbricated scales with a proximal spur, to dorsal regions of the tail, (3) smaller imbricated scales with associated bristle-like structures, to the lateroventral region of the tail, (4) monofilaments, to the head and thorax, (5) clusters of filaments that diverge from a basal plate, to the proximal parts of the limbs (humerus and femur), and (6) clusters of ribbon-shaped structures, to the proximal tibia. All of these structures are most likely epidermal in origin (*contra* Lingham-Soliar 2014). Ongoing analyses of the ultrastructure and chemistry of the different integumentary structures in *Kulindadromeus* will provide a deeper understanding of the anatomical affinity of these structures and, in turn, of the origin and early diversification of epidermal structures, including feathers, in Archosauromorpha. The insights yielded by these ultrastructural and chemical studies will be complemented by our further sedimentological, mineralogical, and taphonomic studies of the Kulinda locality in order to better understand the environmental and diagenetic circumstances leading to exceptional preservation of integumentary features in *Kulindadromeus*.

Acknowledgments The authors want to thank C. Foth and O. Rauhut for inviting them to write this chapter. C. Foth and R. Butler reviewed an earlier version of this chapter and made many constructive and helpful comments.

References

Alifanov VR, Saveliev SV (2014) Two new ornithischian dinosaurs (Hypsilophodontia, Ornithopoda) from the Late Jurassic of Russia. Paleontol J 48(4):414–425

Alifanov VR, Saveliev SV (2015) The most ancient ornithomimosaur (Theropoda, Dinosauria), with cover imprints from the Upper Jurassic of Russia. Paleontol J 49(6):636–650

Bakhurina NN, Unwin DM (1995) A preliminary report on the evidence for 'hair' in *Sordes pilosus*, an Upper Jurassic pterosaur from Middle Asia. In: Sun A, Wang Y (eds) Sixth symposium on mesozoic terrestrial ecosystems and biota. China Ocean Press, Beijing, pp 79–82

Baron MG (2017) *Pisanosaurus mertii* and the Triassic ornithischian crisis: could phylogeny offer a solution?

Hist Biol. http://sci-hub.tw/10.1080/08912963.2017.1410705

Baron MG, Norman DB, Barrett PM (2017) A new hypothesis of dinosaur relationships and early dinosaur evolution. Nature 543:501–506

Barrett PM, Evans DC, Campione NE (2015) Evolution of dinosaur epidermal structures. Biol Lett 11 (6):20150229

Bell P (2012) Standardized terminology and potential taxonomic utility for hadrosaurid skin impressions: a case study for *Saurolophus* from Canada and Mongolia. PLoS One 7(2):e31295

Bjelland S, Volden G, Raa J (1988) A simple essay for determining keratin and collagen degradation in vitro. Arch Dermatol Res 280:50–53

Brown TA, Brown K (2011) Biomolecular archaeology: an introduction. Wiley-Blackwell, Oxford

Buchwitz M, Voigt S (2012) The dorsal appendages of the Triassic reptile *Longisquama insignis*: reconsideration of a controversial integument type. Paläontol Z 86:313–331

Butler RJ, Upchurch P, Norman DB (2008) The phylogeny of the ornithischian dinosaurs. J Syst Pal 6:1–40

Chen PJ, Dong ZM, Zheng SN (1998) An exceptionally well preserved theropod dinosaur from the Yixian Formation of China. Nature 391:147–152

Cincotta A, Pestchevitskaya EB, Sinitsa SM et al (2019) The rise of feathered dinosaurs: *Kulindadromeus zabaikalicus*, the oldest dinosaur with 'feather-like' structures. PeerJ 7:e6239

Cuesta E, Diaz-Martinez I, Ortega F, Sanz JL (2015) Did all theropods have chicken-like feet? First evidence of a non-avian dinosaur podotheca. Cretac Res 56:53–59

Davis PG, Briggs DEG (1995) Fossilization of feathers. Geology 23(9):783–786

Foth C (2012) On the identification of feather structures in stem-line representatives of birds: evidence from fossils and actuopalaeontology. Paläontol Zeitschr 68:91–102

Godefroit P, Demuynck H, Dyke G et al (2013) Reduced plumage and flight ability of a new Jurassic paravian theropod from China. Nat Comm 4:1394

Godefroit P, Sinitsa SM, Dhouailly D et al (2014a) A Jurassic ornithischian dinosaur from Siberia with both feathers and scales. Science 345:451–455

Godefroit P, Sinitsa SM, Dhouailly D et al (2014b) Response to Comment on "A Jurassic ornithischian dinosaur from Siberia with both feathers and scales". Science 346. http://sci-hub.tw/10.1126/science.1260146

Göhlich UB, Chiappe LM (2006) A new carnivorous dinosaur from the Late Jurassic Solnhofen archipelago. Nature 440:329–332

Hu DY, Hou L, Zhang L, Xu X (2009) A pre-*Archaeopteryx* troodontid theropod with long feathers on the metatarsus. Nature 461:640–643

Ji Q, Ji S (1996) On discovery of the earliest bird fossil in China (*Sinosauropteryx* gen. nov.) and the origin of birds. Chin Geol 10(233):30–33

Ji Q, Luo ZX, Yuan CX, Tabrum AR (2006) A swimming mammaliaform from the Middle Jurassic and ecomorphological diversification of early mammals. Science 311:1123–1127

Lingham-Soliar T (2010a) Dinosaur protofeathers: pushing back the origin of feathers into the Middle Triassic? J Ornithol 151:193–200

Lingham-Soliar T (2010b) Response to comments by G. Mayr to my paper "Dinosaur protofeathers: pushing back the origin of feathers into the Middle Triassic?". J Ornithol 151:519–521

Lingham-Soliar T (2014) Comment on "A Jurassic ornithischian dinosaur from Siberia with both feathers and scales". Science 346. http://sci-hub.tw/10.1126/science.1259983

Lingham-Soliar T, Plodowski G (2007) Taphonomic evidence for high-speed adapted fins in thunniform ichthyosaurs. Naturwissenschaften 2007:65–70

Lingham-Soliar T, Wesley-Smith J (2008) First investigation of the collagen D-band ultrastructure in fossilized vertebrate integument. Proc R Soc Lond B 275:2207–2212

Lingham-Soliar T, Feduccia A, Wang X (2007) A new Chinese specimen indicates that 'protofeathers' in the early Cretaceous theropod dinosaur *Sinosauropteryx* are degraded collagen fibers. Proc R Soc Lond B 274:1823–1829

Lucas AM, Stettenheim PR (1972) Avian anatomy, integument. US Department of Agriculture, Washington, DC

Lyman RL (1994) Vertebrate taphonomy. Cambridge University Press, Cambridge

Mayr G (2010) Response to Lingham-Soliar: dinosaur protofeathers: pushing back the origin of feathers into the Middle Triassic? J Ornithol 151:523–524

Mayr G, Peters DS, Plodowski G, Vogel O (2002) Bristle-like integumentary structures at the tail of the horned *Psittacosaurus*. Naturwissenschaften 89:361–365

Mayr G, Pittman M, Saitta E, Kaye TG, Vinther J (2016) Structure and homology of Psittacosaurus tail bristles. Palaeontology 59:793–802

Prum RO (1999) Development and evolutionary origin of feathers. J Exp Zool B 285:291–306

Prum RO (2010) Moulting tail feathers in a juvenile oviraptorosaur. Nature 468:E1

Prum RO, Brush AH (2002) The evolutionary origin and diversification of feathers. Q Rev Biol 77:261–295

Rauhut OWM, Foth C, Tischlinger H, Norell MA (2012) Exceptionally preserved juvenile megalosauroid theropod dinosaur with filamentous integument from the Late Jurassic of Germany. Proc Natl Acad Sci USA 109(29):11746–11751

Saveliev SV, Alifanov VR (2014) A new type of skin derivatives in ornithischian dinosaurs from the Late Jurassic of Transbaikalia. Dokl Akad Nauk 456 (2):251–252

Sereno PC (2012) Taxonomy, morphology, masticatory function and phylogeny of heterodontosaurid dinosaurs. ZooKeys 226:1–225

Sinitsa SM (2011) Jurassic dinosaurs of Transbaikalia. In: Sinitsa SM (ed) Environmental cooperative studies in the cross-border ecological region: Russia, China, and Mongolia. Inst. Min. Res., Ecol. Cryol. SB RAS, Chita, pp 173–176

Sinitsa SM, Starukhina S (1986) New data and problems in stratigraphy and palaeontology of the Upper Mesozoic in East Transbaikalia. In: Novye Dannye po Geologii Zabaikal'ya. Min. Geol. RSFSR, Moscow, pp 46–51

Vinther J, Nicholls R, Lautenschlager S et al (2016) 3D camouflage in an ornithischian dinosaur. Curr Biol 26:2456–2462

Wang X, Zhou Z, Zhang F, Xu X (2002) A nearly completely articulated rhamphorhynchoid pterosaur with exceptionally well-preserved wing membranes and 'hairs' from Inner Mongolia, northeast China. Chin Sci Bull 47:226–230

Witmer LM (2009) Fuzzy origins for feathers. Nature 458:293–295

Xu X (2006a) Scales, feathers and dinosaurs. Nature 440:287–288

Xu X (2006b) Feathered dinosaurs from China and the evolution of major avian characters. Integr Zool 1:4–11

Xu X, Zhou Z, Prum RO (2001) Branched integumental structures in *Sinornithosaurus* and the origin of feathers. Nature 410:200–203

Xu X, Zhou Z, Wang X et al (2003) Four-winged dinosaurs from China. Nature 421:335–340

Xu X, Zheng X, You H (2009) A new type of feathers in a nonavian theropod and the early evolution of feathers. Proc Natl Acad Sci USA 106(3):832–834

Xu X, Zheng X, You H (2010) Exceptional dinosaur fossils show ontogenetic development of early feathers. Nature 29:1338–1341

Yang Z, Jiang B, Mc Namara ME et al (2019) Pterosaur integumentary structures with complex feather-like branching. Nat Ecol Evol 3:24–30

Zhang F, Zhou Z, Xu X, Wang X, Sullivan C (2008) A bizarre Jurassic maniraptoran from China with elongate ribbon-like feathers. Nature 455:1105–1108

Zheng XT, You HL, Xu X, Dong ZM (2009) An Early Cretaceous heterodontosaurid dinosaur with filamentous integumentary structures. Nature 458:333–336

5 Filamentous Integuments in Nonavialan Theropods and Their Kin: Advances and Future Perspectives for Understanding the Evolution of Feathers

Xing Xu

5.1 Introduction

Among various vertebrate integumentary appendages, feathers are the most complex ones with a diversity of modern forms (Lingham-Soliar 2015; Lucas and Stettenheim 1972). How the complexity of various feather forms and their diverse functions originated and evolved has intrigued scholars from different disciplines (Brush 1972; Brush 2000; Chuong et al. 2003; Feo et al. 2015; Harris et al. 2005; Prum 1999; Prum and Brush 2002; Vinther et al. 2009; Xu and Guo 2009; Yu et al. 2002). Insights from avian integumentary anatomy and embryology, functional morphology, developmental biology, biochemistry, and some other disciplines have greatly improved our understanding of the origin and early evolution of feathers over the last few decades (Brusatte et al. 2014; Chuong et al. 2000, 2003; Xu and Guo 2009; Xu et al. 2014), but the most significant recent advances have been made in the field of paleontology, as a result of the discoveries of numerous nonavialan dinosaur fossils preserving feather-like structures or feathers from the Jurassic and Cretaceous beds of northeastern China (Godefroit et al. 2013a; Hu et al. 2009; Norell and Xu 2005; Xu and Guo 2009), Myanmar, Germany, Canada, and Russia (Godefroit et al. 2014; Göhlich and Chiappe 2006; Rauhut et al. 2012; Xing et al. 2016; Zelenitsky et al. 2012). The present chapter reviews the research on feathers or feather-like integumentary appendages in those recently uncovered, globally distributed fossils; highlights and discusses remaining controversies and problems in feather evolution; and proposes future directions in this important research area.

5.2 Filamentous Integuments Among Nonavialan Dinosaurs

Filamentous integuments of diverse morphologies are now known to be present in numerous nonavialan dinosaurs from several ornithischian groups, many nonavialan theropod groups (Norell and Xu 2005; Xu and Guo 2009; Xu et al. 2014), and even some pterosaurs as well (Kellner et al. 2010) (Figs. 5.1, 5.2, and 5.3). The morphologies of these structures range from single filaments to a range of more complex structures, including pennaceous feathers (Barrett et al. 2015; Xu and Guo 2009). Most of these filamentous integumental structures have been identified as early feathers, but their relationships to various modern feather morphologies are not

X. Xu (✉)
Key Laboratory of Vertebrate Evolution and Human Origins, Institute of Vertebrate Paleontology and Paleoanthropology, Chinese Academy of Sciences, Beijing, China

CAS Center for Excellence in Life and Paleoenvironment, Beijing, China
e-mail: xuxing@ivpp.ac.cn

C. Foth, O. W. M. Rauhut (eds.), *The Evolution of Feathers*, Fascinating Life Sciences,
https://doi.org/10.1007/978-3-030-27223-4_5

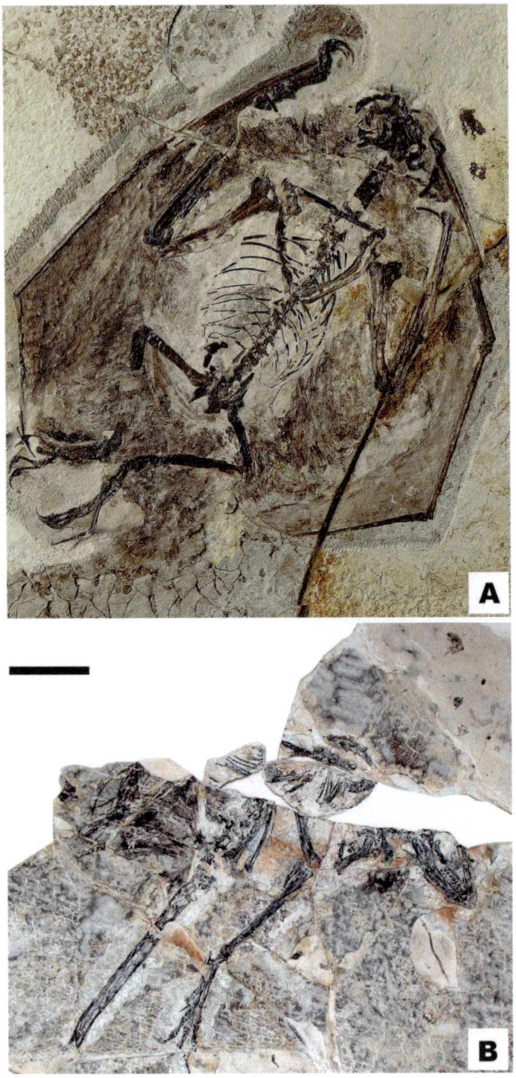

Fig. 5.1 Selected non-theropod avemetatarsalian fossils preserving filamentous integuments. (**a**) The anurognathid pterosaur *Jeholopterus*; (**b**) the heterodontosaurid ornithischian *Tianyulong*. Scale bars, 50 mm

well known because many of their morphological details are often either poorly preserved or have yet to be recovered (Barrett et al. 2015; Xu and Guo 2009). For example, barbules on the pennaceous morphotypes presumably are present in the pennaceous feathers of basal pennaraptorans (Foth et al. 2014), but their morphologies and even their presence have yet to be documented. Other morphotypes present in nonavialan dinosaurs are similar to the bristles and filoplumes of modern birds (Xu and Guo 2009; Xu et al. 2015), but these identifications cannot be confirmed given the lack of preservation of fine morphological features. More controversial is whether some filamentous integuments such as the mono-filamentous integuments seen in some ornithischians are feathers (Barrett et al. 2015) or some kind of feather homolog. Besides the overall poor documentation of the fine morphological details of the fossils of filamentous integument among nonavialan dinosaurs, one of the other lesser resolved issues in these early fossils includes the pattern of distribution of various filamentous structures across the body of these nonavialan dinosaurs and the functions of the diverse filamentous integuments (Xu and Guo 2009).

Nevertheless, at least 13 types of filamentous integumental structures have been reported among nonavialan dinosaurs (Table 5.1) (Xu and Guo 2009; Xu et al. 2010b). The common morphotypes include slender monofilamentous feathers (SMFIs), broad monofilamentous feathers (BMFIs), basally joining filamentous feathers (BJFFs), symmetrical open-vaned feathers (SOVFs), symmetrical close-vaned feathers (SCVFs), and asymmetrical close-vaned feathers (ASVFs). Other than those morphotypes, there are two other unusual forms. One is a proximally ribbon-like close-vaned feather (PRCVF), and the other is a rachis-dominant close-vaned feather (RDCVF). The PRCVFs have been argued to represent sheathed pin feathers (Foth 2012; Prum 2010), but some of their morphological features are not consistent with this identification (Xu et al. 2010a). It is possible that the long ribbon-like proximal portion represents a long calamus, and the PRCVFs are thus a type of calamus-dominant pennaceous feathers.

Several additional morphotypes seem to be present among the known fossil specimens, but are poorly represented in the fossil record. There are five morphotypes in this category: the basally joining shafted filamentous feathers (BJSFFs), the basally joining membranous-based filamentous feathers (BJMBFFs), the radially branched shafted filamentous feathers (RBSFFs), the bilaterally branched filamentous feathers (BBFFs), and the basally joining branched filamentous

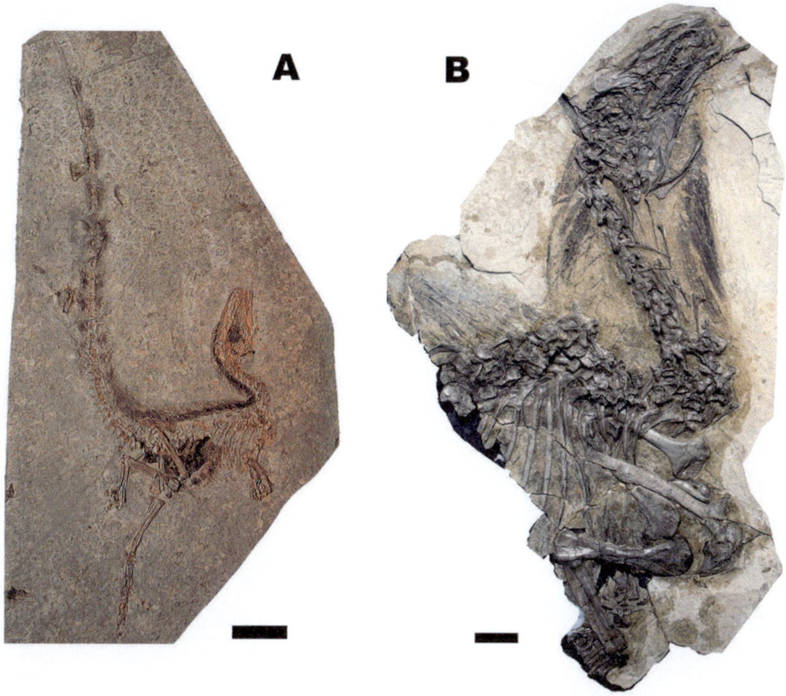

Fig. 5.2 Selected non-pennaraptoran theropod fossils preserving filamentous integuments. (**a**) The compsognathid theropod *Sinosauropteryx*; (**b**) the therizinosauroid theropod *Beipiaosaurus*. Scale bars, 50 mm

feathers (BJBFFs). These less common morphotypes have been identified in some specimens of feathered dinosaurs (Xu and Guo 2009; Xu et al. 2015; Zhang et al. 2008), but the possibility that they are just preservation artifacts cannot be completely excluded.

SMFIs are known in several ornithischians such as the ceratopsian ornithischian *Psittacosaurus* (Mayr et al. 2002), the heterodontosaurid ornithischian *Tianyulong* (Zheng et al. 2009), and the ornithopod ornithischian *Kulindadromeus* (Godefroit et al. 2014), and some theropods such as the megalosaurid theropod *Sciurumimus* (Rauhut et al. 2012), possibly the ornithomimid theropod *Ornithomimus* (Zelenitsky et al. 2012), and possibly the likely compsognathid theropod *Juravenator* (Chiappe and Göhlich 2010). BMFIs are restricted to theropods and are known in the tyrannosauroid theoropod *Yutyrannus* (Xu et al. 2012), possibly the ornithomimid theropod *Ornithomimus*[1] (Zelenitsky et al. 2012), and the therizinosauroid theropod *Beipiosaurus* (Xu et al. 2009). BJFFs are known in the ornithopod ornithischian *Kulindadromeus,* the dromaeosaurid theropod *Sinornithosaurus* (Xu et al. 2001), possibly the scansoriopterygid theropods *Epidexipteryx* (Zhang et al. 2008) and *Yi* (Xu et al. 2015), and the troodontid theropod *Anchiornis* (Hu et al. 2009). BJSFFs are present in *Sinornithosaurus* (Xu and Guo 2009), and SOVFs occur in a basal maniraptoran theropod (Xing et al. 2016). SCVFs are preserved in the oviraptorosaurian theropods *Protarchaeopteryx*, *Caudipteryx* (Ji et al. 1998; Zhou and Wang 2000; Zhou et al. 2000), and *Incisivosaurus*[2] (Xu et al. 2010b), the anchiornithine theropod *Anchiornis* and its kin (Godefroit et al. 2013a, b; Hu et al. 2009), the troodontid theropod *Jinfengopteryx* (Ji et al. 2005; Norell and Xu 2005; Xu and Norell 2006), and the

[1] The pennaceous feathers inferred to be present in *Ornithomimus* are probably BMFFs based on the preserved morphology.

[2] STM4-1 and STM 22-6 were referred to *Similicaudipteryx* in the original description, but reexamination of the two specimens suggest that it is more proper to assign the two specimens to *Incisivosaurus* because the former closely resemble the latter in some dental features, which are unique to *Incisivosaurus*.

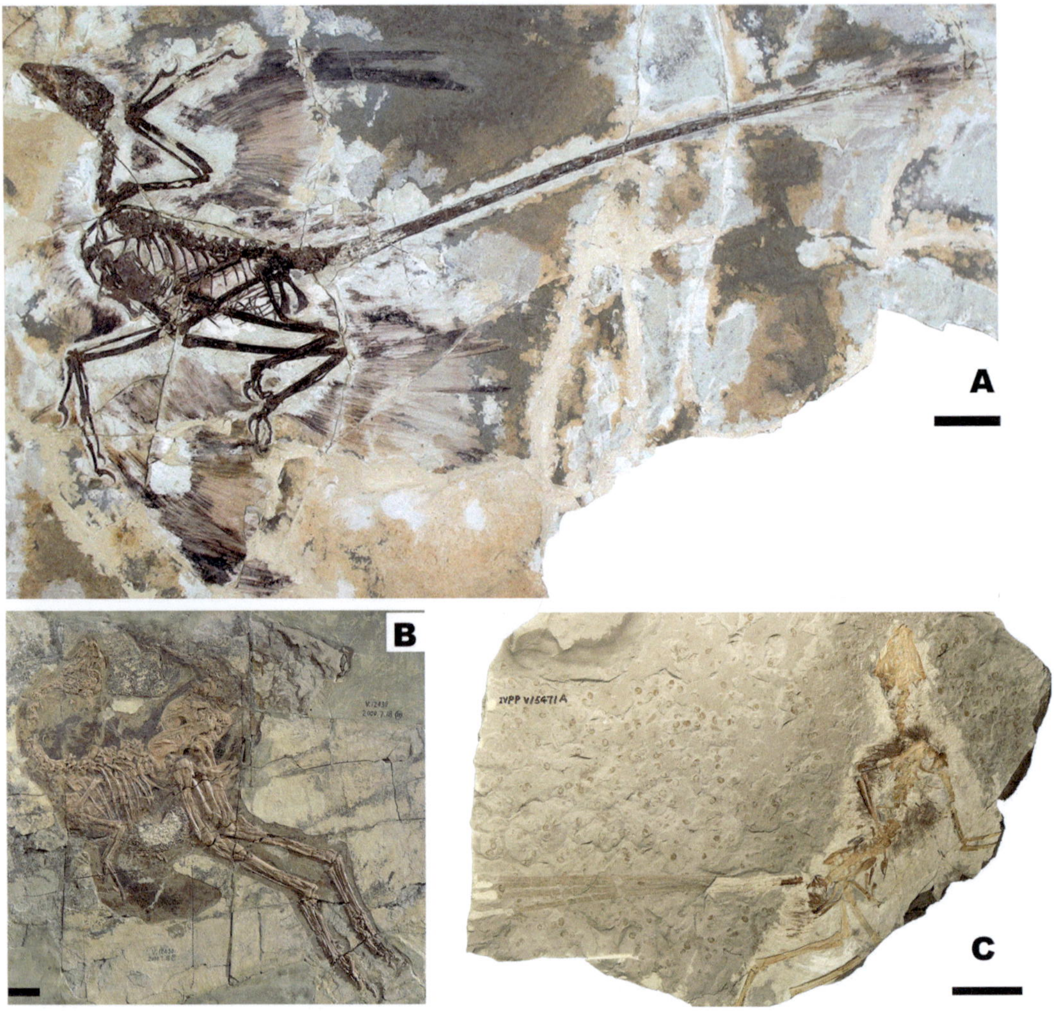

Fig. 5.3 Selected pennaraptoran theropod fossils preserving feathers. (**a**) The droameosaurid theropod *Microraptor*; (**b**) the oviraptorosaurian theropod *Caudipteryx*; (**c**) the scansoriopterygid theropod *Epidexipteryx*. Scale bars, 50 mm

dromaeosaurid theropods *Sinornithosaurus* (Ji et al. 2001) and *Microraptor* (Xu et al. 2003). ASVFs are known in *Microraptor* (Xu et al. 2003), the troodontid *Jianianhualong* (Xu et al. 2017), and the anchiornithine *Caihong* (Hu et al. 2018). PRCVFs are preserved in *Incisivosaurus* (Xu et al. 2010b), and RDCVFs have been found in *Epidexipteryx* (Zhang et al. 2008). RBSFFs occur in *Sinornithosaurus* (Xu et al. 2001), and BJBFFs were reported in *Yi* (Xu et al. 2015). Some filamentous feathers of several nonavialan dinosaurs might be referable to the morphotypes above, but cannot be confidently grouped with them at present. For example, some filamentous feathers of the compsognathid theropod *Sinosauropteryx* appear to have a short quill and long barbs, and are possibly referable to RBSFFs, some filamentous feathers of the basal tyrannosauroid theropod *Dilong* appear to be BBFFs, some filamentous feathers of *Beipiaosaurus* might be RBSFFs, and the extremely short, filamentous elements reported to be associated with the alvarezsauroid theropod *Shuvuuia* (Schweitzer et al. 1999) might be SMFIs. All of these morphotype identifications need confirmation with better preserved fossils.

The patterns of distribution of some of these morphotypes across the dinosaurian body can be reconstructed confidently for some taxa. SMFIs cover most of the body of *Tianyulong* (Zheng et al. 2009) and *Kulindadromeus* (Godefroit

Table 5.1 Major filamentous integumentary structures found among nonavialan dinosaurs

Morphotype	Abbreviation	Definition
Slender monofilamentous integument	SMFI	Single slender filament
Broad monofilamentous integument	BMFI	Single broad filament
Basally joining filamentous feather	BJFF	Several filaments joined at their base without a central shaft
Symmetrical open-vaned feather	SOVF	A slender central shaft and open vanes of similar size on either side
Symmetrical close-vaned feather	SCVF	A thick central shaft and closed vanes of similar size on either side
Asymmetrical close-vaned feather	ASVF	A thick central shaft and closed vanes of different size on either side
Proximally ribbon-like close-vaned feather	PRCVF	A long ribbon-like proximal portion and a pennaceous distal portion
Rachis-dominant close-vaned feather	RDCVF	A hypertrophied rachis and a small/reduced pennaceous portion
Basally joining shafted filamentous feather	BJSFF	Multiple filaments joined at their bases on the distal end of a slender central shaft
Basally joining membranous-based filamentous feather	BJMBFF	Multiple parallel filaments joined at their base to a membranous structure
Radially branched shafted filamentous feather	RBSFF	Multiple filaments that branch radially along the length of a central filament
Bilaterally branched filamentous feather	BBFF	Multiple filaments that branch bilaterally along the length of a central filament
Basally joining branched filamentous feather	BJBFF	Several BJSFFs, RBSFFs, or BBFFs that are joined at their bases

et al. 2014), but they are restricted only to the middle portion of the tail in *Psittacosaurus* (Mayr et al. 2002). BMFIs are distributed over the skull, anterior neck, and posterior tail in *Beipiaosaurus* in a restricted pattern on its body (Xu et al. 2009), and they probably are present along the arm and posterior tail in *Yutyrannus* (Xu et al. 2012). The simply branched feathers such as BJFFs, BJSFFs, RBSFFs, or BBFFs seem to cover much of the body of *Sinosauropteryx* (Currie and Chen 2001) and *Yutyrannus* (Xu et al. 2012). One of the most unusual feather distribution patterns is the extensive feathering on the lower parts of the legs (and feet) of various theropods, including basal birds (Foth et al. 2014; Xu et al. 2003; Zheng et al. 2013). Flight feathers are typically only seen in forelimbs and tail of modern birds, but large SCVFs and ACVFs are present along the lower legs in a number of basal paravians, including basal birds, and in some cases even along the metatarsals (Foth et al. 2014; Xu et al. 2003; Zheng et al. 2013). Extensively feathered feet appear to be a plesiomorphic feature for Coelurosauria (Zheng et al. 2013).

5.3 Current Consensus on the Origin and Early Evolution of Feathers

The discoveries of various filamentous integuments in different nonavialan dinosaurs have provided significant new information on feather evolution (Chen et al. 1998; Ji et al. 1998, 2001; Norell and Xu 2005; Wellnhofer 2004; Witmer 2009a, b; Xu and Guo 2009; Xu et al. 2001, 2009, 2010b; Zhang and Zhou 2006). In combination with recent insights from developmental biology and other morphological aspects of modern feather diversity (Alibardi 2005; Brush 2000; Chuong et al. 2000; Evangelista et al. 2014; Feo et al. 2015; Foth 2011, 2012; Lin et al. 2013a, b; McNamara 2013; McNamara et al. 2013; Moyer et al. 2014; Prum 1999), these new data together have helped to reach some consensus in regard to the overall pattern of feather evolution.

In general, it is accepted now that the first feathers are simple, filamentous structures.

Feather complexity increased incrementally toward the origin of birds, and various feather types, including the highly complex flight feathers, appeared before the evolution of the earliest birds (Xu and Guo 2009; Xu et al. 2014).

Mapping these various filamentous integuments preserved as fossils onto a theropod phylogeny (Fig. 5.4), we can identify several stages in feather evolution: Stage I, the appearance of the filamentous morphology at least at the base of the Tetanurae (e.g., SMFIs in *Sciurumimus)*; Stage II, appearance of the primary branching morphology (e.g., barbs) and planar morphology at the base of the Coelurosauria (e.g., various branched filamentous feathers, including bilaterally branched feathers in the compsognathids and tyrannosauroids); Stage III, appearance of the secondary branching (i.e., plumulaceous barbules) and open vanes at the base of the Maniraptora (e.g., SOVFs in a basal maniraptoran preserved in the amber); Stage IV, appearance of closed vanes (pennaceous barbules) at the base of the Pennaraptora (e.g., SCVFs in several oviraptorosaurians and several deinonychosaurs); and Stage V, appearance of vane asymmetry at the base of Paraves (e.g., ACVFs in basal avialans and several deinonychosaurs). These evolutionary stages are consistent with the evolutionary stages inferred from developmental data derived from living birds, which include five major stages from an undifferentiated cylinder (stage I) to the pennaceous feathers with a closed vane (Prum 1999).

Besides the evolutionary sequence of major morphological features (above), some consensus also has been reached regarding the evolutionary sequence of the various functions of feathers. Locomotion (including flight), insulation, and display represent the three major functions exhibited among modern birds and their feathers, and various other functions also play important roles (Lucas and Stettenheim 1972; Prum and Brush 2002). Although certain morphological features of various feathers can be correlated with certain functions, these correlations are sometimes complex and weak. Nevertheless, functional inferences can be made based on the integumentary features, and they can be further strengthened if relevant osteological data and other lines of data converge on the same functional inference (Xu et al. 2011). For example, a coherent airfoil surface formed by close-vaned feathers may indicate a locomotory function if it is coincidently present in a long-armed theropod, an insulating layer formed by numerous slender filaments suggests an insulatory function, and display function can be achieved by striking colors, distinctive color patterns, or highly specialized shapes of various feathers. These three major functions all have been inferred to be present among nonavialan dinosaurs based on different lines of data (Chen et al. 1998; Evangelista et al. 2014; Feo et al. 2015; Han et al. 2014; Li et al. 2010, 2012, 2014; Xu and Guo 2009; Xu et al. 2014; Zhang et al. 2008).

Most researchers agree that the use of feathers for flight was the last major function that appeared out of the three major functions mentioned above (Ji et al. 1998), but that flight capability might have evolved early in paravian evolution instead of at the base of the Avialae or alternatively flight might have evolved even multiple times independently within the Paraves. This hypothesis suggests that even volant locomotion, a key feature previously used to define birds, is not restricted to birds (Brusatte et al. 2014; Xu and Guo 2009; Xu et al. 2003, 2011), and it is likely that feathers were exapted into a flight function from their initial uses for insulation and display/camouflage (Foth et al. 2014; Li et al. 2014; Xu and Guo 2009; Xu et al. 2009; Zhang et al. 2008, 2010). For example, a display function appeared early in coeluroaurian theropod evolution as indicated by the BMFIs in basal coelurosaurians such as *Yutyrannus* and *Beipiaosaurus* (Xu and Guo 2009; Xu et al. 2010b), and an insulatory function presumably

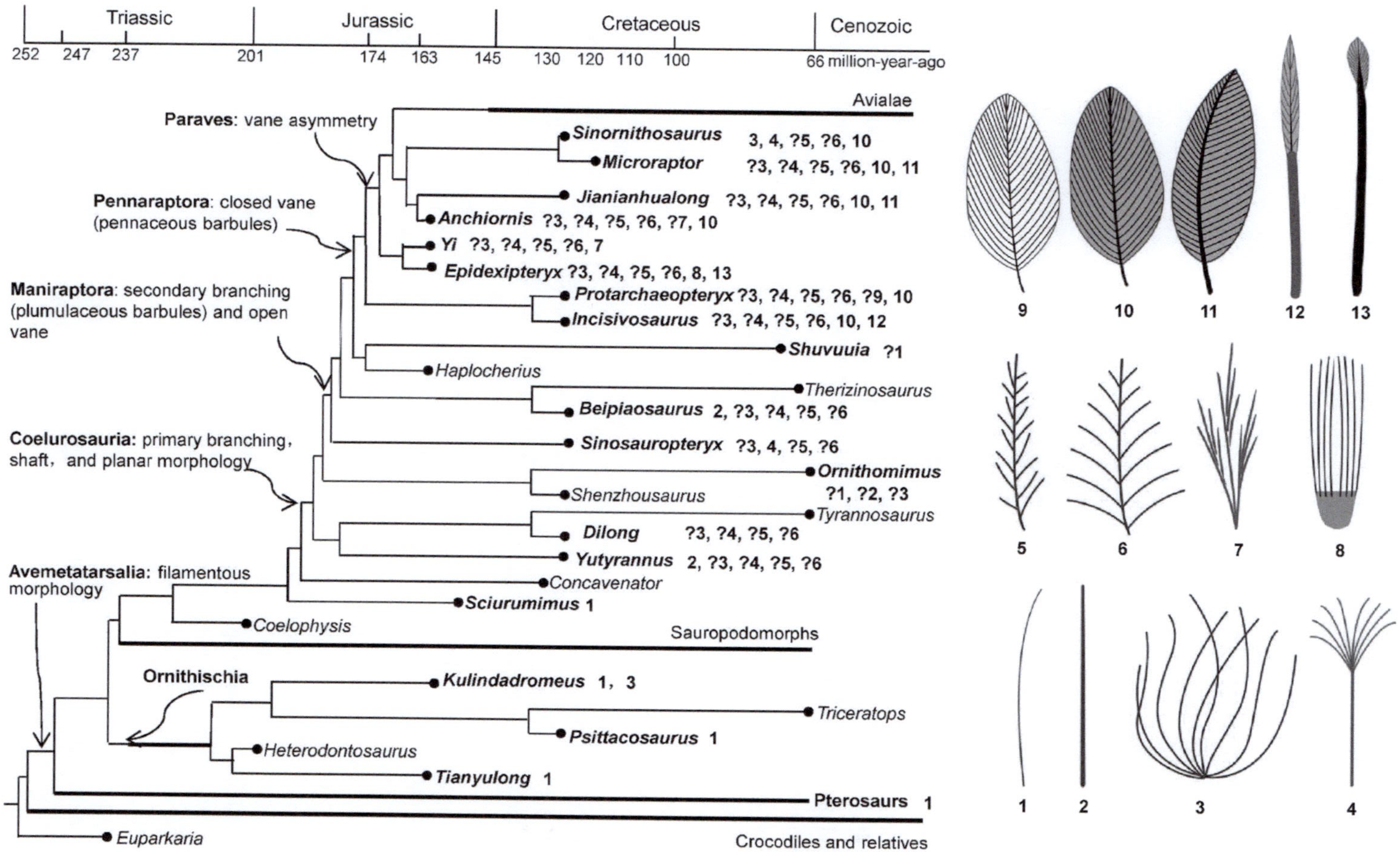

Fig. 5.4 Known feather-like or feather morphotypes across a simplified avemetatarsalian phylogeny. "?" indicates uncertainty regarding the presence of this particular morphotype in this group. 1, SMFF; 2, BMFF; 3, BJFF; 4, BJSFF; 5, RBSFF; 6, BBFF; 7, BJBFF; 8, BJMBFF; 9, SOVF; 10, SCVF; 11, ASVF; 12, PRCVF; 13, RDCVF

appeared early in tetanuran evolution (Rauhut et al. 2012), if not directly tied to the earliest feathers or feather-like structures.

5.4 Remaining Controversial/Unresolved Issues in Feather Evolution

The evolutionary sequence of various feather morphologies occurring among different nonavialan dinosaurs is generally congruent with predictions based on a developmental model (Prum 1999). Along with that model, a clear evolutionary picture for feather origination and evolution is emerging, but there are still many remaining controversial and unresolved issues in feather evolution. For example, at what point in archosaurian phylogeny did the first feathers originate? How did the major microstructural features of feathers evolve? Did early feathers develop ontogenetically in the same way as modern feathers (from a follicle in a feather tract)? Did the major features of feathers appear incrementally in evolution, or did at least some appear simultaneously? And finally, what were the primary functions of various early feathers?

The discoveries of SMFIs in some theropods, several ornithischian groups and in some pterosaurs has led to the proposal of a hypothesis for feather origins in avemetatarsalians (Xu and Guo 2009; Xu et al. 2009; Zheng et al. 2009), suggesting that feathers represent a diagnostic feature of the Avemetatarsalia (a clade including both dinosaurs and pterosaurs). The alternative hypothesis, the theropod hypothesis for feather origin, suggests that feathers evolved only within Theropoda (Barrett et al. 2015). In this latter case, the SMFIs exhibited in some pterosaurs and ornithischians would have nothing to do with modern feathers, and they are just highly specialized integumentary appendages that evolved convergently in a closely related group. Whether the avemetatarsalian or the theropod hypothesis is more acceptable depends much on the identity (homology) of the SMFIs of pterosaurs and ornithischians, and some exceptionally preserved pterosaurian fossils seem to provide strong evidence for the avemetatarsalian hypothesis (Yang et al. 2018).

Even though the avemetatarsalian hypothesis for the origin of feathers can be confirmed (or refuted) with additional specimens and data, the primary function for the first feathers remains unresolved. Clearly the earliest feathers were not used for flight purposes. While an insulatory function is plausible for the earliest and morphologically simplest feathers and feather-like structures, some morphological features seen among early feathers suggest that a display function is equally possible or even more likely. The SMFIs in *Tianyulong* and *Psittacosaurus* are long and rigid, and in the latter taxon have a limited distribution on the body. Furthermore, their density appears to be relatively low compared to that of the presumably insulative filamentous integumentary appendages in modern birds and mammals. Furthermore, the filamentous feathers in basal tetanurans are proportionally thick and fairly stiff in appearance (Currie and Chen 2001; Xu and Guo 2009). This rigidity suggests that display is a viable hypothesis for earliest function of feathers. The primary functions for some other types of feathers such as SCVFs and ACVFs also are debated. The SCVFs have been suggested to have performed various functions such as balancing during locomotion, aiding in the capture of insects, attracting females, or species signals among nonavialan dinosaurs. Among these functions, a display function has been favored more recently for the SCVFs of various basal pennaraptorans (Foth et al. 2014; Li et al. 2010) and even for the tail ACVFs of *Microraptor* (Li et al. 2012). While SCVFs in some basal paravians such as *Anchiornis* have been suggested to have functioned primarily for display (Foth et al. 2014; Li et al. 2010), they could have also been used in aerodynamic locomotion as indicated by coherent airfoil surfaces formed by the SCVFs, consistent with many other features that imply an aerodynamic function in basal paravians, including long and robust arms, flight-adapted cerebral features, and significantly improved metabolism (Balanoff et al. 2013; Dececchi and Larrson 2013; Xu et al. 2014).

It is worthy of mention that a display function is particularly favored by melanosome data derived from several nonavialan theropod fossils (Li et al. 2010, 2012; Zhang et al. 2010). One study even suggests that melanosomes in wing feathers of *Archaeopteryx* might have strengthened *Archaeopteryx*'s feathers for flight (Carney et al. 2012). While the discoveries of fossilized melanosomes opened a new window into the Mesozoic ecosystems (Vinther et al. 2008), some other studies have cautioned the identification of microbodies preserved in fossils as fossilized melanosomes (Lindgren et al. 2015; Moyer et al. 2014). Nevertheless, no evidence has been presented so far to reject the identification of microbodies preserved in fossils as melanosomes. Fossilized melanosomes will continue to provide important data for us to understand various aspects of extinct animals and test different hypotheses regarding the primary functions and even biomechanics of early feathers.

Some uncertainties pertaining to feather evolution are the result of a lack of knowledge about the fine morphological features of various feathers or feather-like structures seen in nonavialan dinosaurs. For example, pennaceous barbules presumably are present in all close-vaned feathers of nonavialan theropods, but to date no direct evidence has been presented. The presence and morphologies of other types of barbules, such as plumulaceous barbules, also are poorly known among various feathers in nonavialan dinosaurs, resulting in a difficulty for the reconstruction of the evolutionary sequence of the rachis and related barbules. Compared to feathered dinosaur fossils from lacustrine deposits, feathers preserved in amber display much finer details which can be comparable to those in modern feathers (McKellar et al. 2011; Vincent 2008; Xing et al. 2016), but in most cases, these feathers are difficult to be placed phylogenetically, and thus their evolutionary implications are less than other larger phylogenetically constrained body fossils.

The lack of morphological details in some specimens also results in a key question remaining unresolved: the homologous relationships between some feather-like structures or various structures seen in the fossil feathers of nonavialan dinosaurs and those of modern feathers and their structures (Saitta et al. 2018). For example, it is not known whether the SMFIs and/or BMFIs are homologous to some types of modern feathers (e.g., the quill feathers in cassowaries (Prum 2005)) or to some parts of modern feathers such as the barbs or rachis. It is not known whether the membranous base of the BJMBFFs represents a type of structure absent among extant birds, or if it is homologous to certain parts of modern feathers (e.g., a large calamus), and we also do not know whether the ribbon-like portion of the PRCVFs are homologous to the tubular feather sheath of emergent feathers (Prum 2010) or to a long calamus of a unusual type of feather.

The developmental data from modern feathers have greatly improved our understanding of feather origin and diversification (Chu et al. 2014; Chuong et al. 2000, 2003, 2012; Jiang et al. 2011; Lin et al. 2013a, b; O'Connor et al. 2012; Prum 2005; Prum and Brush 2002; Widelitz et al. 2003; Wu et al. 2004; Yu et al. 2002; Yue et al. 2005, 2006, 2012) and in particular, recent data from fossil feathers (Norell and Xu 2005; Xu and Guo 2009; Xu et al. 2014) overall support the developmental model of the feather origin and evolution proposed by Prum (1999). However, the variable feathering in the tails of the tetanuran theropod *Juravennator*, the ceratopsian ornithischian *Psittacosaurus*, and the early-diverging ornithischian *Kulindadromeus*, and in other parts of the body of some other dinosaurian taxa (e.g., the lower legs of the basal paravians, and the arms and tail of *Incisivosaurus* during different ontogenetic stages) suggests that there are some developmental differences between early feathers and modern feathers, and these early feathers seem to be less constrained and more flexible in some developmental features than modern feathers (Xu et al. 2010b). Consequently, while it is important to use developmental data from modern feathers to reconstruct the evolutionary history of feathers, caution should be paid on such practices as the developmental features of feathers also evolved.

A better understanding of the origin and evolution of feathers requires a multidisciplinary approach. While new data from neontological morphology, developmental biology, and even comparative genomics have improved our understanding of the morphogenesis, development, and functions of feathers (Alibardi 2005; Chuong et al. 2000; Feo et al. 2015; Prum 2005; Seki et al. 2017), and additionally have provided broad insights into the evolution of feathers. However, it is important to emphasize here that the data from fossil specimens play a key role in reconstructing the evolutionary history of feathers by providing diverse morphological, developmental, phylogenetic, temporal, and other data not available from living species. Consequently, an increase in data from fossil feathers, derived from either the investigation of new fossil specimens or the application of new analytical or other techniques, will provide the final say in the ongoing research to develop an evolutionary model for feathers from their dinosaurian (or avemetatarsalian) origins to their modern diversity.

Acknowledgments The author thanks Thomas Stidham for commenting on the manuscript; Richard Prum, Zhonghe Zhou, Derek Briggs, and Julia Clarke for discussions; and Yi Liu, Lida Xing, and Hailong Zang for illustrations. This study is supported by the National Natural Science Foundation of China (41120124002 and 41688103) and the Strategic Priority Research Program of the Chinese Academy of Sciences (XDB18030504).

References

Alibardi L (2005) Cell structure of developing barbs and barbules in downfeathers of the chick: Central role of barb ridge morphogenesis for the evolution of feathers. J Submicrosc Cytol Pathol 37:19–41

Balanoff AM, Bever GS, Rowe TB, Norell1 MA (2013) Evolutionary origins of the avian brain. Nature 501:93–96

Barrett PM, Evans DC, Campione NE (2015) Evolution of dinosaur epidermal structures. Biol Lett 11:20150229. https://doi.org/10.1098/rsbl.2015.0229

Brusatte S, Lloyd G, Wang S, Norell M (2014) Gradual assembly of avian body plan culminated in rapid rates of evolution across the dinosaur-bird transition. Curr Biol 24:1–7

Brush A (1972) Correlation of protein electrophoretic pattern with morphology of normal and mutant feathers. Biochem Genet 7:87–93

Brush AH (2000) Evolving a protofeather and feather diversity. Am Zool 40:631–639

Carney R, Vinther J, Shawkey MD, d'Alba L, Ackermann J (2012) New evidence on the colour and nature of the isolated *Archaeopteryx* feather. Nat Commun 3:637. https://doi.org/10.1038/ncomms1642

Chen PJ, Dong ZM, Zhen SN (1998) An exceptionally well-preserved theropod dinosaur from the Yixian Formation of China. Nature 391:147–152

Chiappe LM, Göhlich UB (2010) Anatomy of *Juravenator starki* (Theropoda: Coelurosauria) from the Late Jurassic of Germany. Neues Jahrbuch für Geologie und Paläontologie - Abhandlungen 258:257–296

Chu Q et al (2014) Dkk2/Frzb in the dermal papillae regulates feather regeneration. Dev Biol 387:167–178

Chuong C-M, Chodankar R, Widelitz RB, Jiang TX (2000) Evo-Devo of Feathers and Scales: building complex epithelial appendages. Curr Opin Genet Dev 10:449–456

Chuong C-M et al (2003) Adaptation to the sky: defining the feather with integument fossils from Mesozoic China and experimental evidence from molecular laboratories. J Exp Zool (Mol Dev Evol) 298B:42–56

Chuong C-M, Randall V, Widelitz R, Wu P, Jiang T (2012) Physiological regeneration of skin appendages and implications for regenerative medicine. Physiology (Bethesda) 27:61–72

Currie PJ, Chen P-J (2001) Anatomy of *Sinosauropteryx prima* from Liaoning, northeastern China. Can J Earth Sci 38:1705–1727

Dececchi A, Larrson HCE (2013) Body and limb size dissociation at the origin of birds: Uncoupling allometric constraints across a macroevolutionary transition. Evolution 67:2741–2752

Evangelista D et al (2014) Aerodynamic characteristics of a feathered dinosaur measured using physical models—effects of form on static stability and control effectiveness. PLoS One 9:e85203. https://doi.org/10.1371/journal.pone.0085203

Feo TJ, Field DJ, Prum RO (2015) Barb geometry of asymmetrical feathers reveals a transitional morphology in the evolution of avian flight. Proc R Soc B Biol Sci 282:20142864. https://doi.org/10.1098/rspb.2014.2864

Foth C (2011) The morphology of neoptile feathers: ancestral state reconstruction and its phylogenetic implications. J Morphol 272:387–403

Foth C (2012) On the identification of feather structures in stem-line representatives of birds: evidence from fossils and actuopalaeontology. Palaontol Z 86:91–102

Foth C, Tischlinger H, Rauhut OWM (2014) New specimen of *Archaeopteryx* provides insights into the evolution of pennaceous feathers. Nature 511:79–82

Godefroit P, Cau A, Hu D-Y, Escuillié F, Wu W, Dyke G (2013a) A Jurassic avialan dinosaur from China

resolves the early phylogenetic history of birds. Nature 498:359–362. https://doi.org/10.1038/nature12168
Godefroit P, Demuynck H, Dyke G, Hu D-Y, Escuillié F, Claeys P (2013b) Reduced plumage and flight ability of a new Jurassic paravian theropod from China. Nat Commun 4:1394. https://doi.org/10.1038/ncomms2389
Godefroit P et al (2014) A Jurassic ornithischian dinosaur from Siberia with both feathers and scales. Science 345:451–455
Göhlich UB, Chiappe LM (2006) A new carnivorous dinosaur from the Late Jurassic Solnhofen archipelago. Nature 440:329–332
Han G et al (2014) A new raptorial dinosaur with exceptionally long feathering provides insights into dromaeosaurid flight performance. Nat Commun 5:4382. https://doi.org/10.1038/ncomms5382
Harris MP, Williamson S, Fallon JF, Meinhardt H, Prum RO (2005) Molecular evidence for an activator-inhibitor mechanism in development of embryonic feather branching. Proc Natl Acad Sci USA 102:11734–11739
Hu D-Y, Hou L-H, Zhang LJ, Xu X (2009) A pre-*Archaeopteryx* troodontid from China with long feathers on the metatarsus. Nature 461:640–643
Hu D-Y et al (2018) A bony-crested Jurassic dinosaur with evidence of iridescent plumage highlights complexity in early paravian evolution. Nat Commun 9:217. https://doi.org/10.1038/s41467-017-02515-y
Ji Q, Currie PJ, Norell MA, Ji S-A (1998) Two feathered dinosaur from China. Nature 393:753–761
Ji Q, Norell MA, Gao K-Q, Ji S-A, Ren D (2001) The distribution of integumentary structures in a feathered dinosaur. Nature 410:1084–1088
Ji Q, Ji SA, Lu JC, You HL, Chen W, Liu YQ, Liu YX (2005) First avialian bird from China. Geol Bull China 24:197–210
Jiang TX, Tuan TL, Wu P, Widelitz RB, Chuong CM (2011) From buds to follicles: Matrix metalloproteinases in developmental tissue remodeling during feather morphogenesis. Differentiation 81:307–314
Kellner A, Wang XL, Tischlinger H, Campos DA, Hone D, Meng X (2010) The soft tissue of *Jeholopterus* (Pterosauria, Anurognathidae) and the structure of the pterosaur wing membrane. Proc R Soc B Biol Sci 277:321–330. https://doi.org/10.1098/rspb.2009.0846
Li QG et al (2010) Plumage color patterns of an extinct dinosaur. Science 327:1369–1372
Li QG et al (2012) Reconstruction of *Microraptor* and the evolution of iridescent plumage. Science 335:1215–1219
Li QG et al (2014) Melanosome evolution indicates a key physiological shift within feathered dinosaurs. Nature 507:350–353
Lin S et al (2013a) Topology of feather melanocyte progenitor niche allows complex pigment patterns to emerge. Science 340:1442–1445
Lin S et al (2013b) Feather regeneration as a model for organogenesis. Dev Growth Differ 55:139–148
Lindgren J et al (2015) Molecular composition and ultrastructure of Jurassic paravian feathers. Sci Rep 5:13520
Lingham-Soliar T (2015) The vertebrate integument volume 2: structure, design and function. Springer, New York
Lucas AM, Stettenheim PR (1972) Avian anatomy: integument. United States Department of Agriculture, Washington, DC
Mayr G, Peters DS, Plodowski G, Vogel O (2002) Bristle-like integumentary structures at the tail of the horned dinosaur *Psittacosaurus*. Naturwissenschaften 89:361–365
McKellar RC, Chatterton BDE, Wolfe AP, Currie PJ (2011) A diverse assemblage of Late Cretaceous dinosaur and bird feathers from Canadian Amber. Science 333:1619–1622. https://doi.org/10.1126/science.1203344
McNamara ME (2013) The taphonomy of colour in fossil insects and feathers. Palaeontology 56:557–575
McNamara ME, Briggs DEG, Orr PJ, Field DJ, Wang Z (2013) Experimental maturation of feathers: implications for reconstructions of fossil feather colour. Biol Lett 9:20130184
Moyer AE, Zheng W, Johnson EA, Lamanna M, Li D, Lacovara KJ, Schweitzer MH (2014) Melanosomes or microbes: testing an alternative hypothesis for the origin of microbodies in fossil feathers. Sci Rep 4:4233
Norell M, Xu X (2005) Feathered dinosaurs. Annu Rev Earth Planet Sci 33:277–299
O'Connor J, Chiappe L, Chuong C-M, Bottjer D, You H-L (2012) Homology and potential cellular and molecular mechanisms for the development of unique feather morphologies in early birds. Geosciences (Basel) 2:157–177
Prum RO (1999) Development and evolutionary origin of feathers. J Exp Zool (Mol Dev Evol) 285:291–306
Prum RO (2005) Evolution of the morphological innovations of feathers. J Exp Zool (Mol Dev Evol) 304B:570–579
Prum RO (2010) Moulting tail feathers in a juvenile oviraptorisaur. Nature 468:E1
Prum RO, Brush AH (2002) The evolutionary origin and diversification of feathers. Q Rev Biol 77:261–295
Rauhut OWM, Foth C, Tischlinger H, Norell MA (2012) Exceptionally preserved juvenile megalosauroid theropod dinosaur with filamentous integument from the Late Jurassic of Germany. Proc Natl Acad Sci USA 109:11746–11751. https://doi.org/10.1073/pnas.1203238109
Saitta ET, Gelernter R, Vinther J (2018) Additional information on the primitive contour and wing feathering of paravian dinosaurs. Palaeontology 61:273–288
Schweitzer MH, Watt JA, Avci R, Knapp L, Chiappe L, Norell M, Marshall M (1999) Beta-keratin specific immunological reactivity in feather-like structures of

the Cretaceous alvarezsaurid, *Shuvuuia deserti*. J Exp Zool (Mol Dev Evol) 285:146–157
Seki R et al (2017) Functional roles of Aves class specific cis-regulatory elements on macroevolution of bird-specific features. Nat Commun 8:14229
Vincent P (2008) Early evolution of feathers: fossil evidence from Cretaceous amber of France. Proc R Soc B 275(1639):1197–1202
Vinther J, Briggs DEG, Prum RO, Saranathan V (2008) The colour of fossil feathers. Biol Lett 4:522–525
Vinther J, Briggs DEG, Clarke J, Mayr G, Prum RO (2009) Structural coloration in a fossil feather. Biol Lett 6:128–131. https://doi.org/10.1098/rsbl.2009.0524
Wellnhofer P (2004) The plumage of *Archaeopteryx*: feathers of a dinosaur? In: Currie PJ, Koppelhus EB, Shugar MA, Wright JL (eds) Feathered dragons—studies on the transition from dinosaurs to birds. Indiana University Press, Bloomington, pp 282–300
Widelitz RB et al (2003) Molecular biology of feather morphogenesis: a testable model for Evo-Devo research. J Exp Zool (Mol Dev Evol) 298B:109–122
Witmer L (2009a) Feathered dinosaurs in a tangle. Nature 461:601–602
Witmer LM (2009b) Fuzzy origins for feathers. Nature 458:293–294
Wu P et al (2004) Evo-Devo of amniote integuments and appendages. Int J Dev Biol 48:249–270
Xing L et al (2016) A feathered dinosaur tail with primitive plumage trapped in mid-Cretaceous amber. Curr Biol 26:1–9. https://doi.org/10.1016/j.cub.2016.10.008
Xu X, Guo Y (2009) The origin and early evolution of feathers: insights from recent paleontological and neontological data. Vertebrata PalAsiatica 47:311–329
Xu X, Norell MA (2006) Non-avian dinosaur fossils from the Lower Cretaceous Jehol Group of western Liaoning, China. Geol J 41:419–438
Xu X, Zhou Z-H, Prum RO (2001) Branched integumental structures in *Sinornithosaurus* and the origin of feathers. Nature 410:200–204
Xu X, Zhou Z-H, Wang X-L, Kuang X-W, Zhang F-C, Du X-K (2003) Four-winged dinosaurs from China. Nature 421:335–340
Xu X, Zheng XT, You HL (2009) A new feather type in a nonavian theropod and the early evolution of feathers. Proc Natl Acad Sci USA 106:832–834
Xu X, Zheng X, You H (2010a) Reply to "Moultingtail feathers inajuvenileoviraptorisaur". Nature 468:E2. https://doi.org/10.1038/nature09481
Xu X, Zheng XT, You HL (2010b) Exceptional dinosaur fossils show ontogenetic development of early feathers. Nature 464:1338–1341
Xu X, You H, Du K, Han F (2011) An *Archaeopteryx*-like theropod from China and the origin of Avialae. Nature 475:465–470
Xu X et al (2012) A gigantic feathered dinosaur from the Lower Cretaceous of China. Nature 484:92–95
Xu X, Zhou ZH, Dudley R, Mackem S, Chuong CM, Erickson GM, Varricchio DD (2014) An integrative approach to understanding bird origins. Science 346:1253293. https://doi.org/10.1126/science.1253293
Xu X et al (2015) A bizarre Jurassic maniraptoran theropod with preserved evidence of membranous wings. Nature 521:70–73. https://doi.org/10.1038/nature14423
Xu X et al (2017) Mosaic evolution in an asymmetrically feathered troodontid dinosaur with transitional features. Nat Commun 14972:1–12
Yang Z, Jiang B, McNamara ME, Kearns SL, Pittman M, Kaye TG, Orr PJ, Xu X, Benton MJ (2018) Pterosaur integumentary structures with complex feather-like branching. Nat Ecol Evol 3:24–30
Yu M-K, Wu P, Widelitz RB, Chuong C-M (2002) The morphogenesis of feathers. Nature 420:308–312
Yue Z, Jiang T, Widelitz R, Chuong C (2005) Mapping stem cell activities in the feather follicle. Nature 438:1026–1029
Yue Z, Jiang T, Widelitz R, Chuong C (2006) Wnt3a gradient converts radial to bilateral feather symmetry via topological arrangement of epithelia. Proc Natl Acad Sci USA 103:951–955
Yue Z, Jiang T, Wu P, Widelitz R, Chuong C (2012) Sprouty / FGF signaling regulates the proximal-distal fate of feathers. Dev Biol 372:45–54
Zelenitsky DK, Therrien F, Erickson GM, DeBuhr CL, Kobayashi Y, Eberth DA, Hadfield F (2012) Feathered non-avian dinosaurs from North America provide insight into wing origins. Science 338:510–514. https://doi.org/10.1126/science.1225376
Zhang FC, Zhou ZH (2006) Feathers and 'feather-like' integumentary structures in Liaoning birds and dinosaurs. Geol J 41:395–404
Zhang FC, Zhou ZH, Xu X, Wang XL, Sullivan C (2008) A bizarre Jurassic maniraptoran from China with elongate ribbon-like feathers. Nature 455:1105–1108
Zhang FC et al. (2010) Fossilized melanosomes and the colour of Cretaceous dinosaurs and birds. Nature 463:1075–1078. https://doi.org/10.1038/nature08740
Zheng XT, You HL, Xu X, Dong ZM (2009) A heterodontosaurid dinosaur from the Early Cretaceous of China with filamentous integumentary structures. Nature 458:333–336
Zheng XT et al (2013) Hind wings in basal birds and the evolution of leg feathers. Science 339:1309–1312
Zhou Z-H, Wang X-L (2000) A new species of *Caudipteryx* from the Yixian Formation of Liaoning, northeast China. Vertebrata PalAsiatica 38:111–127
Zhou Z-H, Wang X-L, Zhang F-C, Xu X (2000) Important features of *Caudipteryx*-evidence from two nearly complete new specimens. Vertebrata PalAsiatica 38:241–254

Two of a Feather: A Comparison of the Preserved Integument in the Juvenile Theropod Dinosaurs *Sciurumimus* and *Juravenator* from the Kimmeridgian Torleite Formation of Southern Germany

6

Christian Foth, Carolin Haug, Joachim T. Haug, Helmut Tischlinger, and Oliver W. M. Rauhut

6.1 Introduction

The discoveries of numerous theropod dinosaurs from the Upper Jurassic and Lower Cretaceous of China preserved with filamentous integumentary structures in various stages of morphological complexity provided new insights into the early evolution of feathers. These findings indicated that ancestral feathers were filamentous and did not originate in relation to flight, but for other purposes (Prum and Brush 2002; Norell and Xu 2005; Xu and Guo 2009; Foth et al. 2014; Xu et al. 2014). The feathers from these localities are usually preserved as black, carbonized traces related to melanosome preservation (Li et al. 2010, 2012, 2014; Zhang et al. 2010), which for the first time allowed conclusions about ancestral colour patterns in extinct animals (see Vinther et al. 2008).

In the last decade, two early juvenile theropod dinosaurs from the Upper Jurassic limestones of Bavaria (Germany) were described, *Juravenator starki* (Göhlich and Chiappe 2006) and *Sciurumimus albersdoerferi* (Rauhut et al. 2012), both of which were preserved with large traces of soft tissue remains, including skin and feathers (Fig. 6.1). In contrast to the feathered theropods from China, the integumentary structures of both theropods were preserved in the form of apatite (see Frey and Martill 1998; Viohl and Zapp 2006; see below), resulting in a strong autofluorescence signal under UV light with respect to the surrounding matrix. In the present study, the integumentary soft tissues of *J. starki* and *S. albersdoerferi* are investigated with the help of autofluorescence techniques using two different wavelengths (UVA and cyan) and different light filter adjustments. This approach allows corroborating morphological observation on soft tissues due to different reactions on light colours and filter adjustments.

C. Foth (✉)
Department of Geosciences, Université de Fribourg, Fribourg, Switzerland

SMNS Staatliches Museum für Naturkunde Stuttgart, Stuttgart, Germany

C. Haug · J. T. Haug
LMU München, Biocenter, Planegg-Martinsried, Germany

GeoBio-Center of the LMU München, Munich, Germany

H. Tischlinger
Stammham, Germany

O. W. M. Rauhut
GeoBio-Center of the LMU München, Munich, Germany

SNSB, Bayerische Staatssammlung für Paläontologie und Geologie, Munich, Germany

Department for Earth and Environmental Sciences, LMU München, Munich, Germany

C. Foth, O. W. M. Rauhut (eds.), *The Evolution of Feathers*, Fascinating Life Sciences,
https://doi.org/10.1007/978-3-030-27223-4_6

Fig. 6.1 Overview of both juvenile theropods under UV light. (**a**) *Juravenator starki*. (**b**) *Sciurumimus albersdoerferi*. Areas with fossilized integument structures are marked by dotted circles. Scale bar 5 cm

These results in turn are used for taphonomic revision of the preservation of the integumentary structures. Finally, the morphology of the integuments of *J. starki* and *S. albersdoerferi* is compared with that of other feathered theropods and, after a short discussion on the phylogenetic position of *J. starki*, integrated into a phylogenetic-based overview of early feather evolution in theropods.

6.2 General Comments on Soft Tissue Preservation in Plattenkalks of Painten and Schamhaupten

The Upper Jurassic well bedded to laminated lithographic limestones of the southern Franconian Alb (the often so-called Solnhofener Plattenkalks or plattenkalks of the Solnhofen Archipelago) have a great potential for soft tissue (e.g. integuments, inner organs and muscles) preservation, including impressions (Wellnhofer 2009; Foth et al. 2014), melanosome preservation (Tischlinger 1998; Carney et al. 2012) and apatite preservation (Frey and Martill 1998; Briggs 2003; Viohl and Zapp 2006), which differ in their chemical compounds, respectively. The feather impressions of various *Archaeopteryx lithographica* specimens, for instance, are covered by a thin layer rich in phosphorus and sulphur, which might be organic in origin (Bergmann et al. 2010). The isolated *A. lithographica* feather, which is preserved in the form of melanosomes, is rich in copper, nickel and sulphur (Carney et al. 2012; Manning et al. 2013), whereas apatite preservation is, unsurprisingly, rich in calcium and phosphorus (Briggs and Kear 1994; Briggs and Wilby 1996; Briggs 2003). These different chemical compounds lead to different fluorescent characteristics, inasmuch as apatite shows a strong luminous signal under certain wavelengths such as UV light or cyan light (autofluorescence), while the melanosomes and organic cover of the impressions appear to be darker than the surrounding matrix (Frey and Martill 1998; Tischlinger and Frey 2002; Tischlinger and Unwin 2004; Tischlinger 2005, 2009).

Based on the strong luminescence under UV light (see Göhlich and Chiappe 2006; Göhlich et al. 2006; Chiappe and Göhlich 2010; Rauhut et al. 2012), the soft tissues of both *Sciurumimus albersdoerferi* and *Juravenator starki* are mainly preserved in the form of apatite, although Göhlich and Chiappe additionally reported tubercle scales and ventral fibres in the proximal tail region of *J. starki*, which are preserved as visible marks (see below), and in *S. albersdoerferi*, parts of the filaments can be traced as impressions under low-angle light (Rauhut et al. 2012: Fig. S1). According to Briggs and Kear (1994), apatite mineralization occurs mainly under anoxic conditions when the pH drops between 6 and 7, inducing the precipitation of calcium phosphate in the system. This, however, requires a high concentration of phosphate from the very first, which can, for example, originate from the organism itself due to decay of organic matter and dissolution of biogenic apatite or from external sources like microbial mats, in which bacteria can induce apatite precipitation (Hirschler et al. 1990; Briggs et al. 1993; Wilby et al. 1996). These experimental results are consistent with the supposed deposition process for the bindstones of the Painten and Schamhaupten localities, which were both formed under hypersaline and anoxic conditions by microbial mats (predominantly by cyanobacteria) (Link and Fürsich 2001; Viohl and Zapp 2006, 2007). Besides, the Schamhaupten bindstones are strongly silicified due to a higher abundance of representatives of Radiolaria and Silicispongia (Viohl and Zapp 2006, 2007). The almost complete and articulated preservation of *J. starki* and *S. albersdoerferi* indicates that both specimens sank to the sea bottom rapidly after death and did not float for a longer period in open water (see Reisdorf and Wuttke 2012). Thus, the anoxic conditions at the bottom and the overgrowth by microbial mats prevented further decay and finally induced soft tissue mineralization (Hirschler et al. 1990; Briggs and Kears 1993; Wilby et al. 1996). However, in contrast to bone, the main compounds of integumentary structures (i.e. keratin, collagen and colour pigments) are not very rich in phosphorus (Kühn 1974; Murayama et al. 1986; Schweitzer et al. 1999a; McGraw 2003), while various examples of fossilized integumentary structures show a strong phosphate signal (e.g. Schweitzer et al. 1999a; Tischlinger and Unwin 2004; Benton et al. 2008; Bergmann et al. 2010; Tischlinger and Rauhut 2015). This fact may favour a predominant external source of phosphate for apatite formation in the soft tissues of *J. starki* and *S. albersdoerferi*. On the other hand, integumentary structures often work as toxin storage organs (e.g. Reichholf 1996; Metcheva et al. 2006; Dumbacher et al. 2009), accumulating various biogenic and toxic elements, including phosphorus (Marlow and Caldwell 1934; Kelsall and Calaprice 1972; Liu et al. 2006; Metcheva et al. 2006), so that apatite formation still can partly source from the specimen itself.

6.3 Material and Methods

6.3.1 Taxon Sampling

The holotypes of *Sciurumimus albersdoerferi* (BMMS BK 11) and *Juravenator starki* (JME Sch 200) represent both early juvenile theropod dinosaurs from the Torleite Formation (upper Kimmeridgian) of the Jurassic limestones of Bavaria (see Niebuhr and Pürner 2014), in which *J. starki* was found in strongly silicified, laminated bindstones (a type of autochthonous limestone; see Embry and Klovan 1971) located in the Schamhaupten area (Göhlich et al. 2006; Viohl and Zapp 2006, 2007; Chiappe and Göhlich 2010; Viohl 2015a, b), while *S. albersdoerferi* comes from the thin-bedded micritic bindstones of the Painten area (Rauhut et al. 2012; Albersdörfer and Häckel 2015; Viohl 2015a).

Although both specimens resemble each other in general features and body proportions, it could be demonstrated that *J. starki* and *S. albersdoerferi* are taxonomically distinct from each other at the species level (Rauhut et al. 2012). In a phylogenetic assessment, *S. albersdoerferi* was found close to the base of tetanurans, most likely representing a juvenile megalosauroid (Rauhut et al. 2012). In contrast, *J. starki* was originally described as a basal coelurosaur (Göhlich and Chiappe 2006; Chiappe and Göhlich 2010), which was supported by several phylogenetic analyses (e.g. Butler and Upchurch 2007; Rauhut et al. 2012; Loewen et al. 2013; Brusatte et al. 2014; Choiniere et al. 2014). However, Rauhut et al. (2012) already argued that this phylogenetic position may result from the early juvenile stage of *J. starki* (see discussion).

6.3.2 Autofluorescence Methods

During the past 15 years, frequent investigations under ultraviolet (UV) and cyan radiation have shown that fossils from the Upper Jurassic limestones of southern Germany are often autofluorescent (e.g. Tischlinger 2002, 2005; Tischlinger and Unwin 2004; Haug et al. 2009, 2011a, b; Haug and Haug 2011; Frey and Tischlinger 2012; Tischlinger and Arratia 2013), i.e. the objects emit light in a longer certain wavelength after they were exposed to more energetic light. Due to an increased contrast between structure and matrix, this characteristic allows a more precise investigation of morphological details of both hard and mineralized soft tissues, including delicate structures that are poorly or not discernible in visible light. Thus, this technique can be used to establish compressed morphological structures more clearly and to differentiate them from each other, but also from cracks, preparation artefacts and the underlying matrix.

The UV investigations of *Juravenator starki* and *Sciurumimus albersdoerferi* were done with the help of powerful UVA lamps with intensities between 4000 and more than 50,000 μW per 10 mm^2 and a wavelength of 365–366 nm. Because sometimes essential details of morphological structures are poorly or even not visible to the naked eye or under a microscope, they can still be documented with the help of UV light photography, using different filter techniques for the camera, which allow a selective visualization of peculiar fine structures. In most cases, a variety of different colour-correction filters are necessary, because each type of limestone, but also fossilized hard and soft tissues, reacts differently to different wavelengths. Thus, the right combination of filters and exposure times (1 second to several minutes) is needed to highlight the area of interest, depending on the nature of the fossil material and the magnification, intensity and incident angle of the UV lamps. In consequence, the optimum filtering and exposure time must be tested in a series of experiments (Tischlinger 2002; Frey and Tischlinger 2012). The UV photos were taken with analogue cameras (Pentax ME and MX) on slide film (Kodak Professional Elite Chrome ISO 100/21° and Kodak Elite Chrome Extra Colour 100) and with diverse digital cameras (Panasonic Lumix) partly in combination with a Raynox M-250 macro lens.

As stated above, we used a second autofluorescence method (cyan-red) for corroboration of the morphological observations on the integumentary structures. The investigation with cyan-red fluorescence was performed with a Canon EOS Rebel T3i camera and a Canon Macro Photo Lens MP-E 65 mm, in which the camera lens was equipped with a red filter foil taken from cyan-red stereo spectacles. The complementary cyan filter foil was attached to the light sources (three LED torches). After the objects of interest were excited by the cyan light, only the emitted light was able to pass through the complementary red filter, while light with other wavelengths was mostly blocked (Haug et al. 2011a; Haug and Haug 2011).

After documentation, the digital photos were processed in Adobe Photoshop CS3, including brightness adjustment and contrast enhancement. For the cyan-red light investigation, the photos were modified by deleting the blue and green colour channels and transformed afterwards into grey scale to increase the contrast between fluorescent and nonfluorescent structures.

6.4 Results

6.4.1 The Integumentary Structures of *Juravenator starki*

The soft tissues of *Juravenator starki* were first described by Göhlich and Chiappe (2006), Göhlich et al. (2006) and Chiappe and Göhlich (2010). Due to the fact that the bindstones from Schamhaupten are strongly silicified (Viohl and Zapp 2006, 2007), the preparation of the soft tissues was much more difficult in *J. starki*, and the quality of preservation is not as good as in *Sciurumimus albersdoerferi*. The majority of soft tissues in *J. starki* are preserved as apatite, illuminating bright yellow to bright green under filtered UV light (depending on the filter and camera system), and are also visible under cyan light (Figs. 6.2 and 6.3). The remains are restricted to the posterodorsal pelvic region, both tibiae and the tail, in which skin remains can be found in all areas (Fig. 6.2), while feather filaments are restricted to the mid-tail section (Fig. 6.3). Chiappe and Göhlich (2010) further mentioned soft tissue preservation dorsal to the rostrum of the skull, but the identification of this alleged soft tissue is currently impossible because it is covered with glue, illuminating cyan under filtered UV light.

The most conspicuous structures are a large patch of tubercle-like structures along the laterodorsal side of the tail between the 9th and 18th caudal and a series of longitudinal orientated fibre-like structures along the ventral side of the tail between the 12th and 15th caudal (Fig. 6.2a), which are even visible under normal light conditions (see Chiappe and Göhlich 2010). The former structure was originally described as uniformly sized, nonoverlapping, smooth tubercle-like scales, as known from various dinosaur taxa (Osborn 1912; Czerkas 1997; Mayr et al. 2002; Coria and Chiappe 2007; Christiansen and Tschopp 2010; Bell 2012, 2014), while the latter was interpreted as possible remains of tendinal elements of the m. ilioischiocaudalis, remains of a reinforced layer of the vertical septum ventral to the chevrons or bundles of subcutaneous collagen fibres (see Göhlich and Chiappe 2006; Göhlich et al. 2006; Chiappe and Göhlich 2010). On the ventral side of the caudals, between the chevrons, further tubercle-like structures are present, but they are not as well preserved as those on the laterodorsal side. On the basis of the current investigation, performed with a digital camera, both structures illuminate green to brown under filtered UV light similar to the bone and surrounding matrix (Fig. 6.2d). Using a different filter arrangement and an analogue camera, both structures fluoresce in bright blue, while the matrix possesses a darker blue signal, and the bone appears green to brown (Fig. 6.2c). In both cases, the signal is very different from the phosphatized soft tissue remains (see above). Under cyan light, both structures show no signs of autofluorescence, like the surrounding matrix, but in contrast to the bone and phosphatized soft tissues (Fig. 6.2b, e). These findings indicate that the tubercle- and fibre-like structures may chemically be differently composed than both bone and phosphatized soft tissue remains, but also the surrounding matrix. In some areas, both tubercles and fibres are partly covered with a thin layer of phosphatized soft tissues. Consequently, the supposed tubercles are not superficial, but seem to be internal, like the fibre-like structures. Another aspect that should be taken into account is that the tubercle-like structures are not as uniform as originally described. In fact, dorsal to the 12th and 13th caudal, some tubercles converge to short, thick, parallel band-like structures, which are orientated perpendicular with respect to the tail axis. These bands in turn look very similar to the ventral, horizontally orientated fibres, only differing in orientation. In consequence, it is possible that the original interpretation of the structures may not be correct (see discussion).

In contrast, the skin remains from the tail and hindlimb region, which are preserved as apatite, appear smooth and show no signs of scales (Fig. 6.2e–g). Feather filaments are visible on the dorsal and ventral side of the tail. The dorsal feathers are found in a small area around the 18th to 20th caudal (Fig. 6.3a–c), while the ventral filaments are preserved between the 9th and

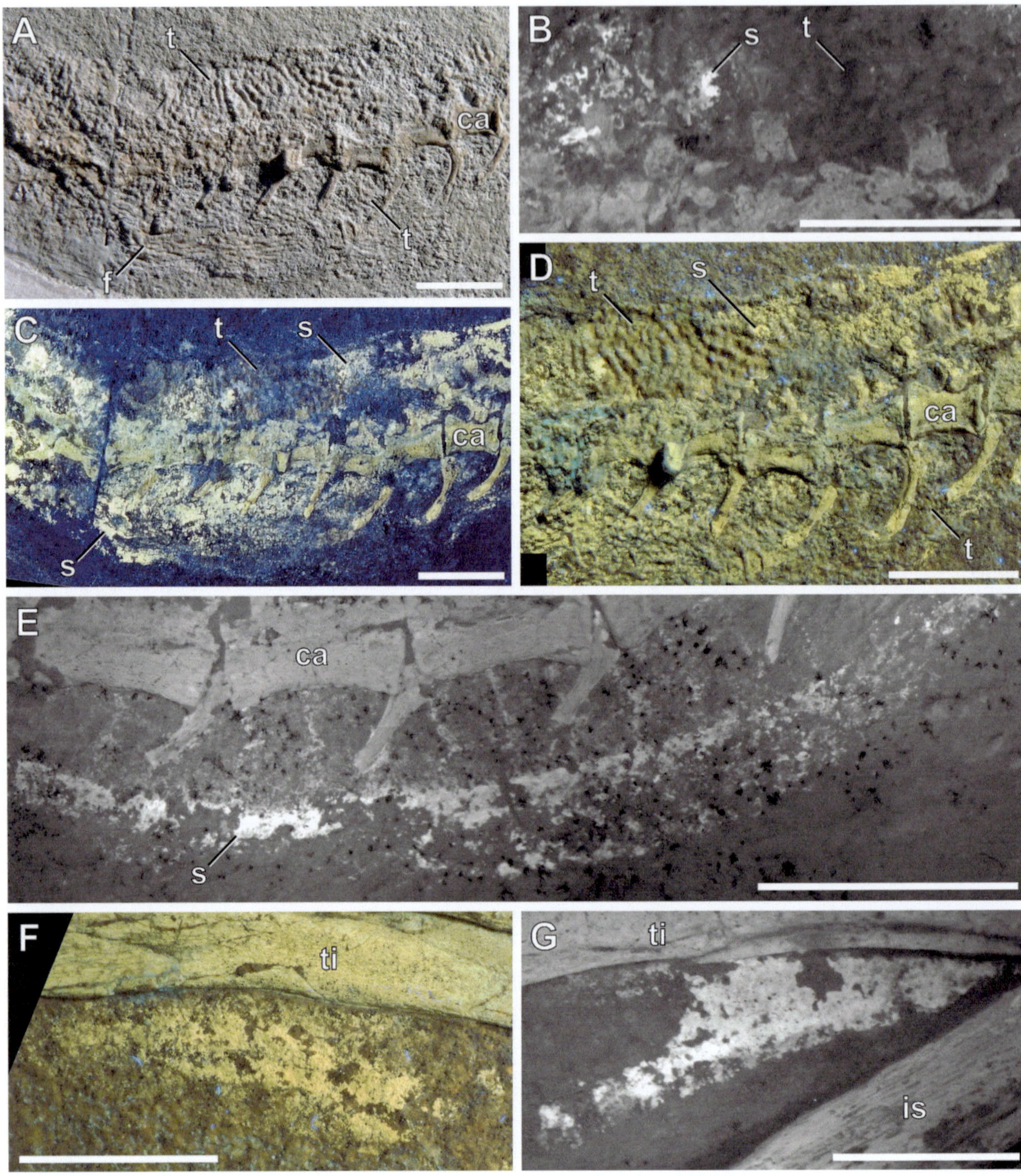

Fig. 6.2 Skin of *Juravenator starki*. (**a–d**) Tubercle and fibre structures of the proximal tail region under (**a**) normal light, (**b**) cyan light, (**c**) filtered UV light taken with an analogue camera and (**d**) filtered UV light taken with a digital camera, which are here interpreted as adipocere. (**e**) Skin remains on the ventral side of the tail. (**f**, **g**) Skin remains of the hindlimb under (**f**) filtered UV light and (**g**) cyan light. *ca* caudal vertebrae, *f* fibres, *is* ischum, *s* skin, *t* tubercles, *ti* tibia

13th caudal vertebrae (Fig. 6.3d). The filaments are parallel oriented and angled 20–45° backwards with respect to the tail axis. The filaments are short, measuring about 4 mm as preserved, and show no sign of branching. However, it should be noted that these filaments could have been longer originally, but parts of these very delicate structures might have been removed during preparation. Due to the quality of preservation, no further observation can be made.

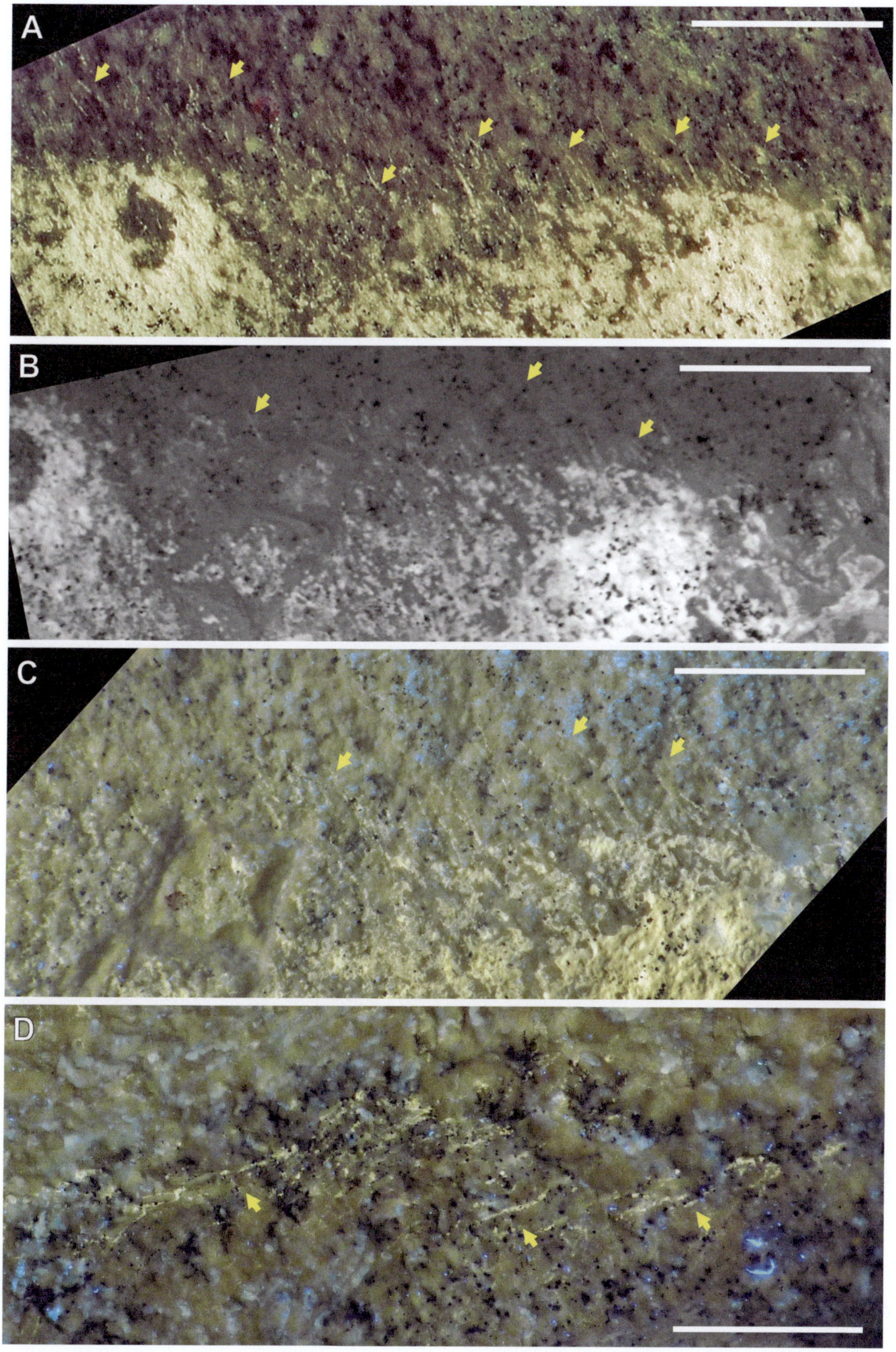

Fig. 6.3 Feather filaments (arrows) of *Juravenator starki*. (**a**, **b**) Filaments on the dorsal side of the tail under (**a**) filtered UV light taken with an analogue camera, (**b**) cyan light and (**c**) filtered UV light taken with a digital camera. (**d**) Filaments on the ventral side of the tail under filtered UV light taken with a digital camera

6.4.2 The Integumentary Structures of *Sciurumimus albersdoerferi*

The holotype of *Sciurumimus albersdoerferi* shows both skin and feather preservation in various parts of the body (Figs. 6.4 and 6.5). Under filtered UV light, the skin remains fluoresce in a bright yellow, which can be well distinguished from the reddish bindstone. The skin remains further show a red autofluorescent signal when illuminated with cyan light, while they are not visible under normal light. The skin remains are

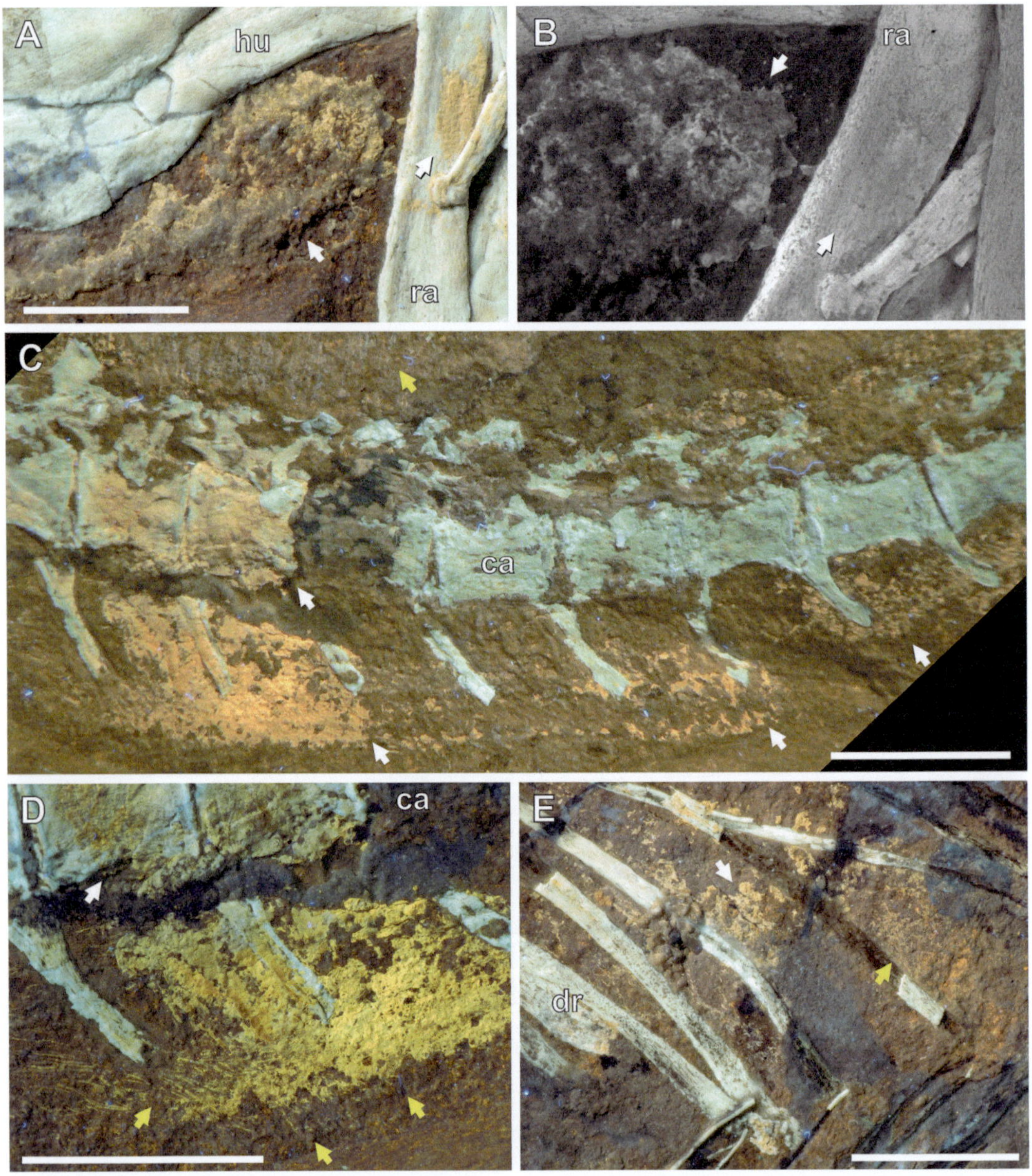

Fig. 6.4 Skin of *Sciurumimus albersdoerferi*. (**a**, **b**) Skin remains of the forelimb under (**a**) filtered UV light taken with digital camera and (**b**) cyan light. (**c**) Overview of skin remains in the tail region. (**d**) Detail of large patch of skin on the ventral side of the tail. (**e**) Skin remains in the thorax region. White arrows mark skin remains and yellow arrows mark feather filaments. *ca* caudal vertebrae, *dr* dorsal ribs, *hu* humerus, *ra* radius

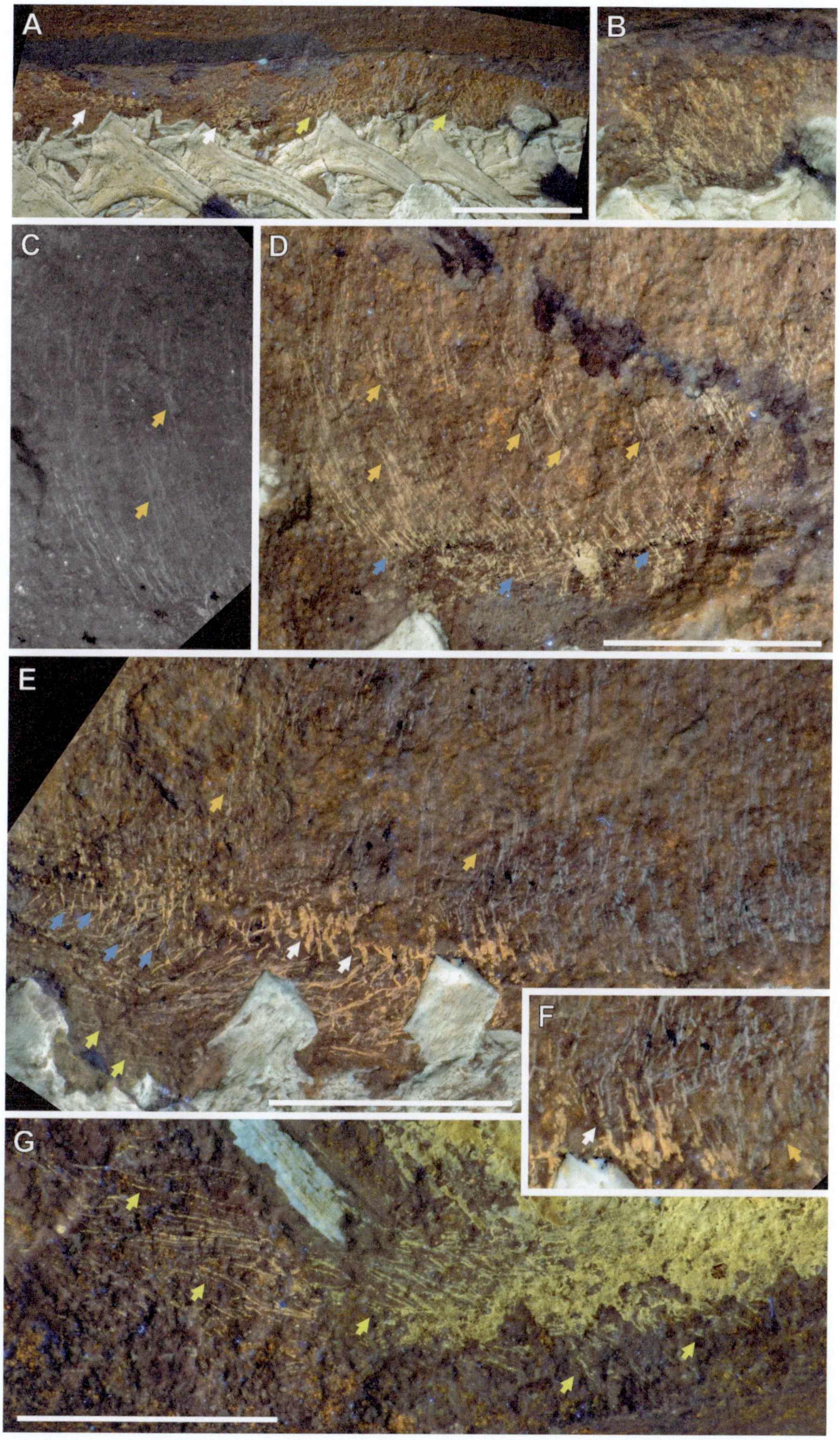

Fig. 6.5 Feather filaments of *Sciurumimus albersdoerferi*. (**a**, **b**) Skin and filaments on the dorsal side of the thorax under filtered UV light taken with digital camera. (**a**) Overview of the integument structures. (**b**) Details of the filaments from the dorsal region. (**c–f**) Integument structure of the dorsal side in the proximal tail

preserved in the region of the scapula (dorsal to the acromion process of the left scapula), forelimb (along the right humerus, ulna and radius) and hindlimb (left femur and tibia). In the original description, the soft tissues along the posterior edge of the tibia of *Sciurumimus* were interpreted as possible fossilized muscle tissue (Rauhut et al. 2012). However, because the surface of this area is rather smooth and does not show any indication for internal structuring (i.e. fascicle and fibres; see Kellner 1996; Chin et al. 2003; Dal Sasso and Maganuco 2011) under UV and cyan-red light, we reinterpret this patch also as skin remains. Further small patches of skin are preserved in the anterior portion of the back, on the lateral and anteroventral side of the torso as well as ventral to the last sacral vertebra. The best-preserved relicts of skin can be found in the anterior and middle tail section (up to the 25th caudal vertebra), along both the dorsal and lateral side (Fig. 6.4c, d). The skin appears generally smooth and shows no signs of external tubercle-like scales. Within the tail region, the skin is even directly preserved in association with feather filaments (see below).

The best feather remains are present along the back, as well as along the dorsal and lateroventral side of the tail (Fig. 6.5). The tail feathers are traceable until the midportion of the tail (27th caudal). Small patches of feathers are further preserved on the lateral side of the torso as well as in the belly region right in front of the knee joint. In the torso and the anterior tail region, the feather remains fluoresce under yellow-filtered UV light and are visible as red structures when illuminated with cyan light. More posteriorly the tail feathers shine greenish under filtered UV light and are barely visible under cyan light anymore, while the most distal tail feather remains appear yellow again (Fig. 6.5c–e).

The reasons for the different fluorescence signal of the feather filaments in the tail region probably result from different chemical compounds, which could be related to a preserved colour pattern. However, with the current method, this is not testable. Although it is principally possible to trace different chemical compounds with the help of unfiltered UV light due to specific characteristic fluorescence colour signals (Kane and Sell 2001), this does not work with the current approach, which works with multiple UV filters for increasing the contrast between matrix, bones and soft tissues (see above). As the identification via unfiltered UV light is further limited to only those chemical compounds being autofluorescent, the differences in the colour signals of the feathers have to be investigated with more sensitive methods using scanning electron microscopy and different spectroscopy techniques in a future approach (see Field et al. 2013; Egerton et al. 2015).

The general morphology of the feathers of *S. albersdoerferi* is described on the basis of the anterior dorsal tail feathers, which are best exposed. The feathers of this region are long, gently backward curved, stiff filaments, which measure 30–40 mm in length, 0.2 mm in width, and are almost perpendicularly orientated with respect to the axis of the tail (Fig. 6.5c–e). More posteriorly, the filaments become shorter, measuring c. 3 mm in the mid-tail portion. The filaments in the anterodorsal region show no signs of branching distally, although some filaments seem to stick closely together (Fig. 6.5d). Because those filaments are not fused both proximally and distally to the contact area, these structures certainly do not represent branching structures, but artificial attachments due to an overlapping in combination with taphonomic compaction of the single filaments (see Foth 2012). Proximally, the

Fig. 6.5 (continued) region. Long filaments under (**c**) cyan light and (**d**) under filtered UV light taken with digital camera. Orange arrows mark artificial overlaps of the monofilaments. Blue arrows mark the proximal ends of the feather filaments. (**e**) Skin remains (white arrows), short (yellow arrows) and long filaments with yellow and greenish fluorescence (orange arrows). (**f**) Detail of the association between skin (white arrows) and long filaments (orange filaments) indicating possible follicles. (**g**) Details of short filaments (yellow arrows) on the ventral side under filtered UV light taken with digital camera

interpretation of the feather morphology is more difficult. Again, in some areas single filaments seem to join each other, creating the impression that they are fused into a calamus-like structure. However, the problem of overlapping and compaction of single filaments in this portion is probably even more evident due to denser packing of filaments at the base. Furthermore, most feathers are not joined proximally, but still solitary before reaching the border of the skin (Fig. 6.5d, f). Thus, we conclude that the branched morphology at the base is in fact artificial and that the actual morphology of the feathers is monofilamentous over their entire length. More proximal to the feathers described, further parallel-orientated filaments with greenish fluorescence (filtered UV) are preserved, which seem to be shorter and run horizontally with respect to the axis of the tail (Fig. 6.5d, f). However, due to their parallel arrangement, the stiff appearance and the colour, we consider these filaments as short feathers, which originate from the dorsolateral side of the tail.

Further feathers are preserved on the ventral side of the tail, which show their best preservation around the large, ventral patch of skin (Fig. 6.5g). In this region, the filaments are almost horizontally orientated with respect to the axis of the tail and are parallel to each other. The filaments are fine and straight and show no sign of branching. In the way they are preserved, the feathers are short, measuring about 2–3 mm. In the border region of the large ventral skin patch, some feathers extend over the ventral border of the skin.

The dorsal trunk feathers are preserved along the mid and posterior portion of the back (Fig. 6.5a, b). As the original matrix in this area was only several millimetres thin, it was not possible to prepare the complete plumage without risking the destruction of the specimen (Raimund Albersdörfer, pers. comm., 2011). Thus, only the proximal part of the feathers is exposed, and nothing can be said about their actual length. The arrangement of these feathers is less regular than in the tail region, showing a high degree of overlapping and artificial contacts. Like in the tail region, some filaments are closely attached to each other, but no real fusion can be detected. Thus, the dorsal trunk feathers seem to be monofilamentous, too. Further small areas with filaments are preserved on the lateral and posteroventral side of the torso (Fig. 6.4e). The preserved filaments are short (c. 4–5 mm) and very thin and show no signs of branching. Because first bone remains of the specimen were originally found in the field from this area, and the torso had to be carefully reconstructed (Raimund Albersdörfer, pers. comm., 2011), large areas of the torso region are infused with glue in order to repair fissures within the matrix. Thus, it is not clear if those filaments are entirely preserved or only small fractions are visible.

Along the dorsal edge of the tail, the association between skin and feathers can be studied best (Fig. 6.5e). Like in other body regions, the skin illuminates yellow under filtered UV light and is preserved as a thin stripe, horizontally orientated with respect to the tail axis. In contrast, the filaments of this region show a greenish fluorescence under the same light conditions. As preserved, the skin seems to be vertically structured in lateral view, while the proximal ends of the vertically orientated feather filaments pierce in between, probably indicating that the feathers were deeply anchored within the skin (Fig. 6.5f), i.e. that the feathers actually grew out of follicles (see discussions).

Ventral to the skin remains of the dorsal edge of the tail, a different kind of filament-like soft tissue is preserved. In contrast to the feather filaments from this region, these structures are short, thick (notably thicker than the individual feather filaments) and wrinkled and are obliquely orientated with respect to the axis of the tail, with a parallel arrangement. Furthermore, these filaments show a yellow signal under UV light, like the skin tissue (Fig. 6.5g). Therefore, we do not interpret these structures as feather filaments. Instead, it is possible that these structures are relicts of subcutaneous muscle tissue or thick dermal collagen bundles. Although both tissue types can potentially fossilize (Kellner 1996; Chin et al. 2003; Lingham-Soliar and Wesley-Smith 2008; Dal Sasso and Maganuco 2011), it is not possible to identify the true nature of these structures on the basis of the current methods and their preservation.

6.5 Discussion

6.5.1 Identification of Tubercles in the Tail Region of *Juravenator starki*

The present examination of the soft tissue in the tail region of *Juravenator starki* reveals that the original interpretation of the tubercles as scales is questionable on the basis of several observations, including the overall kind of preservation, the morphology, the absence of autofluorescence signal and the fact that the tubercles are partly covered with phosphatized soft tissue. In consequence, the tubercles are unlikely to represent a superficial integumentary structure. Furthermore, it is not clear why these very distinct and robust tubercles are only preserved in a single, restricted area, while phosphatized integumental tissue remains can be found along the entire anterior half of the tail. As an alternative explanation, we suggest that the tubercles of *J. starki* could be the result of adipocere formation. Adipocere (corpse or grave wax) is a decomposition product of subepidermal adipose body tissue containing predominantly saturated fatty acids, which are produced due to the hydrolysis and hydrogenation of neutral fats, especially during subaquatic decay (Forbes et al. 2005). Its formation occurs predominantly under moist, warm, anaerobic conditions, a pH range of approximately 5–9 and the presence of putrefactive bacteria (Forbes et al. 2005; O'Brien and Kuehner 2007). Thus, the same environmental conditions leading to apatite preservation of soft tissues (see above) can also form adipocere. Finally, adipocere precipitates into concretions of calcium carbonate $CaCO_3$ (Berner 1968). In this context, the formation of adipocere has previously been hypothesized to play an important role in fossilization of vertebrates from the limestones of the Solnhofen Archipelago (de Buisonjé 1985; Reisdorf and Wuttke 2012). Similar tubercle-like structures have, for instance, also been preserved in the head region of an undescribed ichthyosaur from the Solnhofen limestones housed in the Bürgermeister-Müller-Museum in Solnhofen (Achim Reisdorf, pers. comm. 2015; see Maisch 2015). Because adipocere is generally soft and crumbly, its morphology is very variable, which includes tubercle-like, segmental, bundle-like or even amorphous granular appearances (see Wiman 1942; Reisdorf and Wuttke 2012). Thus, the fibre-like structures found on the ventral side of the tail of *J. starki* also fall in the morphological spectrum of adipocere.

Following this interpretation, the regional restriction of the tubercles can be explained by the presence of a large fat body in the anterior tail section, as known in extant crocodilians (Frey 1988; Huchzermeyer 2003) and turtles (Walter Joyce, pers. comm. 2015). Thus, the current case demonstrates that the discovery of tubercle-like structures in sediments surrounding dinosaur fossils or even on the bones themselves should not be identified by default as skin remains, but evaluated carefully for taphonomic alternatives. In this context, the conspicuous tubercles found in the mid-section of the tail of the French *Compsognathus corallestris* (Peyer 2006) might also be interpreted as remains of adipocere. To test the hypothesized adiposal origin using geochemical methods, the tubercles of *Juravenator* and *Compsognathus* shall contain a high concentration of $CaCO_3$ with a very low $^{13}C/^{12}C$ ratio in the carbonate, as adipocere and lipid carbon are highly depleted in ^{13}C relative to normal marine calcium carbonate (Sondheimer et al. 1966; Berner 1968).

6.5.2 The Skin Morphology of *Juravenator starki* and *Sciurumimus albersdoerferi*

If the tubercle- and fibre-like structures in *Juravenator starki* represent adipocere, as proposed, all integumentary structures are exclusively preserved in the form of apatite, in which the phosphatized skin remains found in the limb and tail region of *J. starki* and *Sciurumimus albersdoerferi* show no evidence for scales whatsoever. Scaly skin remains from the body region are rarely described in theropod dinosaurs (e.g. Bonaparte et al. 1990; Czerkas 1997; Ortega et al. 2010) as many theropod taxa possessed extensive feather plumage, which covered the whole body (Xu and Guo 2009; Xu et al. 2014). Only in the distal foot region scale preservation is more frequent (Gatesy 2001; Ji et al. 2001; Currie et al. 2003; Kundrát 2004; Cuesta et al. 2015), as this region was probably barely covered with feathers, like in the majority of modern birds (Lucas and Stettenheim 1972). Unfortunately, no integumental structures are preserved in the foot region of *J. starki* and *S. albersdoerferi*, which would allow a comparison.

The absence of scales in *J. starki* and *S. albersdoerferi* could be potentially taphonomic, if (a) this kind of soft tissue mineralization does not lead to scale preservation in general or (b) if the scales of these early juvenile theropods were very small or even not fully developed, prohibiting a proper preservation. However, various embryological studies of modern non-avian reptiles and birds (i.e. foot region) indicate that scales are already morphologically well developed and keratinized during hatching (Hamburger and Hamilton 1951; Dhouailly et al. 1980; Alibardi and Thompson 2000, 2001, 2002; Werneburg et al. 2009), a situation also found in fossil sauropod embryos (Coria and Chiappe 2007). Furthermore, several fossils of Rhynchocephalia and squamates from the limestones of the Solnhofen archipelago with phosphatized skin remains (e.g. *Pleurosaurus goldfussi*) demonstrate that scale preservation was not generally impossible in these deposits, although in smaller species (*Sapheosaurus thiollierei*/*Piocormus laticeps*) skin remains also sometimes appear smooth (Tischlinger and Rauhut 2015). Furthermore, preservation of other keratinous structures is not rare in vertebrates from the Solnhofen Archipelago, and at least parts of the keratinous sheaths of the manual and pedal unguals are also preserved in both *Juravenator* and *Sciurumimus* and are usually well visible under UV light (e.g. Chiappe and Göhlich 2010: Fig. 20; Rauhut et al. 2012: Fig. 1c). If we consider the absence of scales in the phosphatized skin remains of *J. starki* and *S. albersdoerferi* to not be taphonomic, the skin would resemble that of modern birds, where body parts covered with feathers (except of the foot region) are scaleless (Lucas and Stettenheim 1972; McNamara et al. 2018). As mentioned above, the skin remains of both theropods are usually found in association with feather filaments, except for the limb regions. However, based on the current data, it is unclear if scales were originally present and if skin morphology may have changed through ontogeny in these theropods.

6.5.3 The Feather Morphology of *Juravenator starki* and *Sciurumimus albersdoerferi*

The new examination reveals that in *Juravenator starki* both the dorsal and ventral sides of the tail were feathered, indicating full plumage coverage of the tail as in other coelurosaurs. Unfortunately, other body parts show no signs of feathers. Because the limestones from the Schamhaupten locality are strongly silicified and extremely hard to prepare, it cannot be ruled out that other areas with soft tissue preservation were accidentally removed during preparation. In contrast, the extension of the plumage can be far better reconstructed in *Sciurumimus albersdoerferi*, indicating that the whole thorax and tail were covered with feathers. The examination further reveals that the anterior tail feathers along the dorsal edge were extremely elongated, while the dorsolateral, lateral and ventral feather filaments were significantly shorter. Thus, the tail of *S. albersdoerferi* was not uniformly bushy, but formed an elongated, dorsal filamentous crest. Similar filamentous crests have been described in the ornithischian dinosaurs *Psittacosaurus* sp. (Mayr et al. 2002, 2016) and *Tianyulong confuciusi* (Zheng et al. 2009). *Ornithomimus* sp. also shows elongated filaments along the dorsal edge of the proximal tail portion (van der Reest et al. 2016).

Skin remains without feather preservation could indicate that tibial regions of both theropods were probably naked, as in *Ornithomimus* sp. (van der Reest et al. 2016), but in contrast to the compsognathid *Sinocalliopteryx gigas* (Ji et al. 2007) and various eumaniraptorians (Zheng et al.

2013; Foth et al. 2014). However, it cannot be ruled out that the absence of feathers in this region may simply be a preservational artefact given that scales are generally expected in unfeathered theropod hindlimbs (see Cuesta et al. 2015).

The feather morphology of *J. starki* and *S. albersdoerferi* was found to be monofilamentous over their entire length. Possible branching patterns in *S. albersdoerferi* were interpreted as taphonomic artefacts caused by the overlap of feathers and the two-dimensional preservation of the fossils itself (see Foth 2012). Thus, the feathers of both theropods represent feathers of stage I according to Prum (1999), or morphotype I and II according to Xu et al. (2010), which were hypothesized to represent the ancestral feather morphologies in both evolutionary scenarios (Fig. 3.6). Similar monofilamentous feathers have been described for the alvarezsaurid *Shuvuuia deserti* (Schweitzer et al. 1999b), the therizinosauroids *Beipiaosaurus inexpectus* (Xu et al. 2009) and *Jianchangosaurus yixianensis* (Pu et al. 2013), two undescribed large-bodied basal coelurosaurs (Xu et al. 2010), and in the manual region of a juvenile *Ornithomimus* sp. (Zelenitsky et al. 2012). Furthermore, monofilamentous filaments (or bristles) were also present in some ornithischian dinosaurs (Mayr et al. 2002, 2016; Zheng et al. 2009; Godefroit et al. 2014), and even pterosaurs (e.g. Kellner et al. 2010), although it is not entirely clear at the moment if these filaments are homologous with theropod feathers (see Barrett et al. 2015). However, in many basal coelurosaurs (see Currie and Chen 2001; Xu et al. 2004, 2012; Ji et al. 2007), the kind of plumage preservation makes the conclusion, if the filaments were still monofilamentous or already branched, representing a more derived feather type, very difficult (see Currie and Chen 2001; Xu and Guo 2009; Xu et al. 2010; Foth 2012; Saitta et al. 2018). Usually, the dense body plumage consists of filaments, which are parallelly oriented to each other over the entire length. Often, single filaments group closely together, especially close to the body. These groupings could be evidence that these filaments were proximally merged, forming a branched feather tuft of multiple barbs emerging from the calamus [radially symmetric branched feather without rhachis—stage II after Prum (1999) and morphotype 3 after Xu et al. (2010)] or joined into a short rhachis [bilaterally symmetric branched feathers with short rhachis—stage IIIa after Prum (1999) and morphotype 5 after Xu et al. (2010)]. However, no certain signs of fusion are identifiable (Fig. 3.5), and the described morphology could alternatively result from taphonomic compaction of overlapping filaments (see Foth 2012). Only for Pennaraptora, branched feather types could be identified without doubt so far (e.g. Ji et al. 1998; Xu et al. 2001, 2015; Foth et al. 2014), although van der Reest et al. (2016) report the presence of branched body feathers containing a rhachis in a newly discovered *Ornithomimus* sp. specimen. If this identification is correct, the origin of bilaterally symmetric feathers can be pushed back to the base of Maniraptoriformes. However, as stated above, this branching structure might also be a preservation artefact. Therefore, more discoveries and closer investigations of plumage morphologies of basal coelurosaurs are necessary to test ideas about feather branching evolution, and autofluorescence techniques could be a key to establish the data needed for this discussion. Besides the UV and cyan-red autofluorescence techniques, a further illumination method has been recently established, using laser-stimulated fluorescence (LSF), which will help in studying soft tissues in fossils in more detail (Kaye et al. 2015; Mayr et al. 2016; Wang et al. 2017).

In the dorsal tail region of *S. albersdoerferi*, skin and feathers were found in direct association with each other, in which the skin remains appear perpendicular to the long axis of the body, creating the impression that the feathers were anchored deep within the skin. If this interpretation is correct, this morphology indicates that the feathers grow out of follicles like modern feathers or mammal hairs (Lucas and Stettenheim 1972; Sengel 1976). However, due to the compaction of the specimen and possible preparation artefacts in this region, the described morphology is still arguable. On the other hand, the outer edge of the skin is the only region in a two-dimensional compressed fossil, where such association could potentially be preserved. Furthermore, the different fluorescence of skin and feathers in this region supports the view that both structures are distinct units and that the filaments pierce into the skin and are not just simple superficial outgrowths (Fig. 6.6e, f). Thus, *S. albersdoerferi* might present evidence for the hypothesis that follicles were associated with the origin of feathers, as suggested by Prum (1999) and Prum

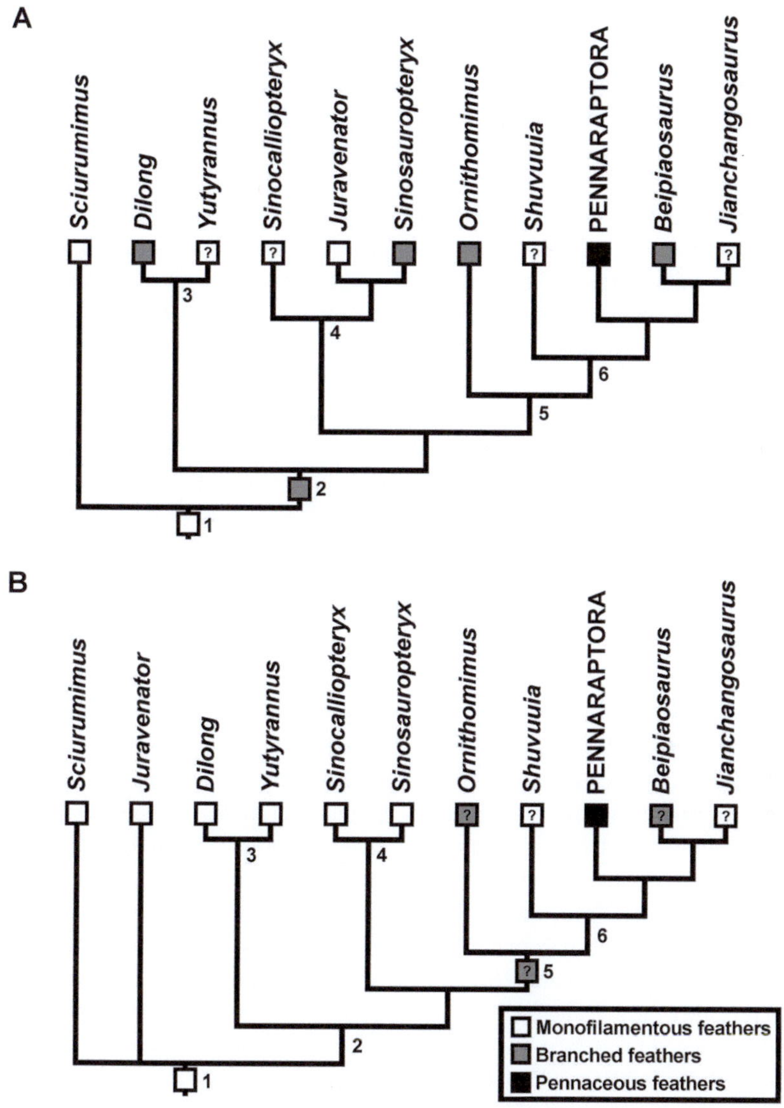

Fig. 6.6 Origin of monofilamentous and branched feather types within theropods with two alternative phylogenetic positions of *Juravenator starki*. (**a**) Traditional interpretation with *J. starki* as compsognathid. Filaments of *Dilong paradoxus*, *Sinosauropteryx prima*, *Ornithomimus* sp. and *Beipiaosaurus inexpectus* are interpreted as branched filaments. In this view, branched feathers are an apomorphic character of Coelurosauria, while several taxa show a secondary simplification of their filaments. (**b**) Alternative interpretation with *J. starki* as a basal tetanuran. The branching patterns of filaments in non-maniraptoriformes are interpreted as overlapping artefacts, so that the feathers in these theropods are actually monofilamentous. In *Ornithomimus* sp. and *Beipiaosaurus inexpectus* simple branched feathers might be present, indicating a potential origin of this feather type in Maniraptoriformes. However, if the assumed branched feathers of these two theropods turn out to be artificial (see above), monofilaments evolved quickly into pennaceous feathers within Maniraptora. **1** Tetanurae, **2** Coelurosauria, **3** Tyrannosauroidea, **4** Compsognathidae, **5** Maniraptoriformes, **6** Maniraptora

and Brush (2002). Furthermore, a very similar situation has been also found in the ornithischian dinosaur *Psittacosaurus* sp. (Mayr et al. 2002), where several bristles penetrate the skin. If these observations are not artificial (sensu Mayr et al. 2016), the presence of follicles could be a possible support for the homology of ornithischian filaments with theropod feathers (see Brusatte et al. 2010; Zheng et al. 2009; Rauhut et al. 2012; Godefroit et al. 2014; contra Barrett et al. 2015; Mayr et al. 2016).

Based on the early ontogenetic stage of both *J. starki* and *S. albersdoerferi*, the feather remains found represent a juvenile plumage. The presence of filamentous feathers indicates that juvenile theropods probably used their feathers for body insulation. However, the presence of a conspicuous dorsal filamentous crest in *S. albersdoerferi* might further indicate a signal or camouflage function. Because of the phylogenetic position of *S. albersdoerferi* close to the base of tetanurans and also the size of this early post-hatchling individual in comparison with other basal tetanuran hatchlings and embryos (e.g. Rauhut and Fechner 2005; Araújo et al. 2013), it has to be assumed that adult representatives of this species were medium- to large-bodied theropods (e.g. Britt 1991; Zhao and Currie 1993; Rauhut 2005; Sadleir et al. 2008). As the increase of body size impacts the need of body insulation for thermoregulation, it is possible that adult individuals lost large parts of their body plumage and were secondarily partly or fully naked (see van der Reest et al. 2016), as it is also the case in many large mammals, such as rhinos or elephants. However, the presence of probably fully feathered, medium- to large-bodied theropods from the Lower Cretaceous of China (Xu et al. 2010, 2012) indicates that plumage reduction in large-bodied theropods during ontogeny is not obligatory.

6.5.4 The Phylogenetic Position of *Juravenator starki*

Before the new observations on the integumentary structures of *Juravenator starki* and *Sciurumimus albersdoerferi* can be included into an evolutionary framework, the phylogenetic position of both theropods has to be discussed. As mentioned above, Rauhut et al. (2012) demonstrated that *S. albersdoerferi* is a basal tetanuran and most probably a juvenile megalosauroid, although the exact phylogenetic position is difficult to establish, due to our still very poor knowledge of ontogenetic changes in basal tetanurans. In contrast, *J. starki* was originally described as basal coelurosaur (Göhlich and Chiappe 2006; Chiappe and Göhlich 2010). Indeed, *J. starki* possesses a number of characters present in basal coelurosaurs, e.g. an elongate anterior process of the maxillary body, which is longer than high; a maxillary fenestra that is situated posterior to rostral border of the antorbital fossa; an enlarged, round orbit, maxillary teeth without anterior serrations; axial epipophyses that do not extend beyond the posterior rim of the postzygapophyses; and long and hair-like cervical ribs (Göhlich and Chiappe 2006; Chiappe and Göhlich 2010).

However, in light of its early juvenile stage, Rauhut et al. (2012) already argued that the phylogenetic position of *J. starki* may be an artefact (see also Tsuihiji et al. 2011). For instance, enlarged, round orbits and maxillary teeth without anterior serrations were found to be juvenile characters of basal tetanurans, which probably evolved in basal coelurosaurs by paedomorphosis (Rauhut et al. 2012; Foth et al. 2016). An elongated anterior process of the maxillary body is present in some compsognathids (Currie and Chen 2001; Hwang et al. 2004) but is also known for megalosauroids (Britt 1991; Sereno et al. 1998; Allain 2002; Rauhut et al. 2012) and some other basal tetanurans (e.g. Zhao and Currie 1993). The slender morphology of the cervical ribs in *J. starki* may be an allometric feature due to its young age. Long cervical ribs are also present in more basal theropods, e.g. in the coelophysid *Coelophysis bauri* (Colbert 1989) or the carcharodontosaurid *Concavenator corcovatus* (Ortega et al. 2010). Due to their gracility, cervical ribs are often incompletely preserved, making a phylogenetic evaluation of this character difficult. Other characters presented by Göhlich and Chiappe (2006) and Chiappe and Göhlich (2010), like the missing mandibular fenestra or the short and low neural spines of the cervical vertebrae, are probably artefacts of deformation. In contrast, *J. starki* possesses several characters that are more typically seen in more basal theropods, like coelophysoids, ceratosaurs or basal tetanurans (Rauhut and Foth 2014, in prep.), such as a medially closed maxillary fenestra (see Megalosauroidea), a short maxillary antorbital fossa anterior to the antorbital fenestra or a supra-acetabular crest that is confluent with the lateral brevis shelf. Thus, an alternative phylogenetic position close to the base of Tetanurae is also plausible. Unfortunately, most of the analyses in which *J. starki* was included (see above) focused on the interrelationships of coelurosaurs and had only a poor sampling of basal theropods. Thus, the actual phylogenetic position of *J. starki* has to be tested in a future approach with a more comprehensive sampling, but this is beyond the scope of the current study.

6.5.5 Insights into Early Feather Evolution in Theropods

The evolutionary implications that can be drawn from the current data are impacted by two major aspects: (1) the phylogenetic position of *Juravenator starki* and (2) the actual morphology of the filamentous feathers in basal coelurosaurs, which has been discussed as monofilamentous or proximally branched (see above). In previous models, branched feathers were believed to be present in basal coelurosaurs (e.g. Prum and Brush 2002; Xu and Guo 2009; Xu et al. 2010, 2014), and *J. starki* was considered as a compsognathid (see Xu and Guo 2009; Barrett et al. 2015). Based on these assumptions (Fig. 6.6a), *Sciurumimus albersdoerferi* would be the phylogenetically most basal feathered theropod, while its monofilamentous feathers would represent the ancestral feather type. The possible anchorage of the filaments within the skin would further imply that follicles were primordial structures for monofilamentous feathers, too. Thus, the reconstructed ancestral stage is in agreement with the hypothesized ancestral feather type proposed by Prum (1999). In basal coelurosaurs, feathers became more complex, evolving proximal branching patterns due to the formation of multiple barbs emerging from the calamus (radially symmetric feathers—stage II after Prum 1999 and morphotype 3 after Xu et al. 2010) or even a short rhachis formation caused by barb fusion at the anterior side of the calamus (bilaterally symmetric feathers—stage IIIa after Prum 1999 and morphotype 5 after Xu et al. 2010). At this stage, the plumage was still filamentous, but contained both monofilamentous and simple-branched feathers (as hypothesized for *Beipiaosaurus inexpectus*). Assuming that the feather types found in the tail region are representative for the entire plumage, *J. starki* would only possess the ancestral feather type, as no branching could be documented. Pennaceous feathers covering the entire body (bilateral symmetric feather with long rhachis and barbs with barbules—stage IV after Prum 1999 and morphotype 8 after Xu et al. 2010) evolved in the clade called Pennaraptora, which includes Oviraptorosauria, Dromaeosauridae, Troodontidae and Avialae (Foth et al. 2014). This whole scenario is very much congruent with the evolutionary model proposed by Prum (1999), which is based on the morphological and developmental hierarchy of feathers (see Lucas and Stettenheim 1972; Prum and Dyck 2003).

However, the reconstructed character evolution might be different if *J. starki* is considered as a basal tetanuran and with the assumption that the filamentous plumages of basal coelurosaurs consisted only of monofilamentous feathers (Fig. 6.6b). In this case, *J. starki* and *S. albersdoerferi* would represent the phylogenetically most basal records for feathered theropods. As in the previous case, monofilamentous feathers growing from a follicle would represent the ancestral feather type, but in contrast to the previous scenario, feather morphology did not change until branched feather types (both radially and bilaterally symmetric feathers, including pennaceous feathers) evolved in the clade Pennaraptora (Foth et al. 2014) or perhaps Maniraptoriformes (van der Reest et al. 2016). This scenario, however, conflicts with the model of Prum (1999), as the proposed transformation of monofilamentous feathers to bilaterally symmetric pennaceous feathers would have skipped the proposed intermediate stage of radially symmetric feathers.

Although morphologically less complex, it has been argued that at least in modern birds radially symmetric feather is secondarily simplified due to the reduction of developmental time (Schaub 1912; Becker 1959; Foth 2011). On the other hand, developmental data of bilaterally symmetric feathers of recent birds show that the rhachis is the fusion product of barbs, merging successively towards each other on the anterior side of the feather anlage (Lucas and Stettenheim 1972; Harris et al. 2002). In consequence, barb formation has to precede rhachis formation in both ontogeny and evolution (Prum 1999; Prum and Brush 2002; Harris et al. 2002). Thus, the only way to solve this conflict without violating the concept of hierarchical organization of feathers would be to assume that both formation processes evolved quasi-contemporaneously. However, two newly discovered anurognathids from the Middle–Late Jurassic Yanliao Group in China (Yang et al. 2018) indicate that the filamentous integuments of ornithodirans could evolve a variety of feather-like branching patterns without a clear rhachis morphology so that more complex branching scenarios need to be considered for the early evolution of feathers.

6.6 Conclusions

The current study revised the integumentary structures of the early juvenile theropod dinosaurs *Juravenator starki* and *Sciurumimus albersdoerferi* from the Upper Jurassic limestones of Bavaria with the help of autofluorescence methods using two different excitation wavelengths (UVA and cyan). The preserved integumentary structures were primarily preserved in the form of phosphate-rich apatite, showing a strong signal of the autofluorescence signal. By this it was possible to describe the integumentary structures of both theropods in much more detail and review possible taphonomic artefacts related to soft tissue preservation. Both theropods possessed monofilamentous feathers and probably smooth skin without scales or with only tiny scales that cannot be distinguished from each other as preserved. In *J. starki*, short feathers could be only traced in the tail region, but their preservation is rather poor compared to *S. albersdoerferi*, which is probably caused by the higher silicification of the limestone matrix from the Schamhaupten area. *S. albersdoerferi*, however, was probably entirely covered with unbranched feathers growing out of follicles. On the dorsal edge of the anterior tail, it had a filamentous crest, which might extend onto the thorax. The tubercle-like structures found in the tail region of *J. starki*, previously described as scales, show no autofluorescence signal and were reinterpreted as remains of adipocere that formed shortly after death and inhibited further decay of specimen. The occurrence of adipocere in a small, restricted area may indicate the former presence of a fat body in the anterior tail region of *J. starki*, as present in recent crocodilians and turtles. This emphasizes the importance of taphonomic considerations in the interpretation of integumentary structures and soft tissues (see Vinther et al. 2008; Foth 2012) and the reconstruction of soft tissue evolution on the basis of fossil organism (Witmer 1995; Sansom et al. 2010; Foth 2014). Furthermore, the current study gives new insights into the early evolution of feathers, although patterns of reconstructed character evolution are sensitive to the phylogenetic position of *J. starki* and the absence or presence of simple branched feather types in basal coelurosaurs. However, current data reveal that *J. starki* and *S. albersdoerferi* might represent the phylogenetically most basal record of feathered theropods. Furthermore, if radially symmetric feathers described for some basal coelurosaurs turn out to be taphonomic artefacts, the transformation of monofilamentous feathers to bilaterally symmetric branched feathers (including pennaceous feathers) might have happened relatively fast and skipped the proposed intermediate proximally branched, radially symmetric stage (see Prum 1999). Due to hierarchical organization of feathers in terms of morphology and development, such transformation would be only possible if barb ridge and rhachis formation evolved quasi contemporaneously. However, given the recent discovery of complex feather-like branching patterns in pterosaur filaments (Yang et al. 2018), it is also possible that the Prum (1999) model may yet be true and that an outstanding gap in the fossil record remains.

Acknowledgements We thank Raimund Albersdörfer and Martina Köbl-Ebert for access to the two specimens and Michael Pittman for critical discussion and reviewing the manuscript. This work was supported by the Volkswagen Foundation under grant I/84 640 (to OR), the Deutsche Forschungsgemeinschaft under grant (FO 1005/2-1) and the Swiss National Science Foundation under grant PZ00P2_174040 (both to CF).

References

Albersdörfer R, Häckel W (2015) Die Kieselplattenkalke von Painten. In: Arratia G, Schultze H-P, Tischlinger H, Viohl G (eds) Solnhofen – Ein Fenster in die Jurazeit. Verlag Dr. Friedrich Pfeil, München, pp 126–133

Alibardi L, Thompson MB (2000) Scale morphogenesis and ultrastructure of dermis during embryonic development in the alligator (*Alligator mississippiensis*, Crocodilia, Reptilia). Acta Zool (Stockholm) 81:325–338

Alibardi L, Thompson MB (2001) Fine structure of the developing epidermis in the embryo of the American alligator (*Alligator mississippiensis*, Crocodilia, Reptilia). J Anat 198:265–282

Alibardi L, Thompson MB (2002) Keratinization and ultrastructure of the epidermis of late embryonic stages in the alligator (*Alligator mississippiensis*). J Anat 201:71–84

Allain R (2002) Discovery of megalosaur (Dinosauria, Theropoda) in the middle Bathonian of Normandy (France) and its implications for the phylogeny of basal Tetanurae. J Vertebr Paleontol 22:548–563

Araújo R, Castanhinha R, Martins R, Mateus O, Hendrickx C, Beckmann F, Schell N, Alves L (2013) Filling the gaps of dinosaur eggshell phylogeny: Late Jurassic Theropod clutch with embryos from Portugal. Sci Rep 3:1924

Barrett PM, Evans DC, Campione NE (2015) Evolution of dinosaur epidermal structures. Biol Lett 11:20150229

Becker R (1959) Die Strukturanalyse der Gefiederfolge von *Megapodius feyc. reinw.* und ihre Beziehungen zu der Nestlingsdune der Hühnervögel. Rev Suisse Zool 66:411–527

Bell PR (2012) Standardized terminology and potential taxonomic utility for hadrosaurid skin impressions: a case study for *Saurolophus* from Canada and Mongolia. PLoS One 7:e31295

Bell PR (2014) A review of hadrosaurid skin impressions. In: Eberth DA, Evans DC (eds) Hadrosaurs. Indiana University Press, Bloomington, pp 572–590

Benton MJ, Zhou Z, Orr PJ, Zhang F, Kearns SL (2008) The remarkable fossils from the Early Cretaceous Jehol Biota of China and how they have changed our knowledge of Mesozoic life. Proc Geol Assoc 119:209–228

Bergmann U, Morton RW, Manning PL, Sellers WI, Farrar S, Huntley KG, Wogelius RA, Larson PL (2010) *Archaeopteryx* feathers and bone chemistry fully revealed via synchrotron imaging. Proc Natl Acad Sci USA 107:9060–9065

Berner RA (1968) Calcium carbonate concretions formed by the decomposition of organic matter. Science 159:195–197

Bonaparte JF, Novas FE, Coria RA (1990) *Carnotaurus sastrei* Bonaparte, the horned, lightly built carnosaur from the Middle Cretaceous of Patagonia. Contrib Sci 416:1–42

Briggs DEG (2003) The role of decay and mineralization in the preservation of soft-bodied fossils. Annu Rev Earth Planet Sci 31:275–301

Briggs DEG, Kear AJ (1993) Fossilization of soft tissue in the laboratory. Science 259:1439–1442

Briggs DEG, Kear AJ (1994) Decay and mineralization of shrimps. PALAIOS 9:431–456

Briggs DEG, Wilby PR (1996) The role of the calcium carbonate-calcium phosphate switch in the mineralization of soft-bodied fossils. J Geol Soc Lond 153:665–668

Briggs DEG, Kear AJ, Martill DM, Wilby PR (1993) Phosphatization of soft-tissue in experiments and fossils. J Geol Soc Lond 150:1035–1038

Britt B (1991) Theropods of Dry Mesa Quarry (Morrison Formation, Late Jurassic), Colorado, with emphasis on the osteology of *Torvosaurus tanneri*. BYU Geol Stud 37:1–72

Brusatte SL, Nesbitt SJ, Irmis RB, Butler RJ, Benton MJ, Norell MA (2010) The origin and early radiation of dinosaurs. Earth Sci Rev 101:68–100

Brusatte SL, Lloyd GT, Wang SC, Norell MA (2014) Gradual assembly of avian body plan culminated in rapid rates of evolution across the dinosaur-bird transition. Curr Biol 24:2386–2392

Butler RJ, Upchurch P (2007) Highly incomplete taxa and the phylogenetic relationships of the theropod dinosaur *Juravenator starki*. J Vertebr Paleontol 27:253–256

Carney RM, Vinther J, Shawkey MD, D'Alba L, Ackermann J (2012) New evidence on the colour and nature of the isolated *Archaeopteryx* feather. Nat Commun 3:637

Chiappe LM, Göhlich UB (2010) Anatomy of *Juravenator starki* (Theropoda: Coelurosauria) from the Late Jurassic of Germany. Neues Jahrb Geol Palaontol Abh 258:257–296

Chin K, Eberth DA, Schweitzer MH, Rando TA, Sloboda WJ, Horner JR (2003) Remarkable preservation of undigested muscle tissue within a Late Cretaceous

tyrannosaurid coprolite from Alberta, Canada. PALAIOS 18:286–294
Choiniere JN, Clark JM, Forster CA, Norell MA, Eberth DA, Erickson GM, Chu H, Xu X (2014) A juvenile specimen of a new coelurosaur (Dinosauria: Theropoda) from the Middle–Late Jurassic Shishugou Formation of Xinjiang, People's Republic of China. J Syst Palaeontol 12:177–215
Christiansen NA, Tschopp E (2010) Exceptional stegosaur integument impressions from the Upper Jurassic Morrison Formation of Wyoming. Swiss J Geosci 103:163–171
Colbert EH (1989) The Triassic dinosaur *Coelophysis*. Mus North Ariz Bull 57:1–160
Coria RA, Chiappe LM (2007) Embryonic skin from Late Cretaceous sauropods (Dinosauria) of Auca Mahuevo, Patagonia, Argentina. J Paleontol 81:1528–1532
Cuesta E, Díaz-Martínez I, Ortega F, Sanz JL (2015) Did all theropods have chicken-like feet? First evidence of a non-avian dinosaur podotheca. Cretac Res 56:53–59
Currie PJ, Chen P (2001) Anatomy of *Sinosauropteryx prima* from Liaoning, northeastern China. Can J Earth Sci 38:1705–1727
Currie PJ, Badamgarav D, Koppelhus EB (2003) The first Late Cretaceous footprints from the Nemegt locality in the Gobi of Mongolia. Ichnos 10:1–13
Czerkas SA (1997) Skin. In: Currie PJ, Padian K (eds) Encyclopedia of dinosaurs. Academic, San Diego, CA, pp 669–675
Dal Sasso C, Maganuco S (2011) *Scipionyx samniticus* (Theropoda: Compsognathidae) from the Lower Cretaceous of Italy. Memorie della Società Italiana di Scienze Naturali e del Museo Civico di Storia Naturale di Milano 37:1–281
De Buisonjé PH (1985) Climatological conditions during deposition of the Solnhofen limestones. In: Hecht MK, Ostrom JH, Viohl G, Wellnhofer P (eds) The beginnings of birds. Freunde des Jura-Museums, Eichstätt, pp 45–65
Dhouailly D, Hardy MH, Sengel P (1980) Formation of feathers on chick foot scales: a stage-dependent morphogenetic response to retinoic acid. J Embryol Exp Morphol 58:63–78
Dumbacher JP, Menon GK, Daly JW (2009) Skin as a toxin storage organ in the endemic New Guinean genus Pitohui. Auk 126:520–530
Egerton VM, Wogelius RA, Norell MA, Edwards NP, Sellers WI, Bergmann U, Sokaras D et al (2015) The mapping and differentiation of biological and environmental elemental signatures in the fossil remains of a 50 million year old bird. J Anal At Spectrom 30:627–634
Embry AF, Klovan E (1971) A late Devonian reef tract on northeastern Banks Island, N.W.T. Bull Can Petrol Geol 19:730–781
Field DJ, D'Alba L, Vinther J, Webb SM, Gearty W, Shawkey MD (2013) Melanin concentration gradients in modern and fossil feathers. PLoS One 8:e59451
Forbes SL, Stuart BH, Dent B (2005) The effect of the burial environment on adipocere formation. Forensic Sci Int 154:24–34
Foth C (2011) The morphology of neoptile feathers: ancestral state reconstruction and its phylogenetic implications. J Morphol 272:387–403
Foth C (2012) On the identification of feather structures in stem-line representatives of birds: evidence from fossils and actuopalaeontology. Paläontol Z 86:91–102
Foth C (2014) Comment on the absence of ossified sternal elements in basal paravian dinosaurs. Proc Natl Acad Sci USA 111:E5334
Foth C, Tischlinger H, Rauhut OWM (2014) New specimen of *Archaeopteryx* provides insights into the evolution of pennaceous feathers. Nature 511:79–82
Foth C, Hedrick BP, Ezcurra MD (2016) Cranial ontogenetic variation in early saurischians and the role of heterochrony in the diversification of predatory dinosaurs. PeerJ 4:e1589
Frey E (1988) Anatomie des Körperstammes von *Alligator mississippiensis* Daudin. Stuttgarter Beiträge zur Naturkunde, A 424:1–106
Frey E, Martill DM (1998) Soft tissue preservation in a specimen of *Pterodactylus kochi* (Wagner) from the Upper Jurassic of Germany. Neues Jahrb Geol Palaontol Abh 210:421–441
Frey E, Tischlinger H (2012) The Late Jurassic pterosaur *Rhamphorhynchus*, a frequent victim of the ganoid fish *Aspidorhynchus*? PLoS One 7:e31945
Gatesy SM (2001) Skin impressions of Triassic theropods as records of foot movement. Bull Mus Comp Zool 156:137–149
Godefroit P, Sinitsa SM, Dhouailly D, Bolotsky YL, Sizov AV, McNamara ME, Benton MJ, Spagna P (2014) A Jurassic ornithischian dinosaur from Siberia with both feathers and scales. Science 345:451–455
Göhlich UB, Chiappe LM (2006) A new carnivorous dinosaur from the Late Jurassic Solnhofen archipelago. Nature 440:329–332
Göhlich UB, Tischlinger H, Chiappe LM (2006) *Juravenator starki* (Reptilia, Theropoda), ein neuer Raubdinosaurier aus dem Oberjura der Südlichen Frankenalb (Süddeutschland): Skelettanatomie und Weichteilbefunde. Archaeopteryx 24:1–26
Hamburger V, Hamilton HL (1951) A series of normal stages in the development of the chick embryo. J Morphol 88:49–92
Harris MP, Fallon JF, Prum RO (2002) Shh-Bmp2 signaling module and the evolutionary origin and diversification of feathers. J Exp Zool B Mol Dev Evol 294:160–176
Haug JT, Haug C (2011) Fossilien unter langwelligem Licht: Grün-Orange-Fluoreszenz an makroskopischen Objekten. Archaeopteryx 29:20–23

Haug C, Haug JT, Waloszek D, Maas A, Frattigiani R, Liebau S (2009) New method to document fossils from lithographic limestones of southern Germany and Lebanon. Palaeontol Electron 12:3.6T

Haug JT, Haug C, Kutschera V, Mayer G, Maas A, Liebau S, Castellani C, Wolfram U, Clarkson ENK, Waloszek D (2011a) Autofluorescence imaging, an excellent tool for comparative morphology. J Microsc 244:259–272

Haug JT, Haug C, Waloszek D, Schweigert G (2011b) The importance of lithographic limestones for revealing ontogenies in fossil crustaceans. Swiss J Geosci 104: S85–S98

Hirschler A, Lucas J, Hubert J-C (1990) Bacterial involvement in apatite genesis. FEMS Microbiol Ecol 73:211–220

Huchzermeyer FW (2003) Crocodiles. CABI, Wallingford

Hwang SH, Norell MA, Ji Q, Gao K (2004) A large compsognathid from the Early Cretaceous Yixian Formation of China. J Syst Palaeontol 2:13–30

Ji Q, Currie PJ, Norell MA, Ji S (1998) Two feathered dinosaurs from northeastern China. Nature 393:753–761

Ji Q, Norell MA, Gao K, Ji S, Ren D (2001) The distribution of integumentary structures in a feathered dinosaur. Nature 410:1084–1088

Ji S, Ji Q, Lü J, Yuan C (2007) A new giant compsognathid dinosaur with long filamentous integuments from Lower Cretaceous of northeastern China. Acta Geol Sin 81:8–15

Kane R, Sell H (2001) Revolution in lamps: a chronicle of 50 years of progress. The Fairmont Press, Lilburn

Kaye TG, Falk AR, Pittman M, Sereno PC, Martin LD, Burnham DA et al (2015) Laser-stimulated fluorescence in paleontology. PLoS One 10:e0125923

Kellner AWA (1996) Fossilized theropod soft tissue. Nature 379:32

Kellner AWA, Wang X, Tischlinger H, Campos D, Hone D, Meng X (2010) The soft tissue of *Jeholopterus* (Pterosauria, Anurognathidae, Batrachognathinae) and the structure of the pterosaur wing membrane. Proc R Soc B Biol Sci 277:321–329

Kelsall JP, Calaprice JR (1972) Chemical content of waterfowl plumage as a potential diagnostic tool. J Wildl Manag 36:1088–1097

Kühn K (1974) Struktur und Biochemie des Kollagens. Chem unserer Zeit 8:97–103

Kundrát M (2004) When did theropods become feathered? Evidence for pre-*Archaeopteryx* feathery appendages. J Exp Zool B Mol Dev Evol 302B:355–364

Li Q, Gao K, Vinther J, Shawkey MD, Clarke JA, D'Alba L, Meng Q, Briggs DEG, Prum RO (2010) Plumage color patterns of an extinct dinosaur. Science 327:1369–1372

Li Q, Gao K, Meng Q, Clarke JA, Shawkey MD, D'Alba L, Pei R, Ellison M, Norell MA, Vinther J (2012) Reconstruction of *Microraptor* and the evolution of iridescent plumage. Science 335:1215–1219

Li Q, Clarke JA, Gao K, Zhou Z, Meng Q, Li D, D'Alba L, Shawkey MD (2014) Melanosome evolution indicates a key physiological shift within feathered dinosaurs. Nature 507:350–353

Lingham-Soliar T, Wesley-Smith J (2008) First investigation of the collagen D-band ultrastructure in fossilized vertebrate integument. Proc R Soc Lond B 275:2207–2212

Link E, Fürsich FT (2001) Hochauflösende Feinstratigraphie und Mikrofaziesanalyse der Oberjura Plattenkalke von Painten, Südliche Frankenalb. Archaeopteryx 19:71–88

Liu X, Zhao S, Sun L, Yin X, Xie Z, Luo H, Wang Y (2006) P and trace metal contents in biomaterials, soils, sediments and plants in colony of red-footed booby (Sula sula) in the Dongdao Island of South China Sea. Chemosphere 65:707–715

Loewen MA, Irmis RB, Sertich JJW, Currie PJ, Sampson SD (2013) Tyrant dinosaur evolution tracks the rise and fall of Late Cretaceous oceans. PLoS One 8: e79420. https://doi.org/10.1371/journal.pone.0079420

Lucas AM, Stettenheim PR (1972) Avian anatomy: integument, part I & II. U.S. Government Printing Office, Washington, DC

Maisch MW (2015) Fischechsen (Ichthyosauria). In: Arratia G, Schultze HP, Tischlinger H, Viohl G (eds) Solnhofen – Ein Fenster in die Jurazeit. Verlag Dr. Friedrich Pfeil, München, pp 422–430

Manning PL, Edwards NP, Wogelius RA, Bergmann U, Barden HE, Larson PL, Schwarz-Wings D et al (2013) Synchrotron-based chemical imaging reveals plumage patterns in a 150 million year old early bird. J Anal At Spectrom 28:1024–1030

Marlow HW, Caldwell MJ (1934) A chemical and X-ray study of "flightless" feathers. J Hered 25:265–268

Mayr G, Peters DS, Plodowski G, Vogel O (2002) Bristle-like integumentary structures at the tail of the horned dinosaur *Psittacosaurus*. Naturwissenschaften 89:361–365

Mayr G, Pittman M, Saitta E, Kaye TG, Vinther J (2016) Structure and homology of *Psittacosaurus* tail bristles. Palaeontology 59:793–802

McGraw KJ (2003) Melanins, metals, and mate quality. Oikos 102:402–406

McNamara ME, Zhang F, Kearns SL, Orr PJ, Toulouse A, Foley T et al (2018) Fossilized skin reveals coevolution with feathers and metabolism in feathered dinosaurs and early birds. Nat Commun 9:2072

Metcheva R, Yurukova L, Teodorova S, Nikolova E (2006) The penguin feathers as bioindicator of Antarctica environmental state. Sci Total Environ 362AD:259–265

Murayama K, Takahashi R, Yokote Y, Akahane K (1986) The primary structure of feather keratins from duck (*Anas platyrhynchos*) and pigeon (*Columba livia*). Biochim Biophs Acta 873:6–12

Niebuhr B, Pürner T (2014) Lithostratigraphie der Weißjura-Gruppe der Frankenalb (außeralpiner

Oberjura) und der mittel- bis oberjurassischen Reliktvorkommen zwischen Straubing und Passau (Bayern). Schriftenr Dtsch Ges Geowissenschaften 83:5–72

Norell MA, Xu X (2005) Feathered dinosaurs. Annu Rev Earth Planet Sci 33:277–299

O'Brien TG, Kuehner AC (2007) Waxing grave about adipocere: soft tissue change in an aquatic context. J Forensic Sci 52:294–301

Ortega F, Escaso F, Sanz JL (2010) A bizarre, humped Carcharodontosauria (Theropoda) from the Lower Cretaceous of Spain. Nature 467:203–206

Osborn HF (1912) Integument of the iguanodontid dinosaur *Trachodon*. Mem Am Mus Nat Hist 1:33–54

Peyer K (2006) A reconsideration of *Compsognathus* from the Upper Tithonian of Canjuers, southeastern France. J Vertebr Paleontol 26:879–896

Prum RO (1999) Development and evolutionary origin of feathers. J Exp Zool B Mol Dev Evol 285:291–306

Prum RO, Brush AH (2002) The evolutionary origin and diversification of feathers. Q Rev Biol 77:261–295

Prum RO, Dyck J (2003) A hierarchical model of plumage: morphology, development, and evolution. J Exp Zool B Mol Dev Evol 298B:73–90

Pu H, Kobayashi Y, Lü J, Xu L, Wu Y, Chang H, Zhang J, Jia S (2013) An unusual basal therizinosaur dinosaur with an ornithischian dental arrangement from northeastern China. PLoS One 8:e63423

Rauhut OWM (2005) Osteology and relationships of a new theropod dinosaur from the Middle Jurassic of Patagonia. Palaeontology 48:87–110

Rauhut OWM, Fechner R (2005) Early development of the facial region in a non-avian theropod dinosaur. Proc R Soc Lond B 272:1179–1183

Rauhut OWM, Foth C (2014) New information on the theropod dinosaurs from the Late Jurassic lithographic limestones of southern Germany. J Vertebr Paleontol, Program and Abstracts: 212

Rauhut OWM, Foth C, Tischlinger H, Norell MA (2012) Exceptionally preserved juvenile megalosauroid theropod dinosaur with filamentous integument from the Late Jurassic of Germany. Proc Natl Acad Sci USA 109:11746–11751

Reichholf JH (1996) Die Feder, die Mauser und der Ursprung der Vögel. Archaeopteryx 14:27–38

Reisdorf AG, Wuttke M (2012) Re-evaluating Moodie's opisthotonic-posture hypothesis in fossil vertebrates part I: reptiles – the taphonomy of the bipedal dinosaurs *Compsognathus longipes* and *Juravenator starki* from the Solnhofen Archipelago (Jurassic, Germany). Palaeobiodivers Palaeoenviron 92:119–168

Sadleir RW, Barrett PM, Powell HP (2008) The anatomy and systematics of *Eustreptospondylus oxoniensis*, a theropod dinosaur from the Middle Jurassic of Oxfordshire, England. Monogr Palaeontol Soc 627:1–82

Saitta E, Gelernter R, Vinther J (2018) Additional information on the primitive contour and wing feathering of paravian dinosaurs. Palaeontology 61:273–288

Sansom RS, Gabbott SE, Purnell MA (2010) Non-random decay of chordate characters causes bias in fossil interpretation. Nature 463:797–800

Schaub S (1912) Die Nestdunen der Vögel und ihre Bedeutung für die Phylogenie der Feder. Verh Naturforsch Ges Basel 23:131–182

Schweitzer MH, Watt JA, Avci R, Forster CA, Krause DW, Knapp LW, Rogers R, Beech I, Marshall M (1999a) Keratin immunoreactivity in the Late Cretaceous bird *Rahonavis ostromi*. J Vertebr Paleontol 19:712–722

Schweitzer MH, Watt JA, Avci R, Knapp LW, Chiappe LM, Norell MA, Marshall M (1999b) Beta-keratin specific immunological reactivity in feather-like structures of the Cretaceous alvarezsaurid, *Shuvuuia deserti*. J Exp Zool B Mol Dev Evol 285:146–157

Sengel P (1976) Morphogenesis of skin. Cambridge University Press, Cambridge

Sereno PC, Beck AL, Dutheil DB, Gado B, Larsson HCE, Lyon GH, Marcot JD et al (1998) A long-snouted predatory dinosaur from Africa and the evolution of Spinosaurids. Science 282:1298–1302

Sondheimer E, Dence WA, Mattick LR, Silverman SR (1966) Composition of combustible concretions of the alewife, *Alosa pseudoharengus*. Science 152:221–223

Tischlinger H (1998) Erstnachweis von Pigmentfarben bei Plattenkalk-Teleosteern. Archaeopteryx 16:1–18

Tischlinger H (2002) Der Eichstätter *Archaeopteryx* im langwelligen UV-Licht. Archaeopteryx 20:21–38

Tischlinger H (2005) Neue Informationen zum Berliner Exemplar von *Archaeopteryx lithographica* H. v. Meyer 1861. Archaeopteryx 23:33–50

Tischlinger H (2009) Der achte *Archaeopteryx* – das Daitinger Exemplar. Archaeopteryx 27:1–20

Tischlinger H, Arratia G (2013) Ultraviolet light as a tool for investigating Mesozoic fishes, with a focus on the ichthyofauna of the Solnhofen archipelago. In: Arratia G, Schultze HP, Wilson MVH (eds) Mesozoic fishes, vol 5: Global diversity and evolution. Verlag Dr. Friedrich Pfeil, München, pp 549–560

Tischlinger H, Frey E (2002) Ein Rhamphorhynchus (Pterosauria, Reptilia) mit ungewöhnlicher Flughauterhaltung aus dem Solnhofener Plattenkalk. Archaeopteryx 20:1–20

Tischlinger H, Rauhut OWM (2015) Schuppenechsen (Lepidosauria). In: Arratia G, Schultze HP, Tischlinger H, Viohl G (eds) Solnhofen – Ein Fenster in die Jurazeit. Verlag Dr. Friedrich Pfeil, München, pp 431–447

Tischlinger H, Unwin DM (2004) UV-Untersuchungen des Berliner Exemplars von *Archaeopteryx lithographica* H. v. Meyer 1861 und der isolierten *Archaeopteryx*-Feder. Archaeopteryx 22:17–50

Tsuihiji T, Watabe M, Tsogtbaatar K, Tsubamoto T, Barsbold R, Suzuki S et al (2011) Cranial osteology of a juvenile specimens of *Tarbosaurus bataar* (Theropoda, Tyrannosauridae) from the Nemegt Formation (Upper Cretaceous) of Bugin-Tsav, Mongolia. J Vertebr Paleontol 31:497–517

Van der Reest AJ, Wolfe AP, Currie PJ (2016) A densely feathered ornithomimid (Dinosauria: Theropoda) from the Upper Cretaceous Dinosaur Park Formation, Alberta, Canada. Cretac Res 58:108–117

Vinther J, Briggs DEG, Prum RO, Saranathan V (2008) The colour of fossil feathers. Biol Lett 4:522–525

Viohl G (2015a) Die Plattenkalk-Typen der Südlichen Frankenalb. In: Arratia G, Schultze H-P, Tischlinger H, Viohl G (eds) Solnhofen – Ein Fenster in die Jurazeit. Verlag Dr. Friedrich Pfeil, München, pp 72–77

Viohl G (2015b) Die Grabung in Schamhaupten. In: Arratia G, Schultze H-P, Tischlinger H, Viohl G (eds) Solnhofen – Ein Fenster in die Jurazeit. Verlag Dr. Friedrich Pfeil, München, pp 119–125

Viohl G, Zapp M (2006) Die Fossil-Lagestätte Schamhaupten (oberstes Kimmeridgium, Südliche Frankenalb, Bayern). Archaeopteryx 24:27–78

Viohl G, Zapp M (2007) Schamhaupten, an outstanding Fossil-Lagerstätte in a silicified Plattenkalk around the Kimmeridgian-Tithonian boundary (Southern Franconian Alb, Bavaria). Neues Jahrb Geol Palaontol Abh 245:127–142

Wang X, Pittman M, Zheng X, Kaye TG, Falk AR, Hartman S, Xu X (2017) Basal paravian functional anatomy illuminated by high-detail body outline. Nat Commun 8:14576. https://doi.org/10.1038/ncomms14576

Wellnhofer P (2009) Archaeopteryx: the icon of evolution. Verlag Dr. Friedrich Pfeil, München

Werneburg I, Hugi J, Müller J, Sánchez-Villagra MR (2009) Embryogenesis and ossification of *Emydura subglobosa* (Testudines, Pleurodira, Chelidae) and patterns of turtle development. Dev Dyn 238:2770–2786

Wilby PR, Briggs DEG, Bernier P, Gaillard C (1996) Role of microbial mats in the fossilization of soft tissues. Geology 24:787–790

Wiman C (1942) Über ältere und neuere Funde von Leichenwachs. Senckenbergiana 25:1–19

Witmer LM (1995) The extant phylogenetic bracket and the importance of reconstructing soft tissues in fossils. In: Thomason JJ (ed) Functional morphology in vertebrate paleontology. Cambridge University Press, Cambridge, pp 19–33

Xu X, Guo Y (2009) The origin and early evolution of feathers: insights from recent paleontological and neontological data. Vertebrata PalAsiatica 47:311–329

Xu X, Zhou Z, Prum RO (2001) Branched integumental structures in *Sinornithosaurus* and the origin of feathers. Nature 410:200–204

Xu X, Norell MA, Kuang X, Wang X, Zhao Q, Jia C (2004) Basal tyrannosauroids from China and evidence for protofeathers in tyrannosauroids. Nature 431:680–684

Xu X, Zheng X, You H (2009) A new feather type in a nonavian theropod and the early evolution of feathers. Proc Natl Acad Sci USA 106:832–834

Xu X, Zheng X, You H (2010) Exceptional dinosaur fossils show ontogenetic development of early feathers. Nature 464:1338–1341

Xu X, Wang K, Zhang K, Ma Q, Xing L, Sullivan C, Hu D, Cheng S, Wang S (2012) A gigantic feathered dinosaur from the Lower Cretaceous of China. Nature 484:92–95

Xu X, Zhou Z, Dudley R, Mackem S, Chuong C, Erickson GM, Varricchio DJ (2014) An integrative approach to understanding bird origins. Science 346:12532931–12532910

Xu X, Zheng X, Sullivan C, Wang W, Xing L, Wang Y, Zhang X, O'Connor JK, Zhang F, Pan Y (2015) A bizarre Jurassic maniraptoran theropod with preserved evidence of membranous wings. Nature 251:70–73

Yang Z, Jiang B, McNamara ME, Kearns SL, Pittman M, Kaye TG, Orr PJ, Xu X, Benton MJ (2018) Pterosaur integumentary structures with complex feather-like branching. Nat Ecol Evol 3:24–30

Zelenitsky DK, Therrien F, Erickson GM, DeBuhr CL, Kobayashi Y, Eberth DA, Hadfield F (2012) Feathered non-avian dinosaurs from North America provide insight into wing origins. Science 338:510–514

Zhang F, Kearns SL, Orr PJ, Benton MJ, Zhou Z, Johnson D, Xu X, Wang X (2010) Fossilized melanosomes and the colour of Cretaceous dinosaurs and birds. Nature 463:1075–1078

Zhao X, Currie PJ (1993) A large crested theropod from the Jurassic of Xinjiang, People's Republic of China. Can J Earth Sci 30:2027–2036

Zheng X, You H, Xu X, Dong Z (2009) An Early Cretaceous heterodontosaurid dinosaur with filamentous integumentary structures. Nature 458:333–336

Zheng X, Zhou Z, Wang X, Zhang F, Zhang X, Wang Y, Wei G, Wang S, Xu X (2013) Hind wings in basal birds and the evolution of leg feathers. Science 339:1309–1312

Feather Evolution in Pennaraptora 7

Ulysse Lefèvre, Andrea Cau, Dongyu Hu, and Pascal Godefroit

7.1 Introduction

Before the discovery of *Sinosauropteryx*, a small Early Cretaceous coelurosaurian theropod from Liaoning Province in China (Ji and Ji 1996; Chen et al. 1998), feathers were thought to be an exclusively avialian character. After two decades of discoveries, feathers and feather-related structures are known to be quite widespread among dinosaurs based on fossils with a great range of size, osteology, limb proportions, and integumentary coverings (Hopson 2001; Ji et al. 2001). Moreover, feathers and feather-related structures may have been present in the common ancestors of both saurischian and ornithischian dinosaurs (Witmer 2009; Godefroit et al. 2014). Here, we briefly describe the feathers and plumage of pennaraptoran theropods, a monophyletic clade containing *Oviraptor philoceratops*, *Deinonychus anthirrhopus*, and *Passer domesticus*) but not Therizinosauroidea (*Therizinosaurus cheloniformis*). This monophyletic group encompasses five groups of bipedal dinosaurs characterized by a trend in size reduction and the presence of pennaceous feathers (Foth et al. 2014; Lee et al. 2014): oviraptorosauria, scansoriopterygidae (a group of hypothetized bat-winged dinosaurs), Troodontidae, Dromaeosauridae, and Avialae.

The presence of feather-related structures is not limited to pennaraptoran dinosaurs. The first evidence of integumentary structures among theropods was reported in *Pelecanimimus*, a small ornithomimosaur from the Lower Cretaceous of Spain (Pérez-Moreno et al. 1994). The specimen shows preserved impressions of subparallel fibers arranged perpendicular to the bone surface below the neck and around the right humerus and elbow (Pérez-Moreno et al. 1994). Two years later, *Sinosauropteryx*, a small coelurosaurian dinosaur from the Lower Cretaceous of China (Chen et al. 2005), was reported by Ji and Ji (1996). Integumentary structures are preserved in this specimen along the dorsal surface of the neck, the back, the hip, and along the upper and lower margins of the tail (Chen et al.

U. Lefèvre
Department of Geology, University of Liège, Liège, Belgium

Earth and Life History O.D., Royal Belgian Institute of Natural Sciences, Bruxelles, Belgium

A. Cau
Earth, Life and Environmental Sciences Department, Alma Mater Studiorum, Bologna University, Bologna, Italy

Geological and Palaeontological Museum "G. Capellini", Bologna, Italy

D. Hu
Paleontological Institute, Shenyang Normal University, Shenyang, China

Key Laboratory for Evolution of Past Life in Northeast Asia, Ministry of Land and Resources (Shenyang Normal University), Shenyang, China

P. Godefroit (✉)
Earth and Life History O.D., Royal Belgian Institute of Natural Sciences, Bruxelles, Belgium
e-mail: pgodefroit@naturalsciences.be

C. Foth, O. W. M. Rauhut (eds.), *The Evolution of Feathers*, Fascinating Life Sciences,
https://doi.org/10.1007/978-3-030-27223-4_7

1998). The nature of these appendages has been largely discussed (Currie and Chen 2001; Zhou and Zhang 2006; Lingham-Soliar 2012) and their affinities with feathers are still debated. Despite the debate about the relationship between these integumentary structures and feathers, integumentary structures have been reported in a medium-sized therizinosaurid theropod, *Beipiaosaurus*, which shows integumentary appendages close to those of *Sinosauropteryx* (Xu et al. 1999a). Zhou and Zhang (2006) suggested that the finely preserved filaments near the ulna of *Beipiaosaurus* were the precursors to secondary flight feathers. Protofeathers were also reported in *Dilong* and *Yutyrannus*, two basal tyrannosauroids from the Cretaceous of China (Xu et al. 2004, 2012). The filamentous protofeathers of *Dilong* were probably branched as in modern feathers and preserved around the posterior left mandible and around the tail (Xu et al. 2004). Filaments composing the feathers were joined at their bases along a central filament (Xu et al. 2004). In *Yutyrannus*, the filamentous integumentary structures were preserved close to the tail, the pelvis, the pes, the neck, and near a limb bone (identified as a humerus) (Xu et al. 2012). It seems that the feather covering in tyrannosauroids was used for thermoregulation and/or display structures rather than in flight (Xu et al. 2004, 2012).

7.2 Oviraptorosauria

Some plumage characteristics present in birds and, in a wider context, in paravians were inherited from this clade of Cretaceous theropod dinosaurs. The presence of integumentary structures in oviraptorosauran theropods was recently reported in *Caudipteryx*, *Protarchaeopteryx*, and *Similicaudipteryx* (Ji et al. 1998; Xu et al. 2010) (Fig. 7.1). The presence of remiges (i.e., flight feathers of modern birds and paravians) were reported in *Caudipteryx* and *Similicaudipteryx* (Ji et al. 1998; Xu et al. 2010). Fourteen remiges were attached to the hand of *Caudipteryx* (i.e., primaries) (Ji et al. 1998), whereas 10 primary and 12 secondary remiges were preserved along each arm of *Similicaudipteryx* (Xu et al. 2010). No flight feathers were reported in *Protarchaeopteryx* (Ji et al. 1998). In *Caudipteryx* and *Similicaudipteyrx*, the remiges were symmetrical (i.e., both vanes had the same width along the rachis) and elongated (Ji et al. 1998; Xu et al. 2010). Unlike in modern birds, in which the first primaries are the longest ones, the middle primaries were reported to be the longest ones in *Caudipteryx* and *Similicaudipteryx* (Lucas and Stettenheim 1972; Ji et al. 1998; Xu et al. 2010). One of the main characteristics of Oviraptorosauran dinosaurs is the reduction of the tail, which ends with a stiffened portion (even a pygostyle in *Nomingia*) (Osmólska et al. 2004). The reduced tail of *Caudipteryx*, *Protarchaeopteryx*, and *Similicaudipteryx* bears several rectrices (i.e., tail feathers of modern birds) organized as a fan (Persons et al. 2014), which are restricted to last caudal vertebrae (Ji et al. 1998; Xu et al. 2010). These rectrices are symmetrical and their basal portion is plumulaceous as in modern birds (Ji et al. 1998). Barbules have been reported in the distal portion of the rectrices in *Protarchaeopteryx* (Ji et al. 1998). Plumulaceous feathers were found around the chest, the tail, and the femora of *Protarchaeopteryx,* whereas plumulaceous feathers were found around the tail of *Caudipteryx* (Ji et al. 1998). Long branching filamentous structures were found all around the skeleton of *Similicaudipteryx* (Xu et al. 2010).

Although the presence of a simple wing (i.e., formed by a single row of remiges) and a tail fan of feathers have been reported among some oviraptorosaurs, they are seen as flightless (Persons et al. 2014). Their feathering was not adapted to ensure an effective role in flight: the remiges were symmetrical, restricted to the forearm, and only slightly longer than the humerus; the distal remiges were shorter than the proximal ones; the forelimbs were proportionally shorter than that of modern birds; and the rectrices were more implied in sexual/visual display than to produce thrust for flight (Ji et al. 1998; Persons et al. 2014).

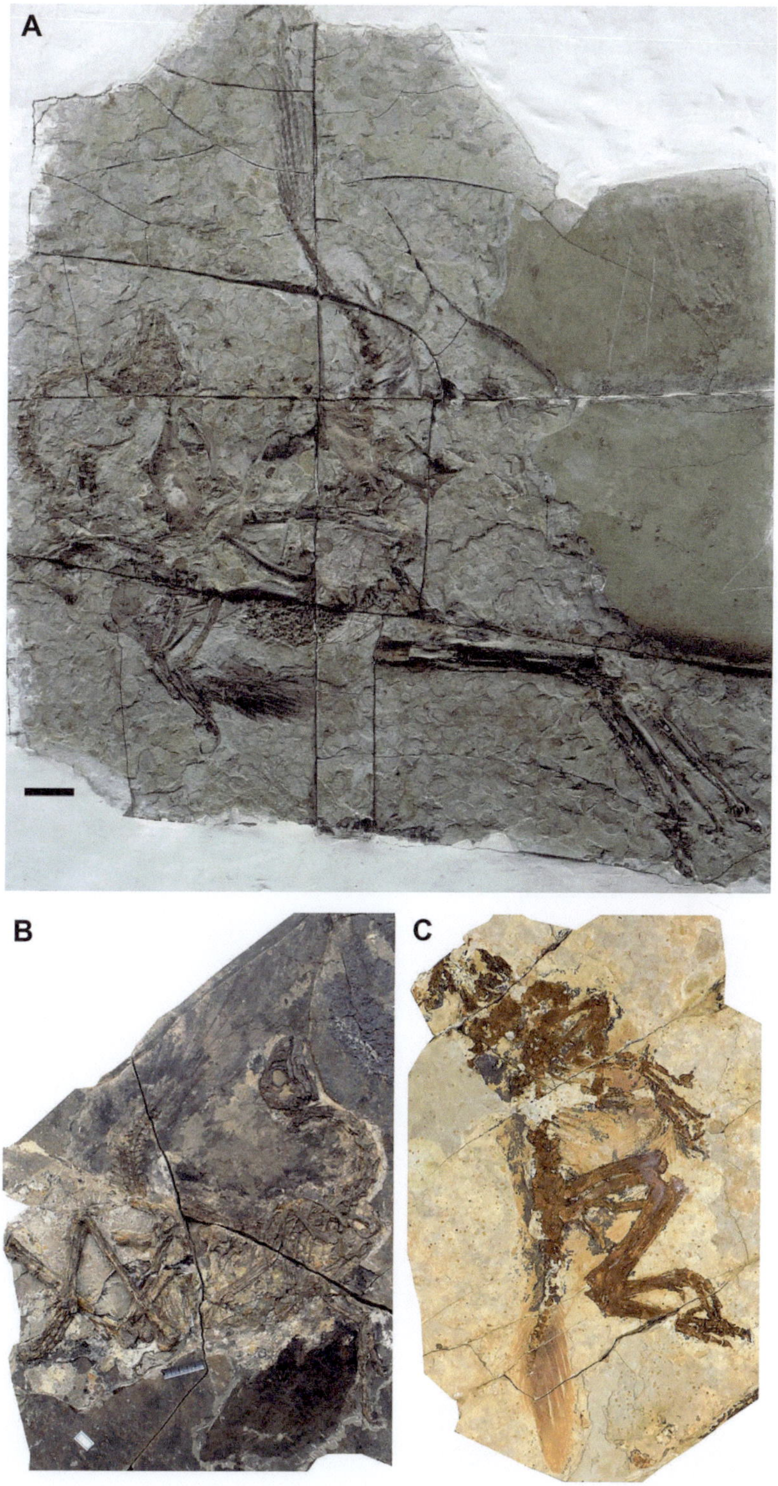

Fig. 7.1 Specimens of oviraptorosaurs preserving feathers. (**a**) NGMC 97-4-A, the holotype of *Caudipteryx* (from Ji et al. 1998). (**b**) STM 22-6, specimen of *Similicaudipteryx* (**c**) STM 4-1, another specimen of *Similicaudipteryx* (from Xu et al. 2010)

7.3 The Strange Case of Scansoriopterygidae

Scansoriopterygidae is a clade of basal paravians that was first characterized by the extreme elongation of the third digit. This is in contrast with birds that retain the plesiomorphic condition of neotheropods in which the second digit is the longest (we follow the "traditional" morphological digital identity homology pattern, in which the digit numbering of the manus of modern birds is I-II-III) (Zhang et al. 2002). The plumage of *Epidendrosaurus*, the first described scansoriopterygid, is limited to faint feather imprints more or less resembling those of *Microraptor* (Zhang et al. 2002). Two types of feather-like structures were subsequently described in *Epidexipteryx*: the dorsal part of the neck, the back, the pectoral region, and the distal part of the hip are covered by filamentous parallel barbs [called non-elongated ribbon-like tail feathers or non-ETF; Zhang et al. (2008a)] comparable to the down-like feathers in some feathered dinosaurs and primitive birds (e.g., *Sinornithosaurus*, *Sinosauropteryx*, *Anchiornis*, *Xiaotingia*) (Xu 2006; Zhang et al. 2008a) (Fig. 7.2). The ten caudal vertebrae of *Epidexipteryx* bear four elongated ribbon-like tail feathers with a broad central element, presumably the rachis, as encountered in Confuciusornithiformes and Enantiornithines (Zhang et al. 2008a, b). As in Confuciusornithiformes, those ornamental feathers may have played a role in visual display (Zhang et al. 2008a). Pennaceous feathers are absent along both the forelimbs and the hindlimbs, so scansoriopterygids were first regarded as secondary flightless paravians (Zhang et al. 2008a). However, the recently discovered *Yi* is characterized by the presence of rod-like structures (i.e., styliform elements) extending from each wrist and delimiting a membranous tissue preserved between the rod-like bones and the manual digits (Xu et al. 2015) (Fig. 7.2b). Those structures are thought to play the same role as the pteroid bone of pterosaurs or the styliform cartilage of flying squirrels and anomalures in supporting a membranous wing (Benett 2007; Xu et al. 2015). Thin stiff filamentous feathers are preserved around the skull, the neck, the forelimbs, and the hindlimbs up to

Fig. 7.2 Selected specimens of Scansoriopterygidae. (**a**) STM 31-2, the holotype of *Yi* characterized by the presence of rod-like structures extending from each wrist (white arrows) (from Xu et al. 2015). (**b**) IVPP V24192, the holotype of *Ambopteryx* (white arrow points to the styliform element) (from Wang et al. 2019)

the metatarsus (Xu et al. 2015). Some feathers in the skull and neck regions present a simple branching pattern with multiple radiating filaments, whereas feathers attached to the limbs are composed of an undifferentiated shaft-like structure with numerous parallel distal filaments (Xu et al. 2015). The presence of a membranous wing in scansoriopterygidae has been recently confirmed with the discovery of *Ambopteryx* (Wang et al. 2019) (Fig. 7.2c). A styliform element is preserved in articulation with the left wrist, and the matrix surrounding the left hand, right forelimb, and abdomen is coated in a brownish continuous layer bearing ripple-like striations (Wang et al. 2019). These elements confirm that the elongated third digit and the styliform element serve as supports for a membranous wing (Xu et al. 2015; Wang et al. 2019).

Initially, the long manus of Scansoriopterygidae was thought to be an adaptation for grasping or grabbing tree trunks rather than for flapping flight (Zhang et al. 2002). It has also been proposed that the elongated third digit of scansoriopterygids serves as a tool for extracting grubs during foraging (Wang et al. 2019). Now this trait is viewed as an adaptation to hold a membranous wing (Xu et al. 2015; Wang et al. 2019). The apparent lack on pennaceous feathers along the forelimbs in Scansoriopterygidae is therefore tentatively viewed as a secondary loss within this clade to allow the presence of a bat wing.

7.4 Anchiornithidae

This new informal group of small-bodied paravian theropods from the Upper Jurassic of China is characterized by extensive feathering. Depending on the phylogenetic analysis, anchiornithids are seen as basal paravians (Lefèvre et al. 2017), basal troodontids (Hu et al. 2009; Turner et al. 2012), or basal deinonychosaurians (Xu et al. 2011, 2015; Hu et al. 2018). The clade includes *Anchiornis*, *Pedopenna*, *Eosinopteryx*, *Aurornis*, *Serikornis*, *Caihong*, and *Ostromia* (Haarlem specimen) (Fig. 7.3). Only the last species comes from Germany, whereas the first five are from the Upper Jurassic of northeastern China (Inner Mongolia and Liaoning provinces). The presence of a feathered wing along the forelimb in Anchiornithidae is a pennaraptoran symplesiomorphy shared with Oviraptorosauria, although it may represent a symplesiomorphy shared with Ornithomimosauria because the presence of feathers is assumed in an adult specimen of *Ornithomimus* (Zelenitsky et al. 2012). Although the forewing of Oviraptorosauria seems to be formed by a single layer of long remiges, the wing of the basal paravians *Anchiornis* and *Serikornis* is formed by multiple layers of indistinct contour and flight feathers (Longrich et al. 2012; pers. obs.). The forewing of *Anchiornis* is formed by short symmetrical feathers (Hu et al. 2009; Pei et al. 2017), but some primary remiges show a slight curvature in the rachis (Saitta et al. 2017). The barbs of the remiges and major coverts of *Anchiornis* are not tight together, especially at their tip (Saitta et al. 2017). Saitta et al. (2017) suggested that the absence of differentiation between proximal and distal barbules results in feathers with open vanes, such as those reported in *Anchiornis*. In *Serikornis*, remiges are short, symmetrical, straplike, and undifferentiated (i.e., there are no significant length differences between remiges and wing coverts) with coverts contributing substantially to the airfoil (Lefèvre et al. 2017). The feathers are devoid of barbules along the forelimb and hindlimb (Lefèvre et al. 2017). The same configuration was observed in the secondary wing feathers of *Eosinopteryx* (Godefroit et al. 2013a). In *Aurornis*, forewing feathers are not preserved, with the exception of bundles of parallel filaments joined together proximally in the proximal third of the tail (Godefroit et al. 2013b). In *Caihong*, the feathers are too densely preserved to be adequately described (Hu et al. 2018). However, the forelimb is covered by vaned feathers with the proximal primaries and the distal secondaries being the longest remiges (Hu et al. 2018). These feathers are 2.4 times as long as the humerus with narrow rachises (Hu et al. 2018). One striking feature is the presence of alular feathers on the first digit of *Caihong* (considering the I-II-III pattern for digit numbering in birds and bird relatives) (Hu et al. 2018).

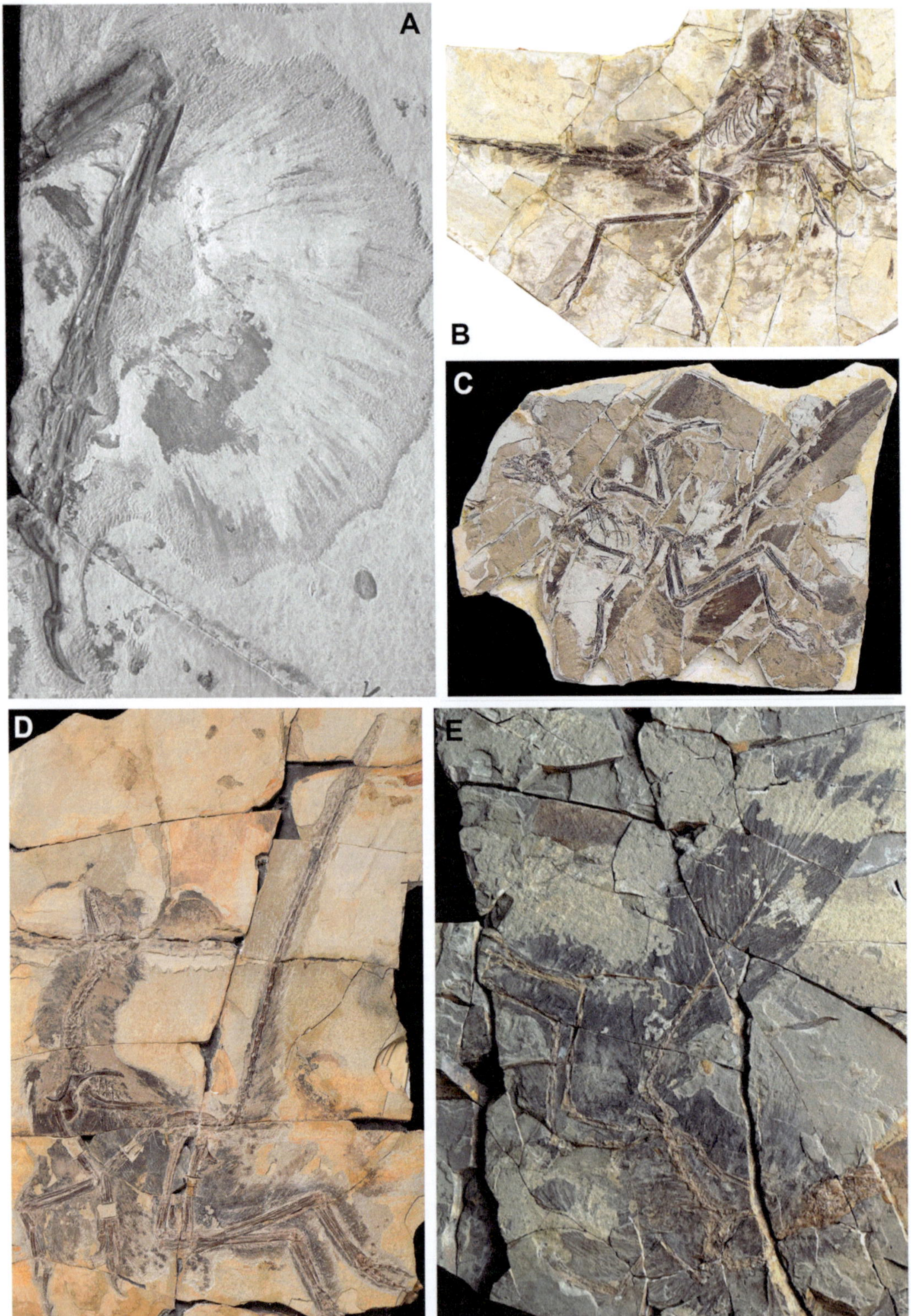

Fig. 7.3 Selected specimens of Anchiornithids with the best preserved plumage. (**a**) IVPP V12721, the holotype of *Pedopenna* (from Xu and Zhang 2005). (**b**) YFGP-T5197, the holotype of *Eosinopteryx* (from Godefroit et al. 2013b). (**c**) YFGP-T5199, a specimen of *Anchiornis* (from Lindgren et al. 2015). (**d**) PMOL-AB00200, the holotype of *Serikornis* (from Lefèvre et al. 2017). (**e**) PMOL-B00175, the holotype of *Caihong* (from Hu et al. 2018)

Unlike in oviraptorosaurs, hindlimb feathers are well developed in Anchiornithidae. Trousers of elongated and symmetrical pennaceous tibial and metatarsal feathers are present in *Pedopenna, Eosinopteryx, Serikornis, Caihong* and *Anchiornis* (Xu and Zhang 2005; Hu et al. 2009, 2018; Godefroit et al. 2013a; Lefèvre et al. 2017; Pei et al. 2017; Saitta et al. 2017). The hindlimb feathers in *Caihong* are reported to be relatively more elongated than those of other paravians (Hu et al. 2018). The anterior side of the legs of *Caihong* is covered by numerous small feathers that extend on the pedal digits (Hu et al. 2018). This trait is also reported in *Anchiornis* and *Serikornis,* where the extensive feathering formed by a bundle of filaments extends up to the end of each toe (Hu et al. 2009; Lefèvre et al. 2017). This character is also seen in some modern birds, with an insulating or protective function (Weidensaul 1995; Hu et al. 2009).

Although elongated rectrices are present along the distal half of the tail in Oviraptorosauria (Ji et al. 1998; Xu et al. 2010) and *Microraptor* (Li et al. 2012), rectrices are absent in *Serikornis, Aurornis*, and *Eosinopteryx*; however, the absence of tail feathering may be a taphonomic artefact of melanosome preservation. In *Caihong*, large pennaceous feathers are attached to all caudal vertebrae together with covert feathers (Hu et al. 2018). The tail feathers in the most anterior part of the tail are less organized than in the middle and posterior regions but are supposed to be pennaceous (Hu et al. 2018). In *Anchiornis*, the proximal portion of the tail appears to be covered with plumulaceous features because no rachises were observed, whereas the middle and distal portions of the tail are associated with elongated pennaceous feathers that appear almost parallel to the caudal column (Hu et al. 2009; Pei et al. 2017). However, at this point, the tail of these basal paravians seems to have no aerodynamic function in increasing the total lift of the animal, nor any display function. The long bony tail was completely covered by elongated but symmetrical pennaceous rectrices in *Anchiornis* (Hu et al. 2009; Pei et al. 2017), resembling the rectrices in *Archaeopteryx* (in this species, the lateral rectrices are asymmetrical) (Foth et al. 2014).

7.5 Troodontidae

This phylogenetic position of this clade of small, lightly built maniraptorans from Cretaceous deposits of Asia and North America has been recently challenged. Pending on the phylogenetic analysis, it is positioned as the sistergroup of Dromaeosauridae (Turner et al. 2012; Lefèvre et al. 2017) in a larger clade named Deinonychosauria or as the sistergroup of Avialae (Godefroit et al. 2013b; Foth and Rauhut 2017). While dromaeosaurids from the Yixian and Jiufotang formations of Liaoning Province are well-documented concerning feather types and plumage, most troodontids appeared to lack integumentary structures; although numerous troodontid specimens have been discovered in the Yixian Formation of Liaoning Province (*Sinovenator, Mei, Sinusonasus*), most have been found in the Lujiatun Bed, which favorable for the three-dimensional preservation of specimens but not for the preservation of integumentary structures. Therefore, the apparent lack of feathers is surely due to an artifact of taphonomy or is caused by an erasing during preparation rather than a true absence. However, the discovery of *Jianianhualong* in the Lower Cretaceous Yixian Formation has greatly improved our understanding of troodontid feathering (Xu et al. 2017) (Fig. 7.4). The body of *Jianianhualong* is covered by long feathers, with the dorsal and sacral ones longer than the cervical feathers (Xu et al. 2017). Large pennaceous feathers are preserved dorsally and ventrally to nearly all caudal vertebrae as in *Jinfengopteryx*, another troodontid from China (Fig. 7.4a), forming a frond-like feathery tail (Xu et al. 2017). These feathers are curved with the distal edge convex and the proximal edge concave (Xu et al. 2017). In one well-exposed tail feather, the trailing vane is about twice as wide as the leading vane, with the blunt distal end wider than the proximal region (Xu et al. 2017). Feathers are also preserved near the humerus, ulna, and the middle portion of the tibia, but few details are preserved (Xu et al. 2017). Integumentary structures are also reported in *Jinfengopteryx*, another presumably

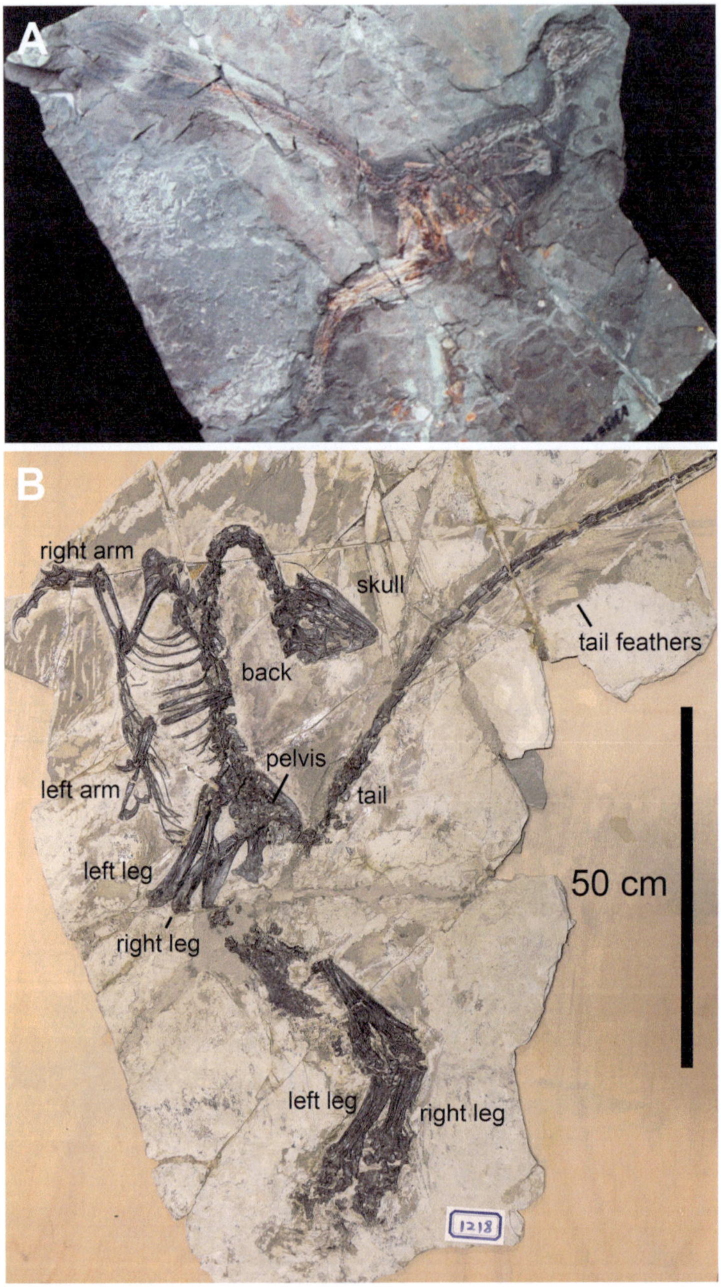

Fig. 7.4 Selected specimens of Troodontidae with preserved plumage. (**a**) CAGS-IG-04-0801, the holotype of *Jinfengopteryx* (from Ji and Ji 2007). (**b**) DLXH 1218, the holotype of *Jianianhualong* (from Xu et al. 2017)

basal troodontid theropod from the Early Cretaceous Huajiying Formation Heibei Province in China (Ji and Ji 2007). This specimen displays two patches of feathers on the right fingers, but no trace of flight feathers were preserved (Ji et al. 2005; Agnolin and Novas 2013). Only long and symmetrically-vaned tail feathers are well preserved and seem to attach to practically all caudal vertebrae (Ji et al. 2005). This configuration of the tail feathers is closer to that of *Anchiornis*,

Archaeopteryx, and *Jianianhualong*, in which pennaceous feathers are attached to all caudal vertebrae, than that of some oviraptorosaurs and *Serikornis*, in which the tail feathers are gathered on the distal caudal vertebrae (Gatesy and Dial 1996; Ji et al. 2005; Xu et al. 2010).

7.6 Dromaeosauridae

Integumentary structures in dromaeosaurids have been well documented (Fig. 7.5) since the description of *Sinornithosaurus*, the first feathered dromaeosaurid discovered by Xu et al. (1999b). The body of this small dinosaur is covered by filamentous feathers that usually reach 40 mm in length around the postcranial elements and that are shorter around the skull (Xu et al. 1999b). No remiges or rectrices with prominent rachis were observed after a second preparation of the specimen (Xu et al. 2001). However, pennaceous feathers resembling the forewing feathers of *Caudipteryx* are attached along the ulnar region of NGMC 91, another specimen tentatively referred to *Sinornithosaurus* (Ji et al. 1998, 2001; Han et al. 2014; Lü and Brusatte 2015). Two additional types of integumentary structures can be observed in *Sinornithosaurus*: (1) filaments that are equal in length and remain parallel as they extend distally from their common point of insertion and (2) filaments that form the margin of a composed structure in a manner similar to a pennaceous feather vane (Xu et al. 2001). The first type resembles the downy feathers of natal down feathers of bird hatchlings (Foth 2011) and the second one resembles of the structure of a pennaceous feather formed by a central rachis from which multiple barbs arise (Lucas and Stettenheim 1972; Xu et al. 2001). Those structures cover the entire body of NGMC 91 except its distal hindlimbs (Ji et al. 2001).

Xu et al. (2000) described large patches of integumentary structures in the dromaeosaurid *Microraptor*, including contour impressions of narrow and remige-like feathers containing rachis-like structures near the femur (Fig. 7.5a). Integumentary structures are extraordinarily well preserved in *Microraptor* (Xu et al. 2003), including two types of feather-like structures (Xu et al. 2003): its body is covered by numerous 25- to 30-mm-long plumulaceous feathers, whereas the feathers around its skull roof are up to 40 mm long, with some of them displaying well-organized pennaceous vanes (Xu et al. 2003). Large pennaceous feathers are present on the distal end of the tail of *Microraptor*, but in a pattern far different from that in *Archaeopteryx* and *Anchiornis*, in which each caudal vertebra bears rectrices (i.e., large vaned pennaceous feathers) (Hu et al. 2009; Foth et al. 2014). In *Microraptor*, the first rectrices attach from the 15th or 18th 202 caudal vertebrae up to the distal end of the tail (Xu et al. 2003). However, they are inserted more proximally than in *Caudipteryx* and *Protarchaeopteryx* (Ji et al. 1998; Xu et al. 2003) and may have acted as stabilizers and additional lifts against the destabilizing effect of the hindwings during the flight of *Microraptor* (Chatterjee and Templin 2007; Alexander et al. 2010). The hindlimbs of *Microraptor* are covered by numerous pennaceous feathers. Tibial feathers are shorter than the 14 large pennaceous metatarsal feathers. Metatarsal feathers are roughly perpendicular to their corresponding bone, with short symmetrical vanes in the proximal portion of the metatarsus and long asymmetrical vanes in the distal region (Xu et al. 2003). This pattern is far from that in basal non-eumaniraptoran paravians, in which all tibial and metatarsal feathers are symmetrical and equal in length. The forelimb feathers of *Microraptor* have a modern bird pattern, with approximately 12 primaries and 18 longer secondaries that are more than twice as long as the humerus (Lucas and Stettenheim 1972); the distal primaries are more parallel to the manus than the proximal ones. The primaries and distal secondaries have asymmetrical vanes, whereas the proximal secondaries have symmetrical vanes (Xu et al. 2003). The presence of an alular feather on the first digit of *Microraptor* and small alular feathers in *Caihong* have been reported by Foth et al. (2014) and Hu et al. (2018). Coverts are present and are variable in size, although they are considerably shorter than the primaries and the

Fig. 7.5 Selected specimens of Dromaeosauridae with best preserved plumage. (**a**) IVPP V13352, the holotype of *Microraptor* (from Xu et al. 2003). (**b**) HG B016, the holotype of *Changyuraptor* (from Han et al. 2014). (**c**) JPM-0008, the holotype of *Zhenyuanlong* (from Lü and Brusatte 2015)

secondaries—a pattern closer to modern birds than to other basal paravians (Xu et al. 2003; Longrich et al. 2012).

Changyuraptor, also from the Yixian Formation of Liaoning Province, is one meter-long dromaeosaurid with extremely long tail feathers (about one third the length of the skeleton), although forelimb feathers are poorly preserved (Han et al. 2014) (Fig. 7.5b). Some of the pennaceous primaries show a shaft that projects from the distal end of the left hand (Han et al. 2014). *Changyuraptor* preserve numerous pennaceous feathers that form an extensive hindwing as in *Anchiornis*, *Xiaotingia*, *Pedopenna*, *Archaeopteryx*, and *Microraptor*. The wing extends up to the distal end of the metatarsus but the feathers do not extend onto the digits, unlike in *Anchiornis* (Hu et al. 2009; Han et al. 2014). Although it is impossible to know whether they are asymmetrical or not, these hindlimb feathers overlap each other, apparently forming an airfoil all along the tibia and the femur (Han et al. 2014). The tail is almost completely covered by long and symmetrical rectrices that project backwards with an acute angle of 20° (Han et al. 2014). As in *Microraptor*, rectrices from the distal one fourth of the tail are more elongated and form a fan composed of at least eight or nine pairs of long feathers (Xu et al. 2003; Han et al. 2014). Those elongated tail feathers may have acted as a display structure, a pitch control structure during the flight, a way to compensate for the weight of the tail during the flight, or a way to reduce descent speed during landing thanks to the passive flexion of the long bony tail (Hone 2012; Han et al. 2014). Filamentous appendages similar to those of *Sinornithosaurus* are found along the ventral side of the neck and dorsal to the pelvis and base of the tail (Ji et al. 2001; Xu et al. 2001; Han et al. 2014).

Feathers are not limited to small-sized dromaeosaurids. Turner et al. (2007) showed that the posterior forearm of *Velociraptor* bears quill knobs for insertion of feathers, as in modern birds. They also suggested that the reduction or absence of quill knobs in larger and more derived dromaeosaurids may reflect the loss of aerodynamic capabilities (Turner et al. 2007). Lü and Brusatte (2015) described *Zhenyuanlong*, a larger-bodied (about 2 m long), short-armed dromaeosaurid from the Yixian Formation of Liaoning Province (Fig. 7.5c). Unlike in *Tianyuraptor*, another large-bodied dromaeosaurid from the Yixian Formation, integumentary structures are finely preserved around the skeleton of *Zhenyuanlong*. The body of *Zhenyuanlong* is covered by numerous small (<30 mm) filaments (Lü and Brusatte 2015). Their structure cannot be accurately determined because of their poor preservation. Large pennaceous feathers form broad wings along the forearms, with short coverts overlapping longer primaries and secondaries, resembling the condition in modern birds (Lucas and Stettenheim 1972; Lü and Brusatte 2015). Coverts from the ulnar region are perpendicularly inserted, whereas those from the metacarpal part are more obliquely inserted and more parallel to the long axis of the hand (Lü and Brusatte 2015). Even though other pennaceous feathers are not as well preserved as the coverts, approximately 10 primaries and 20 secondaries can be identified along the forelimb (Lü and Brusatte 2015). Secondaries are inserted perpendicularly to the long axis of the bones, like the secondary coverts, whereas primaries form an acute angle with the long axis of the manus (Lü and Brusatte 2015), as in *Archaeopteryx*, *Microraptor*, and modern birds (Lucas and Stettenheim 1972; Xu et al. 2003; Foth et al. 2014; Lü and Brusatte 2015). Although the vane arrangement of the forelimb feathers remains unclear due to poor preservation, those feathers were apparently asymmetrical (Lü and Brusatte 2015). The tail of *Zhenyuanlong* was likely covered by large pennaceous feathers, as suggested by the presence of thin, long, and straight rachides projecting roughly 45° posterodorsally from the longest axis of the tail. Small and simple feathers were apparently inserted on the proximal part of the tail, whereas much larger and more complex feathers were inserted distally, as also observed in *Microraptor* (Xu et al. 2003; Lü and Brusatte 2015). The presence or absence of a tail fan in *Zhenyuanlong* cannot be assessed because the distal part of the tail is not preserved. Hindlimb feathers are apparently absent in *Zhenyuanlong*, although it cannot be ruled out that it is an artifact of preservation (Lü and Brusatte 2015). Because of the small size of its forelimbs, *Zhenyuanglong* could not fly or glide. Therefore, the presence of a complex bird-like plumage along the forelimbs of this large dromaeosaurid can be viewed as a probable heritage from a flying or gliding ancestor, secondarily used for sexual or social display (Lü and Brusatte 2015).

7.7 General Pattern of Feather Evolution in Pennaraptoran Theropods

Feathers and feather-related structures are widely distributed among dinosaurs, from small ornithischians to large-bodied theropods such as *Yutyrannus* (Xu et al. 2012; Godefroit et al. 2014). The presence of pennaceous feathers along the forelimb, as observed in anchiornithids (e.g., *Eosinopteryx*, *Anchiornis*, and *Serikornis*) and more derived paravians, was likely inherited from a common ancestor shared with Oviraptorosauria or possibly Ornithomimosauria because quill-knobs have been described in one specimen of *Ornithomimus* (Zelenitsky et al. 2012). We hypothesize that those feathers were secondarily lost in scansoriopterygids to accommodate for the bat wing. Unlike in ornithomimosaurs and oviraptosaurs, the forelimb wing of anchiornitids is multi-layered, formed by several layers of undifferentiated coverts and flight feathers (Longrich et al. 2012).

The undifferentiated forelimb pennaceous feathers in anchiornithids were obviously not adapted for flight: their vanes were short and symmetrical, and their rachis was probably too slender and flexible (Godefroit et al. 2013a, b). The presence of open-vaned feathers in *Anchiornis* also underlines this lack of aerodynamic function (Saitta et al. 2017). Moreover, the forelimbs of anchiornithids were probably proportionally too short and their pectoral girdle lacks adaptations for flapping flight (Godefroit et al. 2013a). It might therefore be hypothesized that those basal paravians principally used their forelimb feathers as visual displays for social and/or sexual behaviors, although we do not rule out that they could use their primitive wings for gliding from tree to tree or prolonging leaps during cursorial locomotion.

Remiges became progressively better adapted for flight in more derived paravians, including some small dromaeosaurids and Avialae thought to be at least good gliders (Xu et al. 2003; Longrich 2006; Foth et al. 2014). Feather specialization in short coverts overlapping longer flight feathers is present both in Dromaeosauridae [e.g., *Microraptor* and *Changyuraptor*; Xu et al. (2003); Han et al. (2014)] and basal Avialae (e.g., *Archaeopteryx* and Enantiornithes; Foth et al. (2014); Xing et al. (2016). Forelimb plumage is unfortunately not preserved in troodontids, so it remains unclear whether feather specialization is a synapomorphy for Eumaniraptora (Dromaeosauridae + Troodontidae + Avialae) or a convergent evolution in Dromaeosauridae and Avialae.

Asymmetry of flight feathers also progressively appeared in dromaosaurids and avialans. As in modern birds, the forelimb feathers of *Microraptor* and *Archaeopteryx* are characterized by well-defined asymmetrical primaries that become less asymmetrical towards the last secondary feather covered by short coverts (Lucas and Stettenheim 1972; Foth et al. 2014; Feo et al. 2015). The development of feather asymmetry on the forearm is thought be an adaptation to the air pressure applied on the feathers during the flight for the reason that this asymmetrical configuration prevents feathers to twist due to pressure (Feo et al. 2015).

The alula (bastard wing) is formed by more or less symmetrical feathers attached to the alular (first) digit, allowing birds to reach a higher angle of attack during low-speed flight or landings (Lee et al. 2015). The alula is apparently not well documented in non-avialan Paraves and in basal avialans [although a patch of pennaceous feathers seems to be attached to the first digit in *Microraptor* and small alular feathers are reported in *Caihong* (Hu et al. 2018)], clearly appearing later within the Ornithothoraces clade (e.g., *Protopteryx* and *Eoalulavis*) (Sanz et al. 1996; Zhang and Zhou 2000; Xu et al. 2003; Lee et al. 2014).

Elongated hindlimb pennaceous feathers (the "tetrapterygian type") appeared in anchiornithids (*Pedopenna*, *Eosinopteryx*, *Serikornis*, *Anchiornis*) and were retained in dromaeosaurids (e.g., *Changyuraptor, Microraptor*). It has been hypothesized that the development of hindlimb wings might be linked with improvement of the aerodynamic capacities of basal paravians (Longrich 2006). However, as already discussed, those basal paravians were likely not adapted for active flight or even for gliding. The development of hindlimb feathers is therefore tentatively regarded as a social and/or sexual display, as it is also the case for the forelimb feathers in those basal paravians (Foth et al. 2014; O'Connor and Chang 2015). In dromaeosaurids (e.g., *Changyuraptor, Microraptor*), it seems that the hindlimb wings could have helped to stabilize the animals during flight (Chatterjee and Templin 2007; Alexander et al. 2010), adding an aerodynamical function to the social and/or sexual display. Hindlimb feathers were also elongated in several basal Avialae. A distal-to-proximal reduction pattern for leg feathers can be subsequently observed in avialan evolution, and extensively scaled feet might have appeared secondarily at an early stage in ornithuromorph evolution (Zheng et al. 2013). However, Foth et al. (2014) suggest that only a tibial plumage is plesiomorphic and the metatarsal extension of feather trousers evolved independently in Microraptorine, Anchiornithidae, and *Sapeornis*.

Pennaceous tail feathers are present in Oviraptorosauria, in some anchiornithids, and in paravians. It might therefore be hypothesized that they were secondarily lost in the basal paravians *Eosinopteryx* and partially in *Serikornis*. The tail plumage has evidently a visual display function in oviraptorosaurs (Persons et al. 2014), as also in many modern birds. Although the long bony tail in non-avalian paravians and basal birds is reduced to a pygostyle in more advanced Avialae, it can bc hypothesized that the elongated pennaceous feathers in non-avialan eumaniraptorans had a similar function to the rectrices in modern birds (Balmford et al. 1993; Thomas 1993; Fitzpatrick 1998). In the same way, the feathered tail of paravians could have helped to act as a stabilizer during flight (Han et al. 2014). Tail feathers may also be used as a pitch control structure or as a way to compensate for the weight of the long bony tail (Han et al. 2014). In *Archaeopteryx*, the lateral rectrices are asymmetrical, whereas the distal ones are long and symmetrical (Foth et al. 2014). The asymmetrical shape of the lateral tail feathers in *Archaeopteryx* tends to indicate that a secondary aerodynamic function took place with the display function in order to increase the total lift of the animal (Foth et al. 2014). In *Caihong*, the tail surface area formed by pennaceous feathers is proportionally larger than that of *Archaeopteryx* (Hu et al. 2018); it can be presumed that this airfoil acts as an additional aerodynamic display to increase the total lift of the animal. The troodontid *Jianianhualong* also shows asymmetrical tail feathers, suggesting that feather asymmetry in rectrices appeared in the ancestors of Troodontidae + Dromaeosauridae + Avialae (Xu et al. 2017). Xu et al. (2017) suggested that feather asymmetry first evolved in the troodontid tail and then appeared in other parts of the body to assist with locomotion on the ground or through the air.

7.8 Conclusion

Two decades of research and discoveries have shown that the feathers and plumage of modern birds have gone through various evolutionary stages. Feather or feather-like morphotypes were already particularly diversified in non-avialan paravians, including filamentous feathers, monofilaments (associated filaments extending distally from a common point of insertion), asymmetrical pennaceous feathers that are nearly identical or identical to those in modern birds, and more exotic feather-like structures that are unknown in modern birds (ribbon-like tail "feathers" in scansoriopterygids). The important diversity of integumentary structures in non-avialan paravians therefore suggests that they probably already fulfilled a wide array of biological roles, including thermoregulation and visual display, but also potentially maneuverability, brooding, and camouflage (Foth et al. 2014). Flight was obviously one of the last adaptations to appear in feather evolution (Xu and Guo 2009). The plumage structures in some basal paravians were obviously not adapted for flight. However, in *Microraptor* and maybe some of its relatives, the large pennaceous feathers along the forelimb, hindlimb, and tail were similar in morphology (large asymmetrical vane, curved rachis) and organization (specialization in coverts, primaries, and secondaries) to those in modern bird wings, supporting the hypothesis that they could have functioned in flight or gliding.

Several aerodynamic innovations and flight-related morphological adaptations were, likely and independently, experimented within the paravian clade close to the origin of birds, including bat-like membranous aerodynamic surfaces in scansoriopterygids (Xu et al. 2015), elongated hindlimb pennaceous feathers (the "tetrapterygian type") in dromaeosaurids, and the important battery of modifications of the scapular girdle and forelimbs allowing flapping flight in Avialae (Senter 2006). The recent discoveries of exquisitely preserved specimens of feathered dinosaurs, particularly in Late Jurassic and Early Cretaceous deposits from Northeastern China, definitely demonstrate that the origin and early evolution of complex feathers and flight abilities in paravian theropods were not linear processes, but rather were more complex than previously thought.

Institutions

CAGS: Chinese Academy Of Geological Sciences, Beijing, China; **DLXH:** Dalian Xinghai Museum, Dalian, China; **HG:** Paleontological Center, Bohai University, Jinzhou City, China; **IVPP:** Institute Of Vertebrate Paeontology And Paleanthropology, Beijing, China; **JPM:** Jinzhou Paleontological Museum, Jinzhou, China; **NGMC:** National Geological Museum Of China, Beijing, China; **PMOL:** Paleontological Museum Of Liaoning, Shenyang, China; **STM:** Shandong Tianyu Museum Of Nature, Shandong, China; **YFGP:** Yizhou Fossil And Geology Park, Yizhou, China.

References

Agnolin FL, Novas FE (2013) Avian ancestors: a review of the phylogenetic relationships of the theropods Unenlagiidae, Microraptoria, *Anchiornis* and Scansoriopterygidae. Springer, Dordrecht

Alexander DE, Gong E-P, Martin LD, Burnham DA, Falk AR (2010) Model tests of gliding with different hindwing configurations in the four-winged dromaeosaurid *Microraptor gui*. Proc Natl Acad Sci USA 107:2972–2976

Balmford A, Jones IL, Thomas ALR (1993) On avian asymmetry: evidence of natural selection for symmetrical tails and wings in birds. Proc R Soc B 252:245–251

Benett SC (2007) Articulation and function of the pteroid bone of pterosaur. J Vertebr Paleontol 27:881–891

Chatterjee S, Templin JR (2007) Biplane wing planform and flight performance of the feathered dinosaur {Microraptor} gui. Proc Natl Acad Sci USA 104:1576–1580

Chen P-J, Dong Z-M, Zhen S-N (1998) An exceptionally well-preserved theropod dinosaur from the Yixian Formation of China. Nature 391:147–152

Chen P, Wang Q, Zhang H, Cao M, Li W, Wu S, Shen Y (2005) Jianshangou Bed of the Yixian Formation in West Liaoning, China. Sci China Ser D Earth Sci 48:298–312

Currie PJ, Chen P (2001) Anatomy of *Sinosauropteryx prima* from Liaoning, northeastern China. Can J Earth Sci 38:1705–1727

Feo TJ, Field DJ, Prum RO (2015) Barb geometry of asymmetrical feathers reveals a transitional morphology in the evolution of avian flight. Proc R Soc Lond B 282

Fitzpatrick S (1998) Birds' tails as signaling devices: markings, shape, length, and feather quality. Am Nat 151:157–173

Foth C (2011) The morphology of neoptile feathers: ancestral state reconstruction and its phylogenetic implications. J Morphol 272:387–403

Foth C, Rauhut WM (2017) Re-evaluation of the Haarlem *Archaeopteryx* and the radiation of maniraptoran theropod dinosaurs. BMC Evol Biol 17:236

Foth C, Tischlinger H, Rauhut OWM (2014) New specimen of *Archaeopteryx* provides insights into the evolution of pennaceous feathers. Nature 511:79–82

Gatesy SM, Dial KP (1996) From frond to fan: *Archaeopteryx* and the evolution of short-tailed birds. Evolution 50:2037–2048

Godefroit P, Demuynck H, Dyke GJ, Hu D, Escuillié F, Claeys P (2013a) Reduced plumage and flight ability of a new Jurassic paravian theropod from China. Nat Commun 4:1394

Godefroit P, Cau A, Hu D, Escuillié F, Wenhao W, Dyke GJ (2013b) A Jurassic avialan dinosaur from China resolves the early phylogenetic history if birds. Nature 498:359–362

Godefroit P, Sinitsa SM, Dhouailly D, Bolotsky YL, Sizov AV, McNamara ME, Benton MJ, Spagna P (2014) A Jurassic ornithischian dinosaur from Siberia with both feathers and scales. Science 345:451–455

Han G, Chiappe LM, Ji S-A, Habib M, Turner AH, Chinsamy AT, Liu X, Han L (2014) A new raptorial

dinosaur with exceptionally long feathering provides insights into dromaeosaurid flight performance. Nat Commun 5:4382
Hone DWE (2012) Variation in the tail length of non-avian dinosaurs. J Vertebr Paleontol 32:1082–1089
Hopson JA (2001) New perspectives on the origin and early evolution of birds. Peabody Museum of Natural History, New Haven, CT
Hu D, Hou L, Zhang L, Xu X (2009) A pre-{*Archaeopteryx*} troodontid theropod from China with long feathers on the metatarsus. Nature 461:640–643
Hu D, Clarke JA, Eliason CM, Qiu R, Li Q, Shawkey MD, Zhao C, D'Alba L, Jiang J, Xu X (2018) A bony-crested Jurassic dinosaur with evidence of iridescent plumage highlights complexity in early paravian evolution. Nat Commun 9:217
Ji Q, Ji SA (1996) On discovery of the earliest bird fossil in China and the origin of birds. Chin Geol 233:30–33
Ji S, Ji Q (2007) {*Jinfengopteryx*} compared to {*Archaeopteryx*}, with comments on the mosaic evolution of long-tailed avialan birds. Acta Geol Sin 81:337–343
Ji Q, Currie PJ, Norell MA, Ji S-A (1998) Two feathered dinosaurs from the northeastern China. Nature 393:753–761
Ji Q, Norell MA, Gao K, Ji S (2001) The distribution of integumentary structures in a feathered dinosaur. Nature 410:1084–1088
Ji Q, Ji S, Lu J, You H, Chen W, Liu Y, Liu Y (2005) First avialan bird from China (*Jinfengopteryx elegans* gen. et sp. nov.). Geol Bull China 24:197–205
Lee MSY, Cau A, Naish D, Dyke GJ (2014) Sustained miniaturization and anatomical innovation in the dinosaurian ancestors of birds. Science 345:562–566
Lee S-I, Kim J, Park H, Jablonski PG, Choi H (2015) The function of the alula in avian flight. Sci Rep 5:9914
Lefèvre U, Cau A, Cincotta A, Hu D, Chinsamy A, Escuillié F, Godefroit P (2017) A new Jurassic theropod from China documents a transitional step in the macrostructure of feathers. Sci Nat 104:1–13
Li Q, Gao K-Q, Meng Q, Clarke JA, Shawkey MD, D'Alba L, Pei R, Ellison M, Norell MA, Vinther J (2012) Reconstruction of *Microraptor* and the evolution of iridescent plumage. Science 335:1215–1219
Lindgren J, Sjövall P, Carney RM, Cincotta A, Uvdal P, Hutcheson SW, Gustafsson O, Lefèvre U, Escuillié F, Heimdal J, Engdahl A, Gren JA, Kear BP, Wakamatsu K, Yans J, Godefroit P (2015) Molecular composition and ultrastructure of Jurassic paravian feathers. Sci Rep 5(1):13520
Lingham-Soliar T (2012) The evolution of the feather: *Sinosauropteryx*, life, death and preservation of an alleged feathered dinosaur. J Ornithol 153:699–711
Longrich N (2006) Structure and function of hindlimb feathers in *Archaeopteryx lithographica*. Paleobiology 32:417–431
Longrich NR, Vinther J, Meng Q, Li Q, Russell AP (2012) Primitive wing feather arrangement in *Archaeopteryx lithographica* and *Anchiornis huxleyi*. Curr Biol 22:2262–2267
Lü J, Brusatte S (2015) A large, short-armed, winged dromaeosaurid (Dinosauria: Theropoda) from the Early Cretaceous of China and its implication for feather evolution. Sci Rep 5:11775
Lucas AM, Stettenheim PR (1972) Avian anatomy: integument, parts I. U.S. Agricultural Research Service, Washington, DC
O'Connor JK, Chang H (2015) Hindlimb feathers in paravians: primarily "wings" or ornaments? Biol Bull 42:616–621
Osmólska H, Currie PJ, Barsbold R (2004) Oviraptorosauria. In: Weishampel DB, Dodson P, Osmólska H (eds) The Dinosauria. University of California Press, Berkeley, CA
Pei R, Li Q, Meng Q, Norell MA, Gao K-Q (2017) New specimens of *Anchiornis huxleyi* (Theropoda: Paraves) from the Late Jurassic of northeastern of China. Bull Am Mus Nat Hist 411:1–66
Pérez-Moreno BP, Sanz JL, Buscalioni AD, Moratalla JJ, Ortega F, Rasskin-Gutman D (1994) A unique multitoothed ornithomimosaur dinosaur from the Lower Cretaceous of Spain. Nature 370:363–367
Persons SW, Currie PJ, Norell MA (2014) Oviraptorosaur tail forms and functions. Acta Palaeontol Pol 59:553–567
Saitta ET, Gelernter R, Vinther J (2017) Additional information on the primitive contour and wing feathering of paravian dinosaurs. Palaeontology 61:273–288
Sanz JL, Chiappe LM, Perez-Moreno BP, Buscalioni AD, Moratalla JJ, Ortega F, Poyato-Ariza FJ (1996) An Early Cretaceous bird from Spain and its implications for the evolution of avian flight. Nature 382:442–445
Senter P (2006) Scapular orientation in theropods and basal birds, and the origin of flapping flight. Acta Palaeontol Pol 51:305–313
Thomas ALR (1993) On the aerodynamics of birds' tail. Philos Trans R Soc B 340:361–380
Turner AH, Makovicky PJ, Norell MA (2007) Feather quill knobs in the dinosaur *Velociraptor*. Science 317:1721
Turner AH, Makovicky PJ, Norell MA (2012) A review of dromaeosaurid systematics and paravian phylogeny. Bull Am Mus Nat Hist 371:1–206
Wang M, O'Connor JK, Xu X, Zhou Z (2019) A new Jurassic scansoriopterygid and the loss of membranous wings in theropod dinosaurs. Nature 569:256–259
Weidensaul S (ed) (1995) Raptors: the birds of prey. Michigan University Press, New York, NY
Witmer LM (2009) Fuzzy origins for feathers. Nature 458:293–295
Xing L, McKellar RC, Wang M, Bai M, O'onnor JK, Benton MJ, Zhang J, Wang Y, Tseng K, Lockley MG, Li G, Zhang W, Xu X (2016) Mummified precocial bird wings in mid-Cretaceous Burmese amber. Nat Commun 7:12089
Xu X (2006) Scales, feathers and dinosaurs. Nature 440:287–288

Xu X, Guo Y (2009) The origin and early evolution of feathers: insights from recent paleontological and neontological data. Vertebrata PalAsiatica 47:311–329

Xu X, Zhang F (2005) A new maniraptoran dinosaur from China with long feathers on the metatarsus. Naturwissenschaften 92(4):173–177

Xu X, Tang Z, Wang X (1999a) A therizinosaurid dinosaur with integumentary structures from China. Nature 399

Xu X, Wang X, Wu X-C (1999b) A dromaeosaurid dinosaur with a filamentous integument from the Yixian Formation of China. Nature 401:262–266

Xu X, Zhou Z, Wang X (2000) The smallest known non-avian theropod dinosaur. Nature 408:705–708

Xu X, Zhou Z-H, Prum RO (2001) Branched integumental structures in *Sinornithosaurus* and the origin of feathers. Nature 410:200–204

Xu X, Zhou Z, Wang X, Kuang X, Zhang F, Du X (2003) Four winged dinosaurs from China. Nature 421:335–340

Xu X, Norell MA, Kuang X, Wang X, Zhao Q, Jia C (2004) Basal tyrannosauroids from China and evidence for protofeathers in tyrannosauroids. Nature 431:680–684

Xu X, Zheng X, You H (2010) Exceptional dinosaur fossils show ontogenetic development of early feathers. Nature 464:1338–1341

Xu X, You H, Du K, Han F (2011) An *Archaeopteryx*-like theropod from China and the origin of Avialae. Nature 475:465–470

Xu X, Wang K, Zhang K, Ma Q, Xing L, Sullivan C, Hu D, Cheng S, Wang S (2012) A gigantic feathered dinosaur from the Lower Cretaceous of China. Nature 484:92–95

Xu X, Zheng X, Sullivan C, Wang X, Xing L, Wang Y, Zhang X, O'Connor JK, Zhang F, Pan Y (2015) A bizarre Jurassic maniraptoran theropod with preserved evidence of membranous wings. Nature 521:70–73

Xu X, Currie P, Pittman M, Xing L, Meng Q, Lü J, Hu D, Yu C (2017) Mosaic evolution in an asymmetrically feathered troodontid dinosaur with transitional features. Nat Commun 8:14972

Zelenitsky DK, Therrien F, Erickson GM, DeBuhr CL, Kobayashi Y, Eberth DA, Hadfield F (2012) Feathered non-avian dinosaurs from North America provide insight into wing origins. Science 338:510–514

Zhang F, Zhou Z (2000) A primitive enantiornithine bird and the origin of feathers. Science 290:1955–1959

Zhang F, Zhou Z, Xu X, Wang X (2002) A juvenile coelurosaurian theropod from China indicates arboreal habits. Naturwissenschaften 89:394–398

Zhang F, Zhou Z, Xu X, Wang X, Sullivan C (2008a) A bizarre Jurassic maniraptoran from China with elongated ribbon-like feathers. Nature 455:1105–1108

Zhang F, Zhou Z, Benton MJ (2008b) A primitive confuciusornithid bird from China and its implications for early avian flight. Sci China Ser Earth Sci 51:625–639

Zheng X, Zhou Z, Wang X, Zhang F, Wang Y, Wei G, Wang S, Xu X (2013) Hind wings in basal birds and the evolution of leg feathers. Science 339: 1309–1312

Zhou Z, Zhang F-C (2006) Origin and early evolution of feathers: evidence from the early cretaceous of China. Acta Zool Sin 52:125–128

The Feathers of the Jurassic Urvogel *Archaeopteryx*

8

Nicholas R. Longrich, Helmut Tischlinger, and Christian Foth

8.1 Introduction

8.1.1 *Archaeopteryx* as a Transitional Fossil

How does evolution produce novel forms and novel functions? Studying the variation of birds under domestication, Darwin (1859) hypothesized that the environment selects upon variants in a population and, over generations, slowly transforms one species to another. He hypothesized that this mechanism, natural selection, not only produced the variety of mockingbirds and finches on the Galápagos Islands but ultimately produced birds themselves and all other species on the planet.

Darwin's hypothesis for a shared ancestry uniting all vertebrates predicted that intermediate forms must have existed in deep time linking one group of organisms, such as birds, to another, such as reptiles. However, no such intermediates were known, either living or fossil. Yet in 1861, just 2 years after the publication of *On the Origin of Species*, Darwin's prediction was borne out by the discovery of the Jurassic *Archaeopteryx* (Owen 1863; Figs. 8.1 and 8.2). In subsequent editions of the Origin, Darwin would cite *Archaeopteryx* as a transitional fossil, and it has literally been a textbook example of evolution ever since.

Archaeopteryx was precisely the kind of creature predicted by Darwin's theory. It combined uniquely avian features such as feathers, wing, and furculum, with primitive features including teeth, clawed fingers, and a long bony tail. Since the first specimen, a total of 11 skeletal specimens have been referred to the genus *Archaeopteryx* with certainty (Wellnhofer 2009; Foth et al. 2014; Rauhut et al. 2018), while the Haarlem and the newly discovered Mühlheim (13th) specimen are considered representing different genera (see Foth and Rauhut 2017; Rauhut et al. 2019). These fossils have helped to bridge the divide between birds and their reptilian ancestors, and they have continued to shed light on, and spur debate about, the origins of birds and their flight. In all these discussions, the plumage has played a central role, helping to understand the affinities of *Archaeopteryx*, its biology, and the evolution of avian flight.

8.1.2 *Archaeopteryx* and the Dinosaurian Ancestry of Birds

While the primitive morphology of *Archaeopteryx* was critical to recognizing the dinosaurian ancestry of birds, it was the feathers that were critical to establishing the avian affinities of

N. R. Longrich (✉)
Department of Biology and Biochemistry, The Milner Centre for Evolution, University of Bath, Bath, UK
e-mail: N.R.Longrich@bath.ac.uk

H. Tischlinger
Stammham, Germany

C. Foth
Departement für Geowissenschaften, Universität Freiburg, Freiburg, Switzerland

C. Foth, O. W. M. Rauhut (eds.), *The Evolution of Feathers*, Fascinating Life Sciences,
https://doi.org/10.1007/978-3-030-27223-4_8

Fig. 8.1 The isolated feather. Discovered in 1861, it was the first specimen to be discovered and the original holotype of *Archaeopteryx lithographica*. Despite dating to the Late Jurassic, it is remarkably similar to the wing feathers of extant birds. It is likely to represent a covert. Scale = 25 mm

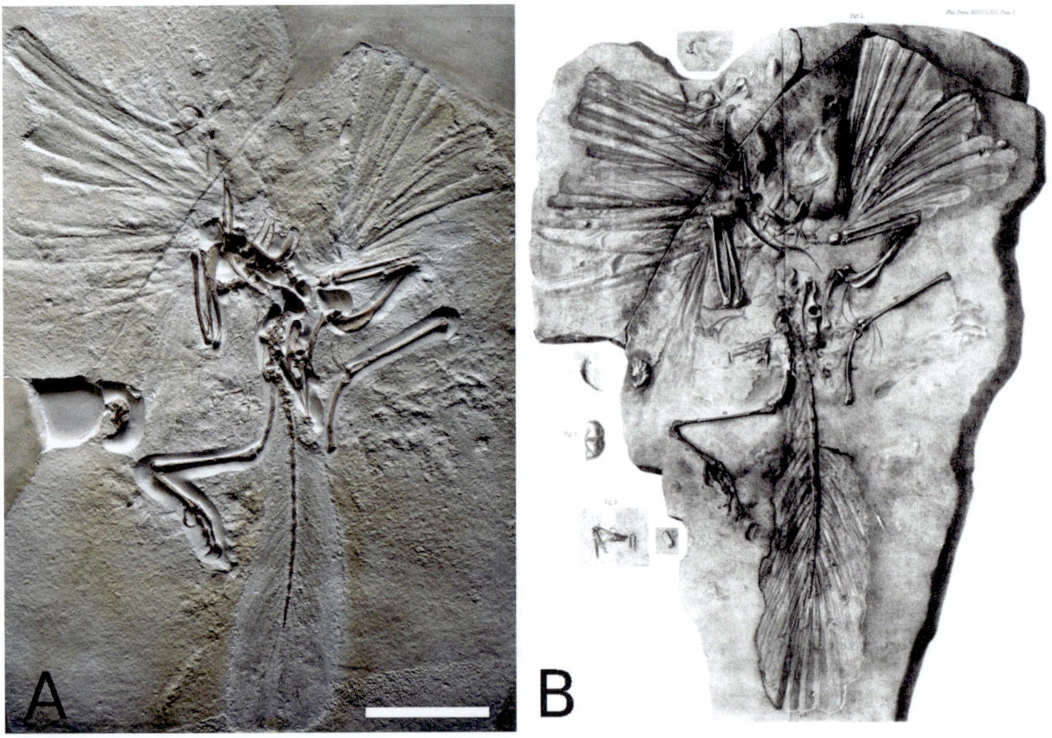

Fig. 8.2 (**a**) The neotype of *Archaeopteryx lithographica* (London specimen). Discovered in 1861, it demonstrated the association of advanced, avian-like plumage with a primitive skeletal structure similar to that of dinosaurs such as *Compsognathus longipes*. (**b**) Owen's illustration of the London *Archaeopteryx*. Scale = 100 mm

Archaeopteryx in the first place. The feathered wings of birds are not just unique to birds but so complex that they are unlikely to have evolved more than once. Because of this, there has never been any serious debate about the avian affinities of *Archaeopteryx*, despite its primitive skeletal morphology. But without the wings, it is unlikely that the transitional nature of *Archaeopteryx* would have been appreciated; as John Ostrom remarked, if not for the feathers, it is probable that *Archaeopteryx* would have been classified as a dinosaur—as indeed happened for the Eichstätt specimen (Wellnhofer 2009), where the feathers are only faintly preserved. It was this combination of avian-like feathers and a primitive skeletal morphology, resembling that of basal coelurosaurs such as *Compsognathus*, which first led Huxley to propose a link between dinosaurs and birds (Huxley 1868, 1870). This idea was later revived and extended by the work of Ostrom (1973, 1976) and the phylogenetic analyses of Gauthier (1986).

Historically, the rarity of early birds meant that the problem of the origins of birds was synonymous with the origins of *Archaeopteryx*. However, recent discoveries have broken up the long branch that once separated crown birds from non-avian theropods such as *Compsognathus* and *Tyrannosaurus*. Such discoveries include feathered coelurosaurs (Ji et al. 1998; Currie and Chen 2001), winged maniraptorans such as *Anchiornis* (Hu et al. 2009) and *Microraptor* (Xu et al. 2003), and basal avialans including *Jeholornis* (Zhou and Zhang 2003a), *Sapeornis* (Zhou and Zhang 2003b), and Confuciusornithidae (Chiappe et al. 1999). Along with the wealth of osteological evidence for a dinosaurian origin of birds, such fossils have provided new and unexpected evidence for the dinosaurian ancestry of birds. These include not only the presence of proto-feathers in non-avialan theropods (Currie and Chen 2001; Xu et al. 2012) but also derived, avian-like features of physiology and behaviour in non-avialan dinosaurs, including rapid growth rates (Erickson et al. 2001, 2004), endothermy (Eagle et al. 2011), avian-like sleeping posture (Xu and Norell 2004), and brooding of eggs (Norell et al. 1995).

As a result of these discoveries, *Archaeopteryx* is no longer quite so central to the problem of the origins and early evolution of birds. Nevertheless, *Archaeopteryx* is still among the oldest representative of Avialae, and its anatomy is well understood due to the exceptional preservation of both the skeleton and feathers. Therefore, it continues to be important to our understanding of the evolution of birds and the relationships between non-avialan coelurosaurs and stem birds (Longrich et al. 2012; Foth et al. 2014) and, in particular, to the origin of flight.

8.1.3 The Flight Ability of *Archaeopteryx* and the Origin of Flight

If the origins of birds have been largely resolved, the origin of avian flight remains controversial. Hypotheses continue to be divided into two general classes, arboreal and cursorial. In the arboreal hypothesis (Norberg 1985a, 1990; Feduccia 1996; Chatterjee and Templin 2003; Longrich 2006), airfoils initially evolve to slow descents, extend leaps, and provide stability and control as animals leap and fall through the forest canopy. Incremental increases in both the surface area and aspect ratio of airfoils improve performance. These include increased lift for slower descents, improved lift/drag ratios with glide angles below 45°, and improved stability, control, and manoeuvrability. Finally, flapping evolves not to generate lift but to generate thrust to compensate for drag, allowing for flatter descents and ultimately level flight. In these scenarios, control and lift evolve first, followed by thrust.

Cursorial hypotheses (Ostrom 1979) and its variants, such as wing-assisted incline running (Dial 2003), envision the wings evolving in the context of running, either along the ground or up inclines. These scenarios tend to emphasize initial use of the wings to produce thrust, followed by the evolution of lift sufficient to support the body in flight. These scenarios are not necessarily mutually exclusive in that an arboreal bird ancestor may have exploited terrestrial environments to some degree and vice versa.

To some degree, biomechanical modelling and modern analogues (or lack thereof) can be used to test the feasibility of these hypotheses. However, they must ultimately be tested against the fossil record. Here, *Archaeopteryx* is critical because it represents one of the earliest and most primitive dinosaurs to use the feathers in an aerodynamic context. It is unlikely to represent a direct ancestor of modern birds, but *Archaeopteryx* exhibits few if any specializations that would rule it out as being an ancestor. It most likely lies at the tip of a short side branch and can therefore be used as a model for the ancestor, providing insight into how it did or did not fly. For example, if flight evolved in a ground-up context, then *Archaeopteryx* should retain adaptations for ground-up takeoff at relatively low speeds. Conversely, a trees-down scenario would predict that *Archaeopteryx* would be well adapted for launching from heights and manoeuvring within vegetation. The morphology of *Archaeopteryx* is therefore important to test hypotheses about the origins and evolution of bird flight.

The purpose of this chapter is not to review all that has been written about the feathers of *Archaeopteryx*, which has already been done (Wellnhofer 2009). Rather, this chapter is meant to sum up our current knowledge: what we know and do not know about feather taphonomy, morphology, arrangement, and, last, their functional and evolutionary implications. Furthermore, the fact that *Archaeopteryx* no longer lies alone at the base of the avian tree, rather than diminishing its importance, allows a deeper understanding of its morphology and biology. Comparisons with other basal forms, particularly *Anchiornis* and *Microraptor*, allow for reciprocal illumination (e.g. Longrich et al. 2012). Hypotheses about feather structure, arrangement, and function in *Archaeopteryx* can be tested against other taxa and vice versa.

Archaeopteryx is an animal that is of exceptional interest to paleontology and the source of no small degree of controversy. Furthermore, while the current work is unlikely to answer all questions or agree with all opinions, it provides an updated summary of our understanding and a starting point for future discussions.

8.2 Preservation, Taphonomy, and Interpretation of the Feathers

All *Archaeopteryx* skeletons are preserved with at least some trace of plumage, except for the 12th (Schamhaupten) specimen. However, the feathers are poorly preserved in most specimens and do not allow for a detailed study of the plumage or feather structure. The exceptions are the Berlin (Wellnhofer 2009; Longrich et al. 2012; Fig. 8.3) and Altmühl (11th) specimens (Foth et al. 2014) which show both the most extensive and best preserved feathering, while the London (Wellnhofer 2009), Munich (Wellnhofer 2009), and Thermopolis (Mayr et al. 2005, 2007) specimens preserve remiges and rectrices. The Maxberg specimen, unfortunately lost since 1991, preserves wing and leg feathers (Wellnhofer 2009), but existing casts and photos do not allow a detailed study of its plumage.

Although the feathers of *Archaeopteryx* are often described as impressions, it is more accurate to refer them as collapsed moulds (Rietschel 1985; Longrich et al. 2012). In the case of the isolated feather, it is preserved as organic residues of feather melanins and keratins (Bergmann et al. 2010; Carney et al. 2012).

8.2.1 Preservational Environment and Burial

All known specimens of *Archaeopteryx* come from the Jurassic lithographic limestones of Bavaria (Wellnhofer 2009) which are lower Tithonian in age (Viohl 1985). Feather preservation is the result of the unusual environmental conditions of the ancient Solnhofen Archipelago.

The lagoons of the Plattenkalk basins lay along the northern edge of the Tethys Sea, located between sponge-microbial mounds and coral patch reefs, which protected them from the open ocean (Viohl 1985; Wellnhofer 2009). The

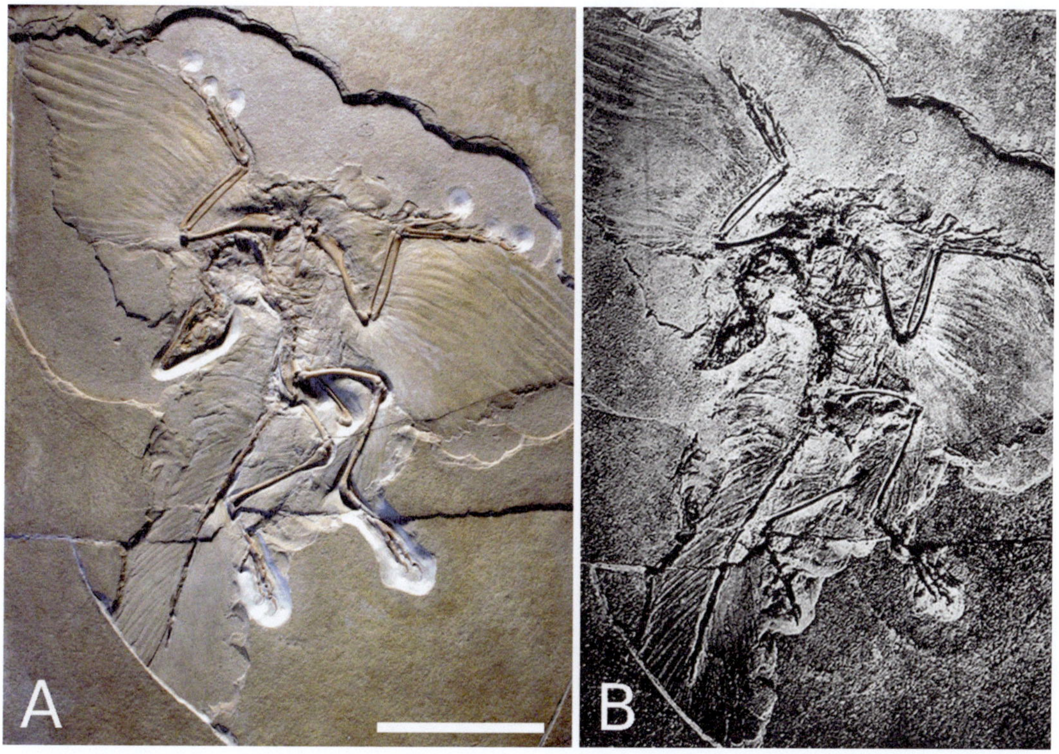

Fig. 8.3 (**a**) Main slab of Berlin specimen of *Archaeopteryx*, which has the most completely preserved plumage of any specimen. (**b**) First photograph of the Berlin specimen (published in 1879) prior to preparation, showing long feathers around the hindlimbs and neck that have since been prepared away. Scale = 100 mm

climate was warm and semiarid (Viohl 1985; Wellnhofer 2009), resulting in high evaporation rates, which in turn led to a hypersaline, anoxic bottom layer (Wellnhofer 2009). These conditions were hostile to animal life, as shown by the absence of bioturbation or epibionts, and would have protected the soft tissues from scavenging or disturbance before and after burial (Wellnhofer 2009; Arratia et al. 2015). Under these conditions, animals sinking to the seafloor could remain undisturbed and were slowly buried under a rain of fine limestone particles, which then moulded the soft tissues.

In some specimens, such as the London, Maxberg, Munich, Thermopolis, and Altmühl specimens, the skeletons exhibit a degree of disarticulation (Wellnhofer 2009) suggesting scavenging and/or decomposition of soft tissues. In such cases, the position of feathers may have been shifted prior to burial, and taphonomic processes must be taken into consideration in interpreting these fossils. However, in some extraordinarily specimens, including the Berlin and Eichstätt specimen, the skeletons are preserved in articulation and with feathers in life position (Wellnhofer 2009). Here, burial probably occurred very soon after death, within days or even hours, with minimal scavenging or decomposition. These specimens are significant because they preserve not only feathers but also the position of the feathers in life, or at least, at the moment of death.

8.2.2 Preservation of Feathers

After sinking to the bottom, the body and feathers were buried by fine particles of calcium carbonate that rained down from the water column (Rietschel 1985; Wellnhofer 2009). Rietschel (1985) proposed a taphonomic model, followed

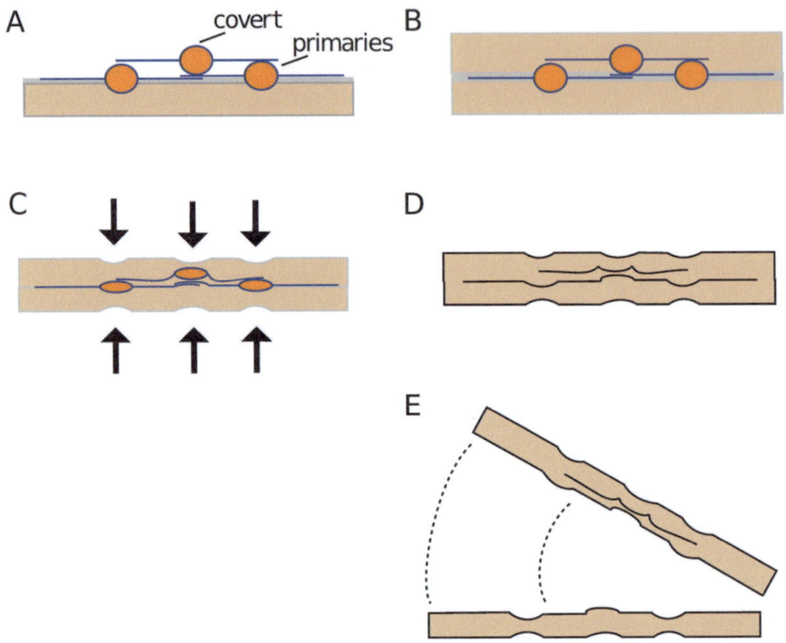

Fig. 8.4 Taphonomic model for the preservation of *Archaeopteryx* feathers, from Longrich et al. (2012) and modified from Rietschel (1985). Primaries and coverts are pressed into the sediment (**a**) and then buried (**b**). Compaction of the feathers (**c**) leads to displacement of sediment above and below the rachis (**d**). When the slab is split apart, a single layer of feathers is revealed, along with 'shaft shadows' resulting from compaction of the sediment (**e**)

here, where the feathers were encased in and moulded by fine sediment particles. Following burial, the soft tissues broke down and the natural moulds collapsed under the sediment's weight. A similar model has been proposed for fossils such as ammonites (de Buisonjé 1985; Viohl 1985). In this model, the moulds collapse vertically under the sediment load, but do not expand laterally because the surrounding sediments are also under pressure. If so, then features such as diameters of the feather shafts should be preserved accurately. Finally, when discovered, the fossils are split along the bedding plane, which may result in the moulds being split open, revealing the morphology of the feathers (Fig. 8.4).

A key implication of this model, in which feathers are preserved as moulds rather than impressions, is that although we can only see a cross section through the wing, the feathers are actually preserved in three dimensions in the matrix, albeit compressed in the vertical dimension. In places, the feathers may be split away to reveal different layers of the wings (Rietschel 1985; Longrich et al. 2012). In the Berlin specimen (Fig. 8.3), for example, the main slab exposes the wing in dorsal view, as the dorsal surface of the skeletal elements is visible. However, although from above, it is actually the ventral surface of the wing that is seen (Rietschel 1985). The situation is analogous to an archaeological site in which the upper stories of a building have been removed and we are looking down at the foundations of the building.

Another implication of this model is that 'shaft shadows', grooves that pass through the remiges (Heilmann 1926), do not represent a taphonomic artefact ('double struck' impressions) but are traces of feathers concealed beneath other feathers (Rietschel 1985; Longrich et al. 2012; Fig. 8.4).

This means that unlike the feathered dinosaurs of the Jehol (Ji et al. 1998; Xu et al. 2003; Norell and Xu 2005) and Daohugou (Hu et al. 2009; Sullivan et al. 2014; Xu et al. 2016) assemblages, where the plumage is preserved as a thin, essentially two-dimensional organic layer (Zhang et al. 2006; Foth 2012), the preservation of the feathers in *Archaeopteryx* makes it possible to examine three-dimensional patterns of overlap and layering of the plumage.

Another implication is that in some cases, poorly preserved traces of feathers may not be the moulds themselves but rather the 'printing' of the collapsed moulds into overlying or underlying layers of sediment. Following Rietschel's (1985) model, collapse of the rachis causes displacement of the overlying and underlying sediment layers. If so, the poorly preserved feathers seen in some specimens, notably the Maxberg, the 9th (Ottmann and Steil specimen, 'chicken wing'), and for the most part the Munich specimen, may not correspond to the feathers themselves but represent artefacts or a palimpsest of the feathers. Similarly, some specimens in which feathers are not visible, including the Solnhofen and Daiting specimens, could preserve feathers deeper in the limestone.

8.2.3 Organic Preservation

Soft tissues also preserve as organic traces in at least some *Archaeopteryx* specimens. In contrast to the skeletal specimens, the single feather is preserved as a thin organic layer. Although previously interpreted as remains of precipitated manganese dioxide (MnO_2) (Griffiths 1996; Wellnhofer 2004) when analysed using backscattered SEM and synchrotron rapid scanning X-ray fluorescence mapping, the traces show no evidence for MnO_2. Instead, they possess high concentrations of organic copper, organic sulphur (Carney et al. 2012), sulphate and nickel (Manning et al. 2013). Sulphur most likely comes from the original feather keratin (Wogelius et al. 2011), while copper is potentially a biomarker for eumelanin pigmentation (Wogelius et al. 2011). Remains of eumelanins are detectable across the feather. Although the preservation of melanosomes has been disputed (Manning et al. 2013; Moyer et al. 2014), the identification of melanosomes is consistent with the regular arrangement of the preserved bodies, their orientation in line with the barbules, and the fact that colour patterns are preserved (Tischlinger and Unwin 2004), with feather tip being darker than the proximal vane (Carney et al. 2012). Similarly, X-ray fluorescence mapping shows that copper, nickel, and sulphur are not uniformly distributed, and instead, the distal tip and outer vane of the feather were darker than inner vane (Manning et al. 2013). Examination of the fossils using synchrotron imaging also reveals the presence of elevated levels of phosphorus in association with the rachis (Bergmann et al. 2010). These traces suggest that organic residues may be more common in *Archaeopteryx* than previously thought.

8.2.4 Feather Nonpreservation

Given the exceptional preservation of soft tissues seen in some Urvogel specimens, particularly the Berlin, London, and Altmühl specimens, it is tempting to conclude that we have a highly complete picture of the plumage. However, this is almost certainly not the case. As seen by comparing specimens (Table 8.1), feather preservation is variable. In the Berlin and Altmühl specimen, remiges, coverts, rectrices, leg feathers, and body contour feathers are preserved in detail. In many specimens, including the London, Munich, Thermopolis, and Eichstätt specimen, only remiges and rectrices were preserved. In others, including the Haarlem (see Foth and Rauhut 2017 for alternative taxonomic classification), Solnhofen, and Ottmann and Steil ('chicken wing') specimen, only poorly preserved traces of the remiges are visible, and in the Schamhaupten (12th) and Mühlheim (13th) specimen, feathers are absent (Rauhut et al. 2018, 2019). In the Daiting specimen, no feathers are visible under normal light, but neck feathers are visible under ultraviolet (Tischlinger 2009).

This emphasizes that even in the Solnhofen, where exceptional soft tissue preservation is relatively common (Arratia et al. 2015), nonpreservation is the rule. Therefore, it is unlikely that a small sample of 11 *Archaeopteryx* specimens has revealed all important aspects of the plumage. We currently lack clear evidence for feathers (or their absence) on the snout, head, alular digit, feet, or toes, for example. However, given that there is no trace of scales in

Table 8.1 Preservation of feathers in the currently known specimens of *Archaeopteryx*, *Ostromia*∗ and *Alcomonavis*∗∗ (numbering of Urvogel specimens in parenthesis)

Specimen	Remiges	Dorsal coverts	Ventral coverts	Leg feathers	Rectrices	Body contour feathers
Feather (0)	–	X	–	–	–	–
London (1)	X	X	–	–	X	–
Berlin (2)	X	X	X	X	X	X
Maxberg (3)	X	–	–	X	–	–
Haarlem∗ (4)	X	–	–	–	–	–
Eichstätt (5)	X	–	–	?	X	–
Solnhofen (6)	X	–	–	–	–	–
Munich (7)	X	–	–	–	X	–
Daiting (8)	–	–	–	–	–	X
Ottmann and Steil (9)	X	–	–	–	–	–
Thermopolis (10)	X	–	–	–	–	–
Altmühl (11)	X	X	X	X	X	–
Schamhaupten (12)	–	–	–	–	–	–
Mühlheim∗∗ (13)	–	–	–	–	–	–

these regions, and because closely related forms from the Jehol and Daohugou bear feathers on these parts of the body, it would be premature to conclude that feathers are absent. Soft tissues need to be reconstructed based on positive evidence, and the absence of evidence for soft tissues should not be taken as strong evidence for their absence.

8.3 Feather Morphology

8.3.1 Feather Structure

The morphology of the individual feathers of *Archaeopteryx* is remarkably similar to modern birds (Lucas and Stettenheim 1972; Fig. 8.1). However, they differ in significant details. In traditional terms, the feathers have a modern 'bauplan'. In modern terms, the similarities in morphology suggest that the developmental genes and processes determining feather morphology were similar to those of crown birds (see Chap. 2). All the observed feathers are variations on this basic bauplan; for purposes of description, we start with the isolated feather.

Proximally, the shaft of the feather is developed as a hollow calamus, which distally gives rise to a central rachis. As preserved, the calamus is rather short, but originally von Meyer (1862) figures and describes faint traces of a longer calamus. The rachis is slightly curved posteriorly, while the vanes show a slight asymmetry (Elzanowski 2002; Wellnhofer 2009). A series of large barbs branch off on either side of the rachis. Proximally, the barbs are separated, but distally they interlock to form closed vanes. In his initial description, von Meyer (1862) identified barbules, which has been confirmed by modern imaging techniques (Carney et al. 2012). Barbule morphology is not readily visible, but the barbs lock together to form vanes, as in the contour and flight feathers of modern birds. Most likely, the hooklets of distal barbules grasped the pennulum of proximal barbules to hold the barbs together as in crown birds (Lucas and Stettenheim 1972).

Studies of the microstructure of the feather likewise suggest that it was structurally similar to modern feathers. Melanosomes, microscopic bodies containing melanin, are visibly aligned parallel to the barbs, as in modern birds (Carney et al. 2012). Melanins serve multiple purposes, including camouflage and display, but also protect the feathers against decay (Goldstein et al. 2004; Gunderson et al. 2008) and provide resistance to wear and damage from ultraviolet light

(Bergman 1982). *Archaeopteryx* would undoubtedly have been subject to all of these selective forces, needing to court mates, conceal itself from prey and predators, and protect its feathers against damage. Based on the morphology, the single feather likely represents a dorsal covert from the primary region, indicated by its asymmetry (Carney et al. 2012).

8.3.2 Ptilosis

In contrast to the scales of dinosaurs and other diapsids, the feathers of *Archaeopteryx* were regularly arranged. The primitive condition in diapsids is for the body scales to be arranged into an irregular or semiregular mosaic over most of the body, but often with larger scales or scutes regularly arranged along the back (Chang et al. 2009). Such a pattern is seen in many non-avian dinosaurs including hadrosaurids (Bell 2014), ceratopsians (Vinther et al. 2016), thyreophorans (Christiansen and Tschopp 2010; Arbour et al. 2014), and sauropods (Chiappe et al. 1998) and more distant relatives of birds including crocodylians (Grigg and Kirshner 2015), turtles, and many lepidosaurs (Pianka and Vitt 2003), as well as on the feet of birds (Lucas and Stettenheim 1972).

By contrast, the arrangement of the feathers of the wings and body in *Archaeopteryx* and crown birds (Lucas and Stettenheim 1972) is far more regular, with feathers regularly spaced. Feathers of birds and eumaniraptorans are spaced in a diagonal grid across the body, which is critical to the function of the feathers. The base of each feather is embedded in skin and fatty tissue and linked to other feathers by an elaborate network of erector and depressor muscles (Lucas and Stettenheim 1972; Homberger and de Silva 2000). These muscles allow feather tracts to be raised and lowered, which is used for thermoregulation, controlling aerodynamic forces, and display (Lucas and Stettenheim 1972; Homberger and de Silva 2000).

Although the arrangement between individual pennaceous feathers of the body cannot be identified in *Archaeopteryx*, the remiges and coverts are arranged in alternating tiers, a feature shared with *Anchiornis* (Longrich et al. 2012; Pei et al. 2017) and crown birds (Lucas and Stettenheim 1972). The alternating rows of feathers seem to simply be an elaboration of the diagonal grid. Most likely, the regular spacing of the feathers across the body evolved first (Christiansen and Bonde 2004) to allow insulator feathers to be raised and lowered synchronously, with this pattern retained and modified to produce the regular arrangement of feathers on the wing and tail.

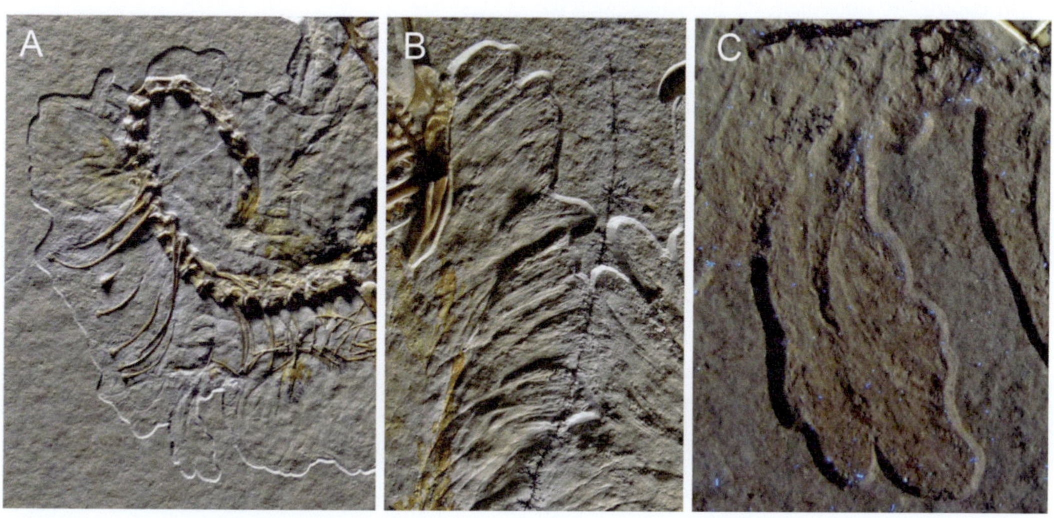

Fig. 8.5 Body feathers of the Altmühl *Archaeopteryx*. (**a**) Feathers of the neck and body; (**b**) close-up of hackle feathers in the neck region; (**c**) close-up of a pair of body features from (**a**)

8.3.3 Body Contour Feathers

Body feathers are preserved in the Berlin and Altmühl specimens. In the Berlin specimen, contour feathers are preserved above the back and anterior to the pectoral girdle (Christiansen and Bonde 2004). In the Altmühl specimen, body feathers run along the neck and surround the thorax (Foth et al. 2014; Fig. 8.5).

Some of the hackle (neck) feathers of the Altmühl specimen are dissociated from the cervical region along with the skull and are instead preserved between the tail and the skull (Fig. 8.5b). They show that the hackle feathers are about half the length of the body feathers and covered the neck at least up to the head. The Berlin specimen shows that body feathers possess a rachis as in the contour feathers of crown birds, but due to poor preservation, it is unclear whether they had closed or open vanes (Christiansen and Bonde 2004). Two isolated feathers exposed in the belly region of the Altmühl specimen clearly show a pennaceous morphology (Fig. 8.5c). They also show a long rachis that curves posteriorly as in crown birds (Lucas and Stettenheim 1972). The vanes are symmetrical, but narrower than in modern birds (Lucas and Stettenheim 1972). In this respect, the body feathers resemble the remiges and rectrices. This is consistent with the idea that differentiation of the various feather types was more poorly developed in early Paraves (see Chap. 5).

In contrast to nonvolant coelurosaurs such as the compsognathid *Sinosauropteryx* (Currie and Chen 2001) and the tyrannosauroid *Yutyrannus* (Xu et al. 2012), body feathers are extremely elongate, with the feathers of the belly region being longer than their associated ribs. Similarly, elongate body feathers occur in *Anchiornis* (Hu et al. 2009), *Microraptor* (Xu et al. 2003), and ornithothoracine birds (see Chap. 9; Lucas and Stettenheim 1972). In modern birds, the length of the feathers means that the body can be covered and effectively insulated by a combination of feathered tracts (pterylae) and unfeathered tracts of the skin (apteria) (Lucas and Stettenheim 1972). Given the existence of elongate body feathers in *Archaeopteryx*, *Anchiornis*, and other Paraves, they may would not have needed to be fully feathered to adequately insulate and streamline the body. It is possible that basal Paraves already possessed pterylae and apteria as hypothesized by Christiansen and Bonde (2004) for *Archaeopteryx*. Although this interpretation is plausible, it is difficult to test on the basis of the available fossils.

8.3.4 Remiges and Coverts

All specimens except for the Daiting and Schamhaupten specimens preserve wing feathers. Details of the wing plumage, including rachis and barbs, are well preserved in the Berlin, London, Munich, Thermopolis, and Altmühl specimens (Table 8.1), with the feathers best seen in the Berlin (Figs. 8.6 and 8.7), London (Fig. 8.2), and Altmühl specimens (Fig. 8.8).

The wing skeleton of these specimens is seen in dorsal view. However, it is not the dorsal but the ventral surface of the wing that appears to be preserved in the Berlin, Munich, and Thermopolis specimens (Heinroth 1923; Helms 1982; Mayr et al. 2007; Wellnhofer 2009; Longrich et al. 2012) as indicated by the exposure of the ventral coverts, i.e. we are looking down at the lower layers of the wing (Rietschel 1985) and the upper layers of the wing have been pulled away. The London specimen appears to preserve the ventral surface of the left wing and the dorsal surface of the right (Longrich et al. 2012).

The situation is less clear for the Altmühl specimen (Fig. 8.8). Because the leading edge vane of the more proximal primaries overlaps and obscures the trailing edge vane of the nearby distal feathers, Foth et al. (2014) hypothesized that the dorsal wing surface is seen (interpretation followed by CF). However, as the preserved primary coverts extend obliquely and have their tips angled inward relative to the primaries, the

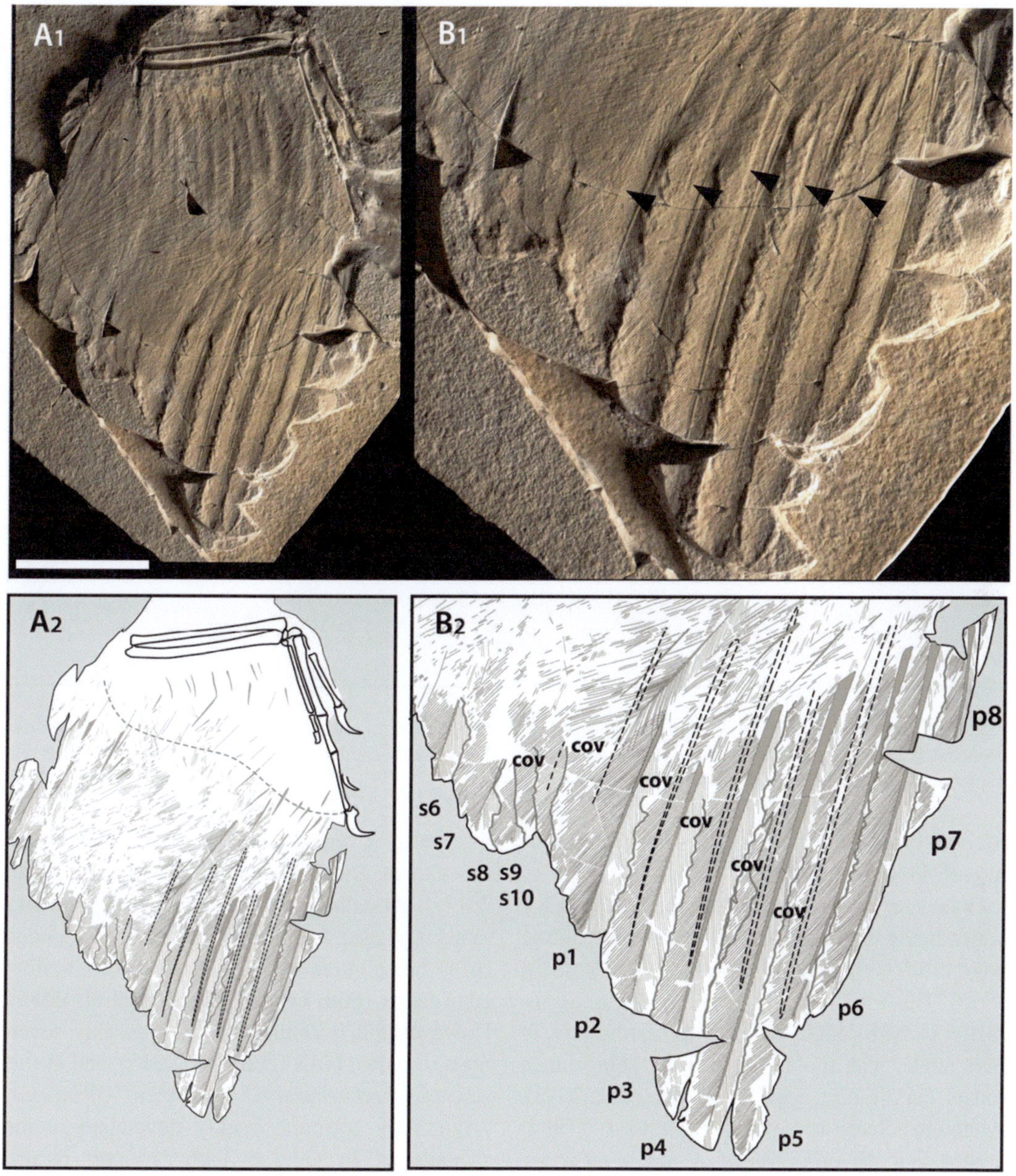

Fig. 8.6 Left wing of the Berlin *Archaeopteryx*, counterslab showing primaries, secondaries, coverts, and impressions of concealed feathers as interpreted by Longrich et al. (2012). Abbreviations: *p8-p1* primaries 1–8; *s6-10* secondaries 6–10, *cov* coverts. Scale = 50 mm

arrangement corresponds more to the situation of the ventral surface in modern bird wings (interpretation followed by NRL). Therefore, a detailed taphonomic study of the wing preservation is necessary for this specimen.

The remiges resemble those of modern birds in many features, but also exhibit primitive features not seen in crown birds. The feather shafts are curved posteriorly in the distal primaries and secondaries. However, at least the distal half of the proximal primaries in the Berlin and Altmühl specimen show the opposite curvature, with the tips curving forward (Norberg 1985b; Longrich et al. 2012; Foth et al. 2014).

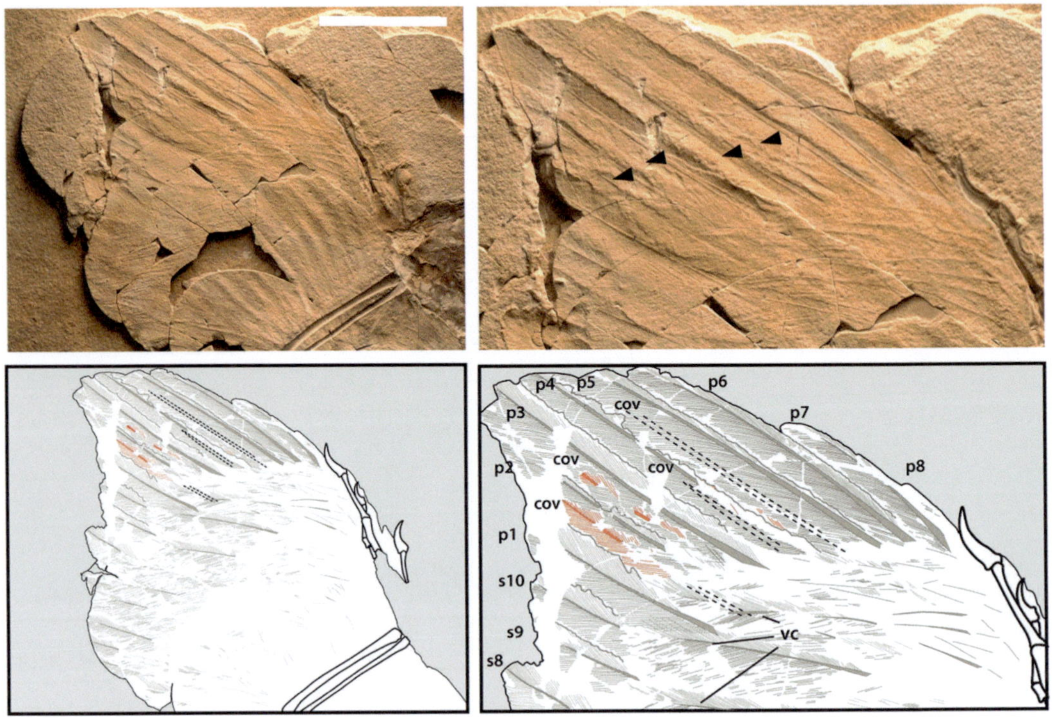

Fig. 8.7 Right wing of the Berlin *Archaeopteryx*, counterslab showing primaries, secondaries coverts and impressions of concealed feathers as interpreted by Longrich et al. (2012). Abbreviations: *p8-p1* primaries 1–8; *s8-10* secondaries 8–10, *cov* coverts. Scale = 50 mm

Modern birds do not exhibit this anterior curvature of the primaries.

Vanes are asymmetrical with the trailing edge vanes being about twice as wide as the leading edge vanes (Feduccia and Tordoff 1979; Norberg 1995). This asymmetry is primarily due to asymmetry in barb length rather than asymmetry in barb angle, with trailing edge barbs being much longer than leading edge barbs (Feo et al. 2015). Although Speakman and Thompson (1994) argued that the primaries of *Archaeopteryx* are symmetrical as in flightless birds (based on the London and Berlin specimen), this is an artefact caused by vanes of neighbouring feathers obscuring the trailing edge, where the fully exposed trailing edges are wide and the feathers are clearly asymmetrical (ratio >3 for outer primaries in the Altmühl specimen) as in volant birds (Norberg 1995; Longrich et al. 2012). This suggests adaptation for aerodynamic function, although it should be noted that some secondary flightless birds retain asymmetric remiges, having a ratio that partly overlaps with that of volant birds (Speakman and Thompson 1994; Wang et al. 2017a). Angulation of the barbs in the primary remiges is also asymmetrical, with leading edge barbs being more strongly angled than trailing edge barbs (Foth et al. 2014; Feo et al. 2015). However, this asymmetry is more weakly developed in crown birds (Feo et al. 2015), and in this respect, *Archaeopteryx* resembles *Anchiornis*. Asymmetry is more weakly developed in the secondaries. In addition, barb curvature is also asymmetrical. Leading edge barbs are straight or weakly curved posteriorly, while trailing edge barbs are strongly curved posteriorly along most of their length, with an inflection at the distal end of the barbs, where the barbs curve towards the tip of the feather. Again, this resembles the condition in modern birds. This asymmetry in curvature is more weakly developed in *Anchiornis* (Longrich et al. 2012).

The rachis is broad proximally, but becomes thinner in their distal 25%. Rachis is robust

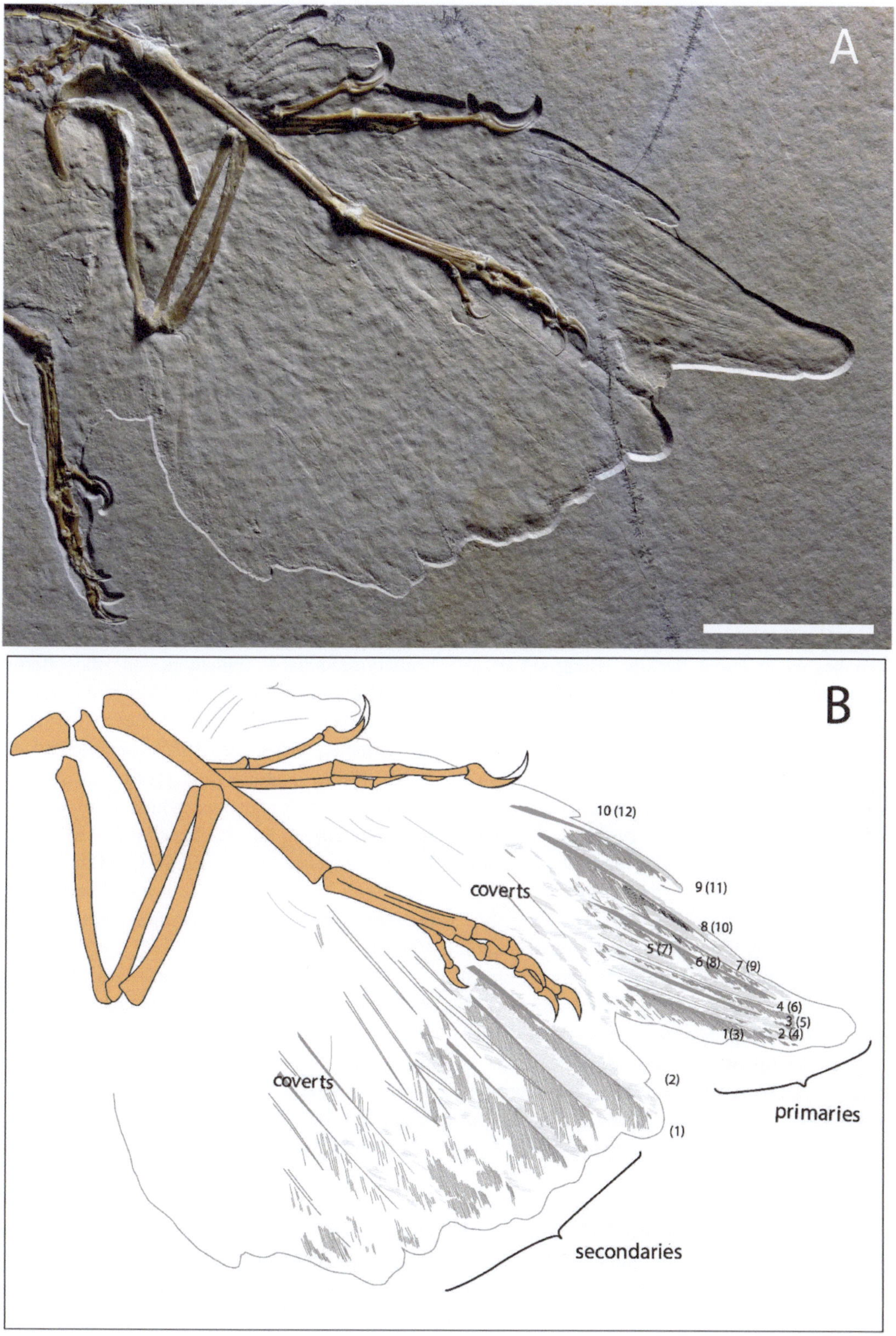

Fig. 8.8 Right wing of the Altmühl specimen of *Archaeopteryx*, showing primaries, secondaries and unidentified coverts. There are at least 10 primaries visible (labelled 1–10 in black), and Foth et al. (2014) identify 12 (labelled 1–12 in grey). Scale = 50 mm

compared to those of body contour feathers, but relatively slender compared to the remiges of crown birds. Although Nudds and Dyke (2010) apparently used incorrect values for both feather diameter and body mass of *Archaeopteryx* in calculating the robusticity of the remiges (Longrich et al. 2012), using more appropriate estimates confirms that the proximal feather shafts are slender compared to volant crown group birds, but similar to those of *Anchiornis* (Longrich et al. 2012). In its distal quarter, however, the relative rachis diameter of the remiges is comparable to modern birds (Foth et al. 2014).

Remiges vary in morphology from distal to proximal, as in modern birds. Distal primary feathers, particularly the four comprising the wing's leading edge, are highly asymmetrical. They have rachis that is strongly curved posteriorly and has strong asymmetrical vanes and strong asymmetry in both barb angulation and curvature. The tips of the distal primary feather are also narrow and acutely pointed. Proximally, the rachis of the primaries becomes straighter and then begins to curve distally (Longrich et al. 2012; Foth et al. 2014). Vanes of proximal remiges are more symmetrical, with broader leading edge vanes, and barb angulation becomes more symmetrical as well. The tips of the proximal primary feathers are much broader and have a blunter end. Secondaries resemble the inner primaries in having relatively broad tips, weakly asymmetrical vanes, and a weak asymmetry in barb angulation.

The wings themselves are relatively short and broad compared to modern birds, with a pointed tip. The leading edge of the wing is formed by four feathers that become successively longer towards the tip of the wing. This design is in contrast to modern birds, where the outermost primary is typically elongate and forms most of the leading edge. From the wingtip, the feathers slowly become shorter again.

The precise number of primaries and secondaries is unclear. Depending on the specimen and interpretation, the number of primaries has been estimated as being anywhere from 8 (Longrich et al. 2012) to 11 (Rietschel 1985; Mayr et al. 2007) or 12 (Heilmann 1926; Norberg 1985b; Stephan 1985; Wellnhofer 2009; Foth et al. 2014). The number of secondaries is at least 10 (Longrich et al. 2012), but others identify 12 (Steiner 1962) to 15 (Stephan 1987).

The Berlin specimen preserves the tips of at least eight primaries (Longrich et al. 2012), indicating a minimum of eight primaries; however, the transition between primaries and secondaries is unclear. The Altmühl specimen has been described as preserving 12 primaries (Foth et al. 2014). There is a distinct gap between the inner and outer remiges in the Altmühl specimen, which is also associated with a shift in feather morphology, however. If this gap represents the transition between primaries and secondaries, there are 10 primaries; however, Foth et al. (2014) interpret this gap originally as an artificial disruption of the wing when the forelimb was disarticulated from the axial skeleton and drifted under the left hindlimb before burial (Fig. 8.8).

The ambiguity in the number of primaries is due not just to the difficulty in identifying the primary-secondary transition. In the Berlin *Archaeopteryx*, four to six additional pairs of wing feathers are preserved, concealed beneath the primaries (Rietschel 1985; Longrich et al. 2012). These feathers are visible as shallow grooves or 'shaft shadows' which correspond to rachis underlying the visible primaries. In places, the primaries have been split away and reveal the rachis and barbs of these feathers, showing that they do in fact correspond to feathers (Rietschel 1985; Longrich et al. 2012).

Although these concealed are typically interpreted as displaced primaries (Rietschel 1985; Foth et al. 2014), several lines of evidence argue against this hypothesis. First, these feathers are not exposed distally on either the right or left wing, and the 'shadows' of the shafts disappear before the edge of the wing suggesting that they are shorter than the primaries (Longrich et al. 2012). Second, the shadow shafts are slightly angled relative to the primary series, rather than lying in parallel with them (Longrich et al. 2012). Third, the feathers alternate with the primaries on both the left and right wings. The regularity of this arrangement is difficult to explain as a

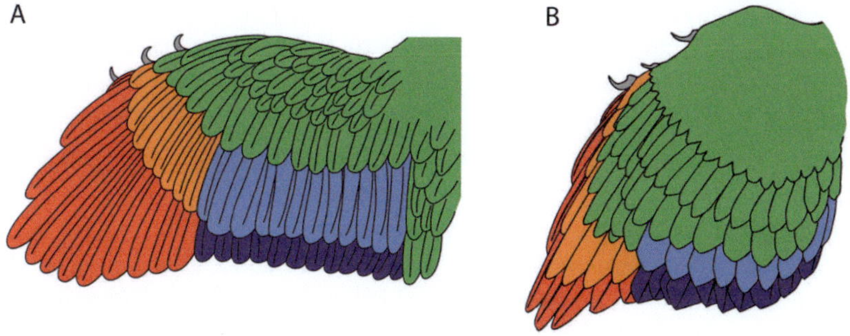

Fig. 8.9 Different reconstructions of the wing of *Archaeopteryx*. (**a**) Original reconstruction, with numerous primaries and short dorsal primary coverts, (**b**) as interpreted by Longrich et al. (2012) with elongate coverts. Red = Primaries; orange = major dorsal coverts, dark blue = secondaries; light blue = major secondary coverts

taphonomic artefact (Longrich et al. 2012). Fourth, the rest of the wing plumage is in place down to individual feathers and barbs. Given that the feather preservation in the Berlin specimen is superior to most other specimens, the arrangement most likely represents the state of the plumage in life (Longrich et al. 2012). If some primaries had been displaced, one would expect to see other feathers and barbs displaced from their life position. Thus, these concealed feathers were interpreted as dorsal coverts (Longrich 2006; Longrich et al. 2012). A similar arrangement can be seen in the London *Archaeopteryx* (Longrich et al. 2012) with feathers that are shorter than the primaries overlying them (Longrich et al. 2012). A similar arrangement to the one proposed for *Archaeopteryx*, with the wing being composed of multiple layers of feathers instead of just the primaries, was subsequently identified in *Anchiornis* (Longrich et al. 2012). The hypothesis of elongate dorsal coverts has not been universally accepted. Foth et al. (2014) have argued that the Altmühl specimen exhibits 12 primaries (see above) and that the concealed feathers in the Berlin specimen represent the 'missing' primaries, as suggested by Rietschel (1985) (Fig. 8.9).

Dorsal secondary coverts appear to be visible on the right wing of the London specimen, where several feathers overlie the secondaries but fail to extend the full chord of the wing. They overlie the primaries and pass obliquely over them, with their tips angled distally. These appear to represent dorsal coverts and extend almost the full width of the wing. They lack the broad, pennaceous vanes of the ventral coverts seen in the Berlin specimen and instead appear to have narrower, closed vanes. The elongation of the dorsal coverts again matches the condition in *Anchiornis*.

Ventral primary coverts are identifiable in the Berlin specimen. They are short distally but become longer proximally. The ventral coverts extend obliquely across the primaries and have their tips angled inward relative to the primaries, as in the primary feathers of crown birds. Ventral secondary coverts are well preserved in the Berlin specimen on the left and right wing (Longrich et al. 2012). As with the ventral primary coverts, they lie with their tips angled inward relative to the associated secondaries. Ventral secondary coverts extend almost the entire chord of the wing, such that only the tips of the secondaries are exposed (Longrich et al. 2012). The ventral coverts have an open pennaceous structure, the barbs interlock at their bases, but distally they are separated, which in the Berlin specimen gives the underside of the wing a shaggy appearance.

The wing of *Archaeopteryx* is usually reconstructed without alular feathers on the first digit (Wellnhofer 2009), which in extant birds increase lift generation at low speeds (Nachtigall and Kempf 1971). Instead, it has been hypothesized that the first digit of the manus might have had an analogous function (Meseguer et al. 2012). However, the Berlin specimen has a possible feather impression distal to the ungual of digit I on the right wing, which may represent an alula or a homologue of the alula. In the Altmühl specimen, the same region is covered by tibial feathers, so that this observation cannot be verified. Although the evidence is weak for *Archaeopteryx*, alular feathers are preserved in other Eumaniraptora, including *Anchiornis* (Longrich et al. 2012), *Microraptor* (Xu et al. 2003), and stem birds such as Enantiornithes (Sanz et al. 1996; Xing et al. 2016). It is therefore likely that *Archaeopteryx* had alular feathers, but if so, new fossils are needed to verify their presence.

The leading edge of the wing may also have been extended by short feathers projecting anterodistally. Traces of these feathers appear to be present in the Berlin *Archaeopteryx* (Longrich et al. 2012) and in the Altmühl specimen. Again these feathers are seen in *Anchiornis* (Longrich et al. 2012) and *Microraptor* (Czerkas et al. 2002). This sharp leading edge is absent in modern birds, where instead the leading edge is covered by small contour feathers that give the leading edge a rounded cross section.

The presence of a propatagium forming the anterior edge between humerus and radius is indicated by a spray of feathers anterior to the deltopectoral crest and the shaft of the humerus of the Berlin specimen. These feathers are similar in position and arrangement to the subhumeral tract of the humerus (Lucas and Stettenheim 1972) and suggest that *Archaeopteryx* had a well-developed propatagium (Martin and Lim 2005). A similar structure can be also found in *Caudipteryx* (Feduccia and Czerkas 2015), *Anchiornis* (Longrich et al. 2012; Wang et al. 2017b), and *Microraptor* (Xu et al. 2003; Feduccia and Czerkas 2015), where an extensive pigmented region anterior to the elbow appears to represent the densely feathered skin of the propatagium.

8.3.5 Hindlimb Feathers

The hindlimb feathers are preserved in the Berlin, Maxberg, and Altmühl specimens and perhaps in the Eichstätt specimen. They are best seen in the Berlin and Altmühl specimen (Figs. 8.10 and 8.11). Unfortunately, the feathers on the main slab of the Berlin specimen were partly destroyed during preparation (Tischlinger and Unwin 2004; Tischlinger 2005). Fortunately, they are still preserved on the counterslab (Fig. 8.10) and cast photos and illustrations of the main slab before

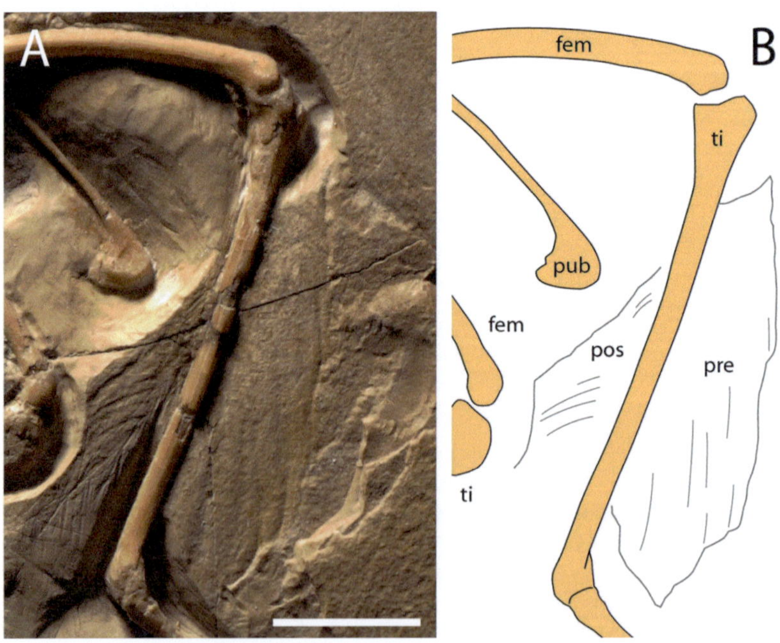

Fig. 8.10 Tibial feathers of the Berlin *Archaeopteryx*. (**a**) Main slab; (**b**) interpretive drawing. Abbreviations: *fem* femur, *ti* tibia, *pos* posttibial feathers, *pre* pretibial feathers. Scale = 25 mm

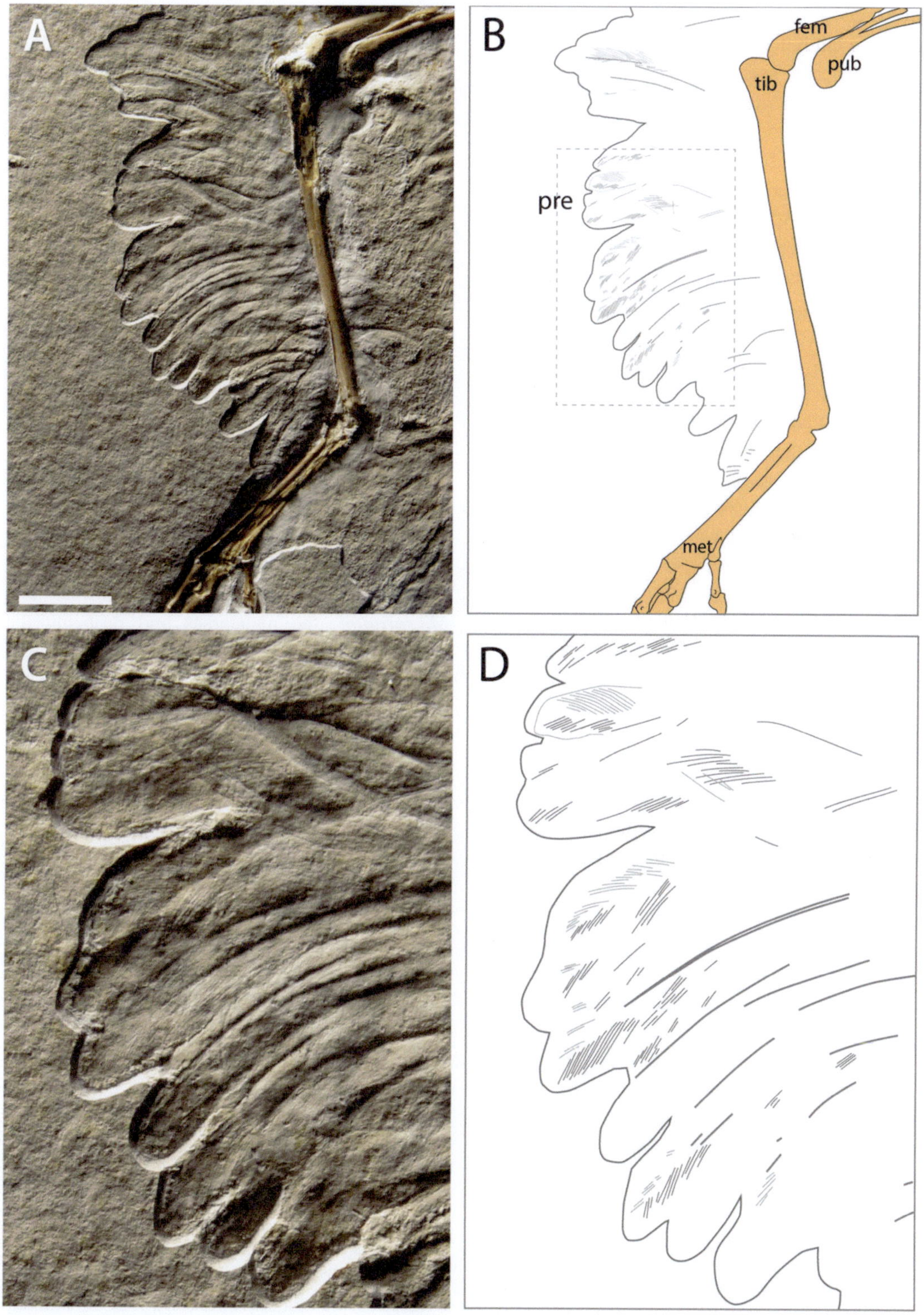

Fig. 8.11 Tibial feathers of the Altmühl *Archaeopteryx*. (**a**) Hindlimb and feathers; (**b**) interpretive drawing; (**c**) close-up of boxed region in (**b**); (**d**) interpretive drawing. Abbreviations: *fem* femur, *ti* tibia, *met* metatarsus, *pre* pretibial feathers. Scale = 25 mm

preparation exist (see Dames 1884; Longrich 2006; Wellnhofer 2009; Fig. 8.3b).

The Berlin specimen has elongate feathers posterior to the femur and tibial feathers that project at an angle from the anterior and posterior margins of the tibia (Longrich 2006). The number of hindlimb feathers is difficult to determine. In the Altmühl specimen, at least 18–20 vaned feathers are preserved anterior to the tibia, with their long axes oriented roughly perpendicular to the bone. Proximally, the tibial feathers measure 51–58% of the tibia length, and distally they decrease to around 37% (Fig. 8.11a). In both specimens, the hindlimb feathers are preserved in the sagittal plane and overlap each other to form a large, fan-shaped structure.

The tibial feathers have a slender rachis that curves posteriorly, as in remiges and rectrices. Vanes are long and slender, again resembling remiges and rectrices (Fig. 8.11b). Longrich (2006) suggested that the vanes were asymmetrical based on the counterslab of the Berlin specimen, but preservation here is poor. The more complete feathers of the Altmühl specimen suggest that the hindlimb feathers are more or less symmetrical (Foth et al. 2014). Barb angulation likewise appears to be relatively symmetrical. However, barb curvature is highly asymmetrical, with both leading and trailing edge barbs curving posteriorly, as in remiges and rectrices, rather than contour feathers.

No feathers are visible on the foot of the Berlin specimen, but a few short feathers project dorsally from the proximal end of the right metatarsus in the Altmühl specimen. In contrast to the long, broad feathers on the tibia, metatarsal feathers appear to have been short and narrow. Proximally, they extend at an angle of approximately 60° relative to the foot; more distally, they project at an angle of around 45°. Elongate, vaned feathers are preserved posterior to the right metatarsus; however, they are parallel it rather than projecting away from it. This suggests that they are not in fact part of the foot feathers and may instead be part of the wing.

The feather arrangement seen in *Archaeopteryx* is similar to that of *Anchiornis* (Hu et al. 2009) and, to a lesser degree, microraptorines including *Microraptor* (Norell et al. 2002; Xu et al. 2003; Pei et al. 2014; Xu and Li 2016) and *Changyuraptor* (Han et al. 2014). In both *Anchiornis* and Microraptorinae, feathers project both from the anterior margin of the tibia and from the posterior margin. The foot bears a short fringe of feathers on its anterior margin and longer, vaned feathers on its posterior margin, which are elaborated in *Microraptor*. The femur appears to bear feathers in *Microraptor* (Norell et al. 2002) and perhaps *Anchiornis* as well (Hu et al. 2009). However, the hindlimb feathers of *Archaeopteryx* are much shorter than either *Anchiornis* or *Microraptor*. *Archaeopteryx* also appears to lack long foot feathers. It is possible that short feathers are attached posterior to the metatarsus and are simply not preserved, but it is unlikely that the elongated feathers seen in *Anchiornis* and *Microraptor* would have escaped preservation. However, given the presence of feathers at the base of the foot, and the fact that the metatarsus is feathered, not scaled, in *Anchiornis* and *Microraptor*, the metatarsus and probably the tops of the toes were probably feathered, rather than scaly.

8.3.6 Rectrices

Archaeopteryx bears a series of rectrices extending along either side of the tail (Fig. 8.12). The number of rectrices is unclear; there are at least 35 pairs of rectrices extending distally from the tip of the bony tail proximally towards the proximal caudals and then to the hips. Rectrices are relatively long at the hips, become shorter at the base of the tail, and then become progressively longer again towards the tip of the tail. Distally, rectrices are longer and more symmetrical (Foth et al. 2014). The distal tail is not preserved except for the Altmühl specimen (Foth et al. 2014). It shows a distinct gap between the last pair of rectrices (Fig. 8.11c). It is unclear if this gap is real, the result of a moult (Foth et al. 2014), or an artefact of taphonomy. The feathers overlap as in modern birds, with distal feathers overlying more proximal feathers. As with the wings (Norberg 1985b), this pattern of overlap causes the asymmetrical feathers to twist up

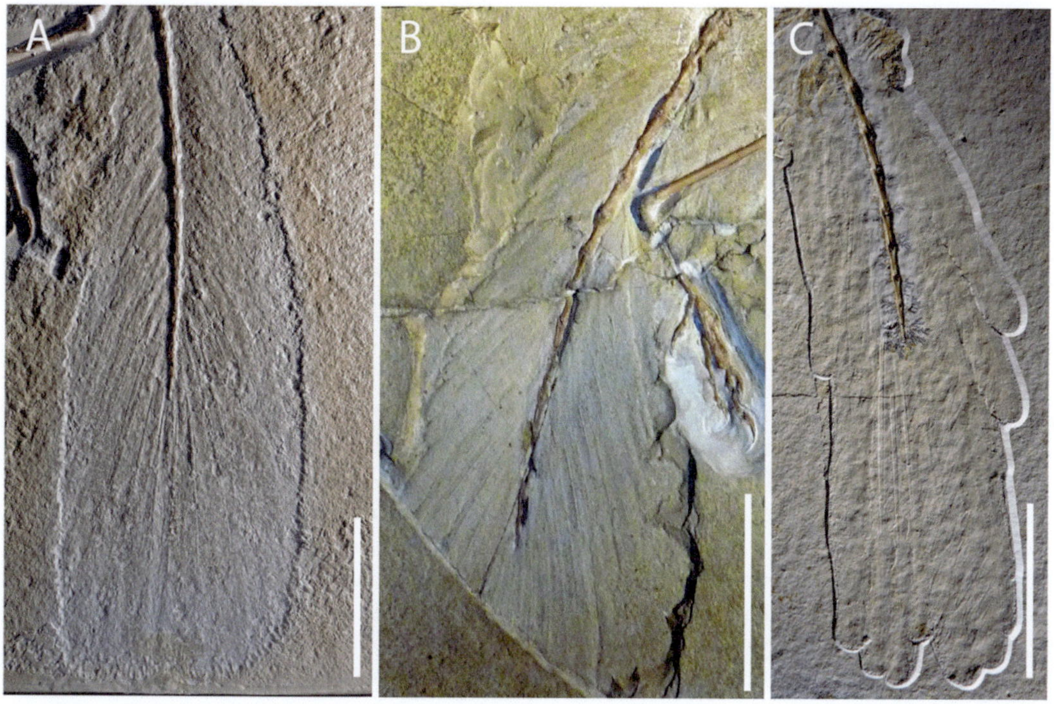

Fig. 8.12 Tail feathers of *Archaeopteryx*. (**a**) London specimen; (**b**) Berlin specimen; (**c**) Altmühl specimen. Scale = 50 mm

against the succeeding feather to form a continuous airfoil surface in flight.

It is unclear whether coverts were present. In crown birds, the tail fan is composed of a single layer of rectrices, with little contribution from coverts. This contrasts with the wings, where coverts overlap the base of the remiges, bracing them and creating a thicker airfoil cross section. Whether coverts have been present in the tail of *Archaeopteryx* is unclear. Several tail feathers in the Altmühl specimen are shorter than the remaining tail feathers and cross them obliquely. It is possible that these represent covert feathers, but it is also possible that they represent moulting feathers.

The morphology of the rectrices is similar to that of the remiges. Rectrices have a broad calamus and an elongate rachis that curves posteriorly in the plane of the feather. Vanes are long, narrow, and in the proximal feathers strongly asymmetrical, with the trailing edge vane being at least 200% the width of the leading edge vane. This is similar to the situation of the flightless troodontid *Jianianhualong* from the Early Cretaceous Jehol group (Xu et al. 2017). The tips of the rectrices form an obtuse angle in proximal feathers and are ovate in the distal rectrices. Barb angulation is symmetrical, but barb curvature is asymmetrical with leading and trailing edge barbs curved towards the trailing edge.

8.4 Discussion

8.4.1 Feather Evolution

In many respects, the feather morphology and arrangement of *Archaeopteryx* is strikingly modern, or seen from another perspective, modern birds are remarkably primitive. Together with *Anchiornis*, *Archaeopteryx* shows that many aspects of feather morphology and arrangement found in modern birds originated in the Jurassic and are retained, within minimal modification, in modern birds (Wellnhofer 2009; Longrich et al. 2012). Conserved features include the basic structure of the pennate feather, including the rachis, barbs, and interlocking barbules, and also the developmental patterns that underlie their growth and moulting. Other conserved features are adaptations for flight, such as curved rachis, vane asymmetry, asymmetry of barb angle and barb curvature, and the overlap of feather vanes to create an airfoil. Last, the basic arrangement of feathers into feather tracts and the associated soft tissues also appear to be ancient. It would not be fair to say that the feathers and their arrangement have not changed; clearly, major changes did occur between *Anchiornis* and *Archaeopteryx* and crown birds in feather structure (Feo et al. 2015) and arrangement (Gatesy and Dial 1996; Longrich et al. 2012). However, these adaptations notwithstanding the feathers of birds represent a remarkable example of stasis. Birds are typically considered highly advanced organisms, and yet in terms of feather morphology, they are every bit as much a 'living fossil' as a crocodile or a coelacanth.

Why the feathers have changed so little is uncertain. It may be that given basic constraints on the mechanical properties of feather proteins and aerodynamics, further refinement simply became impossible. Given the materials available to work with and the basic principles of fluid mechanics, Jurassic feathers were already an optimal or near-optimal solution to creating a wing from elaborated scales. In this respect, avian evolution parallels the evolution of aircraft, where the basic layout of the jet airliner has not changed in half a century, because it represents a mature design that is a near-optimal solution giving existing mechanical and aerodynamic constraints (Longrich et al. 2012). This hypothesis is supported by two observations. First, when birds become secondarily flightless, their feathers often evolve radically different morphologies that differ from those of flying birds (Lüdicke 1974; McGowan 1989; Livezey 2003). Second, feathers that are selected for display, rather than for flight, exhibit a similar departure from the limited range of morphologies seen among flight and contour feathers (Brinckmann 1958; Bleiweiss 1987; Stavenga et al. 2011). Both observations suggest that radical evolution in feather structure is possible, once the selective pressures of flight are removed or at least no longer dominant.

However, existing developmental patterns may also constrain feather morphology. Although the feathers of flightless birds and display feathers do depart from the overall morphology of flight and contour feathers, they are derived from the same basic morphology of a central rachis that branches to create barbs and barbules (see Chap. 2). Once evolved, the developmental processes that underlie feather formation can be modified to produce a diverse but ultimately constrained range of forms. Together, these two forces—stabilizing selection towards an optimal solution and developmental processes that constrain feather variation—probably explain why flight feathers have remained relatively unchanged in the 145 million years since *Archaeopteryx*.

But while the feathering is in many ways similar to those of crown birds, *Archaeopteryx* also shows archaic features that are not seen in living birds. Primitive features of the feathers themselves include relatively slender rachis and the relatively symmetrical barbs. Despite the overall modern appearance of the plumage, *Archaeopteryx* was not a modern bird, and so it is unsurprising that in certain details it is more primitive. Instead, its morphology appears to have been transitional between the older and more primitive *Anchiornis* and more derived pygostylians such as *Confuciusornis*. Primitive features of feather arrangement include the long feathers on the hindlimbs, the development of rectrices along the length of the tail and onto the hips, and the poor differentiation of remiges and coverts. In many of these features, *Archaeopteryx* resembles *Anchiornis* and *Microraptor*, suggesting that these features represent the ancestral condition, rather than specializations.

One of the more striking differences between *Archaeopteryx* and modern birds is the limited differentiation between feather tracts, within feather tracts, and within feathers themselves. Elongated, curved feathers with closed, pennaceous vanes are developed on the hindlimbs and forelimbs and tail in *Archaeopteryx*, whereas in most crown group birds, hindlimb feathers resemble the morphology of body contour feathers (see Lucas and Stettenheim 1972). Similarly, the supposed elongation of the coverts relative to the remiges shows limited differential within the wing feathers. Even within the individual primary feathers, differentiation of barbs is weakly developed, with angulation of anterior and posterior barbs being relatively symmetrical. A similar pattern is seen in *Anchiornis*, except more extreme, with symmetrical feathers that show very little differentiation between primaries and secondaries, remiges and coverts, or forelimbs and hindlimbs (Hu et al. 2009). This suggests that in the early stages of avian evolution, stem birds had yet to fully develop the ability to fully specify feather morphology either within or between feathers. A higher degree of differentiation is seen in Pygostylia. Here, we see highly asymmetrical flight feathers; differentiation of primaries, secondaries, and coverts; reduction of hindlimb feathers; and differentiated rectrices (see Chap. 9). The feather differentiation seen here approaches that of crown Aves, suggesting more refined control of feathering and feather morphogenesis resulting from improvements in regulation of the genes that specify feather morphology.

8.4.2 Feather Function and Behaviour

The structure of the feathers and wings of *Archaeopteryx* suggests that they functioned as efficient airfoils, which served to maintain lift, stability, and control and perhaps to generate thrust. The body feathers probably functioned to insulate and streamline the animal. None of the feathers appear specialized for display, but courtship in birds is heavily dependent on visual cues, and so even if display was not a primary function, the feathers would almost certainly have served a secondary role as display structures. Last, the carefully arranged feathers and barbs strongly indicate that *Archaeopteryx* had evolved grooming behaviours to maintain its plumage.

The flight ability of *Archaeopteryx* has been and remains highly controversial. Although some have proposed that *Archaeopteryx* was incapable of flight, the asymmetry of the feathers as well as their arrangement indicates that the wings and the tail were adapted to function as airfoils (Feduccia and Tordoff 1979; Norberg 1985b). Asymmetry of the distal primaries is seen in all volant birds, but tends to be poorly developed or absent in flightless birds (Feduccia and Tordoff 1979; Speakman and Thompson 1994). Although Speakman and Thompson (1994) argued that the feathers were symmetrical, Norberg (1995) pointed out that the apparent symmetry is a taphonomic artefact and that feather asymmetry in *Archaeopteryx* is comparable to asymmetry in flying birds.

The asymmetry of the feather is an adaptation to the asymmetrical loads experienced in flight. The centre of pressure for an individual feather lies near its leading edge, requiring the main support to be located towards the leading edge (Norberg 1985b). The asymmetry is also important to stabilize the feather in flight; by shifting the main support anterior to the centre of pressure, the feather develops a nose-down pitching moment that causes the trailing edge to twist up against trailing feathers, causing the feathers to automatically assemble into an airfoil (Norberg 1985b).

However, vane asymmetry is only one of several features that differentiate flight feathers from ordinary down, display, and body contour feathers. Flight feathers are characterized by a suite of adaptations for lift generation, which are variably developed depending on their position in the wing. Distal primaries and remiges are oriented at a strong angle relative to the airflow and therefore tend to be under highly asymmetric loads and exhibit corresponding adaptations. The proximal primaries and secondaries are oriented parallel to the airstream and tend to exhibit these features to a lesser degree.

Such features (Table 8.2) include long and robust calami and rachises, posterior curvature of the rachis within the vane, elongate vanes, vane asymmetry, asymmetrical barb angulation and barb curvature, tightly interlocking barbs, and reduction of the downy part of the vane (Lucas and Stettenheim 1972; Norberg 1985b;

Table 8.2 Aerodynamic features of pennaceous feathers (after Longrich 2006) of *Archaeopteryx* from different body regions

	Remiges	Rectrices	Hindlimb
Calamus and rachis slender (0) or robust (1)	1	1	0
Rachis straight (0) or curved in plane of feather (1)	1	1	1
Vanes broad (0) or feathers long and slender (1)	1	1	1
Vanes symmetrical (0) or asymmetrical (1)	1	1	0
Barb angulation symmetrical (0) or asymmetrical (1)	1	1	0
Barb curvature symmetrical (0) or asymmetrical (1)	1	1	1
Barbs loosely interlocked, vane edge ragged (0) or tightly interlocked along their length, vane edge sharp (1)	1	1	1

Longrich 2006; Feo et al. 2015). This suite of features can be considered as functional correlates of an aerodynamic, lift-generating function in feathers.

However, as some secondary flightless birds still possess asymmetric vane ratios falling into the lower range of volant birds, all features cited above as potential adaptations for aerodynamic function, should be still considered with caution in interpreting the flight capacity of extinct stem birds (see Wang et al. 2017a). Asymmetry is consistent with adaptation for aerodynamic function, but adaptation for a function does not necessarily mean the use of that particular function: the asymmetric tail feathers of the Early Cretaceous troodontid *Jianianhualong* show adaptation to flight, but do not function for flight (Xu et al. 2017). Therefore, it is important to consider other features, such as wing loading or skeletal correlates of gliding or flapping flight, in inferring flight ability.

In addition to the shape, the arrangement of the feathers is critical to forming an airfoil. The asymmetry and posterior curvature of the feathers are important in creating a nose-up pitching moment. When arranged in series, they cause distal feathers to twist up against the proximal feather, automatically assembling into an airfoil and stabilizing the wing surface in flight (Norberg 1985b). This arrangement, with proximal feathers overlapping trailing feathers, is seen in the wings and tail of *Archaeopteryx* and perhaps also the legs. All of these features are found, to varying degrees, in the remiges and rectrices of *Archaeopteryx*, arguing for a lift-generating function.

In addition to the structure and arrangement of individual feathers, the shape of the wings and tail in *Archaeopteryx* suggests that they were effective and efficient lift-generating airfoils capable of supporting and steering the bird in flight. To fly, a bird requires relatively low mass relative to wing area, a parameter known as wing loading. Estimates of wing loading in *Archaeopteryx* vary, but they consistently show that it was comparable to modern birds and would have been adequate to support the animal during either gliding or flapping flight (Norberg 1990). Furthermore, to generate lift efficiently, wings must have a relatively long span compared to the chord, a parameter known as aspect ratio, and tapered tips, which help create an elliptical distribution of lift. This shape helps to minimize the wingtip vortices that generate lift-induced drag. Again, *Archaeopteryx* is comparable to modern birds here (Norberg 1990), suggesting well-developed gliding or flapping flight. The tail and perhaps hindlimbs would have also been able to contribute to lift generation (Longrich 2006).

The existence of control structures is also critical, in that flight is of little use if not lethal when it cannot be controlled. The tail of *Archaeopteryx* would have been effective as a delta wing (Gatesy and Dial 1996), which could have functioned as a horizontal stabilizer and elevator to stabilize and control the bird in pitch. In addition, the hindlimb feathers may also have been able to contribute to stability and control (Longrich 2006; Foth et al. 2014) (see below).

Fig. 8.13 Reconstruction of a pair of *Archaeopteryx lithographica* in flight by Carl Buell

There is little doubt that *Archaeopteryx* was capable of aerial locomotion (Fig. 8.13), but whether it was capable of powered flight or not remains open to debate. Nothing in the morphology of the wings and feathers would appear to have prevented flapping flight, and the shape and size of the wings are consistent with the bird's capability of powered flight (Voeten et al. 2018). However, there are no adaptations to the wings that are specific to powered flight: asymmetric, curved feathers are found both in the wings and the tail. The high aspect ratio that is required for powered flight is also effective in improving glide performance. A number of features of the skeleton do suggest adaptation for powered flight, however. The robust furcula and coracoids may have supported the muscles to make a down stroke possible (Olson and Feduccia 1979), while the sharp angle between the coracoids and the scapula would have allowed the pectoralis to both depress and retract the humerus against the lift and thrust produced by flapping flight. The large coracoids would also have increased the length of the pectoralis muscles and therefore the distance over which they could contract, increasing the potential amplitude of a wingbeat. These features may represent adaptations for powered flight, but are also seen in Dromaeosauridae (Paul 2002). Either they do not in fact represent flight adaptations (but were exaptated for flight later on) or flight may have evolved earlier than generally believed.

However, the primitive morphology of the glenoid, the unossified sternum, and the lack of a supracoracoideus pulley (acrocoracoid process) (Wellnhofer 2009) suggest that flapping flight, if it was possible, was rudimentary. In constrast, flap gliding of the sort practiced by petrels and gulls, with glides interspersed with low-amplitude flapping flight using a continuous-vortex gait, was likely possible. Sustained level flight at high speeds, again using low-amplitude flapping and continuous-vortex gait, may have been possible, given that flight is less expensive and flight kinematics are simpler (Rayner 2001). Low-speed flight, requiring more elaborate wing kinematics and high-power output, seems less likely.

The function of the hindlimb feathers remains unclear. The morphology, shape, and arrangement of the feathers suggest that they may not simply be body feathers as traditionally thought. Instead, they appear to be homologous with the elongate feathers seen on the hindlimbs of *Anchiornis* (Hu et al. 2009), *Microraptor* (Xu et al. 2003), and at least some basal pygostylians, including enantiornithes (Zhang and Zhou 2004; Zheng et al. 2013). The arrangement of the feathers in *Archaeopteryx* also indicates that current reconstructions of *Anchiornis* and microraptorines are incorrect in having feathers extending only from the trailing edge of the leg; instead, as in *Archaeopteryx*, feathers of *Anchiornis* (Hu et al. 2009) and microraptorines (Czerkas et al. 2002; Xu and Li 2016) both show that a large fan of feathers extends from the anterior edge of the tibia and the posterior.

As discussed above, flight feathers exhibit a suite of features that are absent in body contour feathers (Lucas and Stettenheim 1972; Longrich 2006), which can be considered functional correlates for flight. The hindlimb feathers of *Archaeopteryx* exhibit some, but not all, of these features (Table 8.2). Features shared with remiges and rectrices feathers include long and narrow vanes, sharp vane edges, posterior curvature of the rachis, and the asymmetry of barb curvature. Yet, the hindlimb feathers lack asymmetry in barb angulation, and the vanes appear to be symmetrical or very weakly asymmetrical. These features vary depending on position in the airfoil; the outer primaries and lateral rectrices invariably show all of these functional correlates, while more proximal remiges and inner rectrices tend to be more symmetrical in terms of vane width, barb angulation, and barb curvature. The weakly developed asymmetry of the vanes of the hindlimb feather therefore does not preclude aerodynamic function, any more than it precludes aerodynamic function for inner secondaries and rectrices.

It is also possible that the hindlimb feathers, although exhibiting features associated with flight, did not actually function for flight and were simply vestiges of the earlier four-winged designs seen in *Anchiornis* and some dromaeosaurids. If so, *Archaeopteryx* might be transitional between the earlier 'four-winged' forms and the later two-winged forms (Prum 2003; Longrich 2006). In contrast, Foth et al. (2014) hypothesized that the homoplastic variation in terms of size, shape, and regional distribution of the hindlimb feathers within Eumaniraptora may indicate an initial display function, which is supported by the presence of complex and gleaming colour patterns within Pennaraptora (e.g. Li et al. 2010, 2012).

Acknowledgements We are indebted to the curators and staff of the Natural History Museum (London), Humboldt

Museum für Naturkunde (Berlin), Jura-Museum (Eichstätt), and Wyoming Dinosaur Center (Thermopolis), especially Raimund Albersdörfer, Martina Köbl-Ebert, Daniela Schwarz, and Burkhard Pohl for access to the specimens in their care, without which this study would not be possible. NRL is grateful to Tony Russell for his patience as a supervisor and mentor and many discussions and also to Philip J. Currie for the discussions. We further thank Jakob Vinther, Oliver Rauhut, and Xu Xing for the discussion and Tom Holtz for his comments on the manuscript. CF is supported by the Swiss National Science Foundation under grant PZ00P2_174040.

References

Arbour VM, Burns ME, Bell PR, Currie PJ (2014) Epidermal and dermal integumentary structures of ankylosaurian dinosaurs. J Morphol 275:39–50

Arratia G, Schultze H-P, Tischlinger H, Viohl G (2015) Solnhofen – Ein Fenster in die Jurazeit. Verlag Dr. Friedrich Pfeil, München

Bell PR (2014) A review of hadrosaurid skin impressions. In: Eberth DA, Evans DC (eds) Hadrosaurs. Indiana University Press, Bloomington, IN, pp 572–590

Bergman G (1982) Why are the wings *Larus fuscus* so dark? Ornis Fenn 59:77–83

Bergmann U, Morton R, Manning P, Sellers W, Farrar S, Huntley K, Wogelius R, Larson P (2010) *Archaeopteryx* feathers and bone chemistry fully revealed via synchrotron imaging. Proc Natl Acad Sci USA 107:9060–9065

Bleiweiss R (1987) Development and evolution of avian racket plumes: fine structure and serial homology of the wire. J Morphol 194:23–39

Brinckmann A (1958) Die Morphologie der Schmuckfeder von *Aix galericulata* L. Rev Suisse Zool 68:485–608

Carney RM, Vinther J, Shawkey MD, D'Alba L, Ackermann J (2012) New evidence on the colour and nature of the isolated *Archaeopteryx* feather. Nat Commun 3:637

Chang C, Wu P, Baker RE, Maini PK, Alibardi L, Chuong C-M (2009) Reptile scale paradigm: Evo-Devo, pattern formation and regeneration. Int J Dev Biol 53:813

Chatterjee S, Templin RJ (2003) The flight of *Archaeopteryx*. Naturwissenschaften 90:27–32

Chiappe LM, Coria RA, Dingus L, Jackson F, Chinsamy A, Fox M (1998) Sauropod dinosaur embryos from the Late Cretaceous of Patagonia. Nature 396:258–261

Chiappe LM, Ji S-A, Ji Q, Norell MA (1999) Anatomy and systematics of Confuciusornithidae (Theropoda: Aves) from the Late Mesozoic of Northeastern China. Bull Am Mus Nat Hist 242:1–89

Christiansen P, Bonde N (2004) Body plumage in *Archaeopteryx*: a review, and new evidence from the Berlin specimen. C R Palevol 3:99–118

Christiansen NA, Tschopp E (2010) Exceptional stegosaur integument impressions from the Upper Jurassic Morrison Formation of Wyoming. Swiss J Geosci 103:163–171

Currie PJ, Chen P-J (2001) Anatomy of *Sinosauropteryx prima* from Liaoning, northeastern China. Can J Earth Sci 38:1705–1727

Czerkas SJ, Zhang D, Li J, Li Y (2002) Flying dromaeosaurs. Dinosaur Museum J 1:98–126

Dames W (1884) Über *Archaeopteryx*. Palaeontologische Abhandlungen 2:119–196

Darwin CR (1859) The origin of species. John Murray, London

de Buisonjé PH (1985) Climatological conditions during deposition of the Solnhofen limestones. In: Hecht MKO, Ostrom JH, Viohl G, Wellnhofer P (eds) The beginnings of birds: proceedings of the international Archaeopteryx conference. Freunde des Jura-Museums Eichstätt, Eichstätt, pp 45–65

Dial KP (2003) Wing-assisted incline running and the evolution of flight. Science 299:402–404

Eagle RA, Tütken T, Martin TS, Tripati AK, Fricke HC, Connely M, Cifelli RL, Eiler JM (2011) Dinosaur body temperatures determined from isotopic (^{13}C-^{18}O) ordering in fossil biominerals. Science 333:443–445

Elzanowski A (2002) Archaeopterygidae (Upper Jurassic of Germany). In: Chiappe LM, Witmer LM (eds) Mesozoic birds: above the heads of dinosaurs. University of California Press, Berkeley, CA, pp 129–159

Erickson GM, Rogers KC, Yerby SA (2001) Dinosaurian growth patterns and rapid avian growth rates. Nature 412:429–433

Erickson GM, Makovicky PJ, Currie PJ, Norell MA, Yerby SA, Brochu CA (2004) Gigantism and comparative life-history parameters of tyrannosaurid dinosaurs. Nature 430:772–775

Feduccia A (1996) The origin and evolution of birds. Yale University Press, New Haven, CT

Feduccia A, Czerkas SA (2015) Testing the neoflightless hypothesis: propatagium reveals flying ancestry of oviraptorosaurs. J Ornithol 156:1067–1074

Feduccia A, Tordoff HB (1979) Feathers of *Archaeopteryx*: asymmetric vanes indicate aerodynamic function. Science 203:1021–1022

Feo TJ, Field DJ, Prum RO (2015) Barb geometry of asymmetrical feathers reveals a transitional morphology in the evolution of avian flight. Proc R Soc Lond B Biol Sci 282:20142864

Foth C (2012) On the identification of feather structures in stem-line representatives of birds: evidence from fossils and actuopalaeontology. Paläontol Z 86:91–102

Foth C, Rauhut OWM (2017) Re-evaluation of the Haarlem *Archaeopteryx* and the radiation of maniraptoran theropod dinosaurs. BMC Evol Biol 17:236

Foth C, Tischlinger H, Rauhut OWM (2014) New specimen of *Archaeopteryx* provides insights into the evolution of pennaceous feathers. Nature 511:79–82

Gatesy SM, Dial KP (1996) From frond to fan: *Archaeopteryx* and the evolution of short-tailed birds. Evolution 50:2037–2048

Gauthier J (1986) Saurischian monophyly and the origin of birds. Memoirs Calif Acad Sci 8:1–55

Goldstein G, Flory KR, Browne BA, Majid S, Ichida JM, Burtt EH Jr, Grubb T Jr (2004) Bacterial degradation of black and white feathers. Auk 121:656–659

Griffiths PJ (1996) The isolated *Archaeopteryx* feather. Archaeopteryx 14:1–26

Grigg G, Kirshner D (2015) Biology and evolution of crocodylians. Cornell University Press, Ithaca, NY

Gunderson AR, Frame AM, Swaddle JP, Forsyth MH (2008) Resistance of melanized feathers to bacterial degradation: is it really so black and white? J Avian Biol 39:539–545

Han G, Chiappe LM, Ji S-A, Habib M, Turner AH, Chinsamy A, Liu X, Han L (2014) A new raptorial dinosaur with exceptionally long feathering provides insights into dromaeosaurid flight performance. Nat Commun 5:4382

Heilmann G (1926) The origin of birds. Witherby, London

Heinroth O (1923) Die Flügel von *Archaeopteryx*. J Ornithol 71:277–283

Helms J (1982) Zur Fossilization der Federn des Urvogels (Berliner Exemplar). Wissenschaftliche Zeitschrift der Humboldt-Universität, mathematisch-naturwissenschaftliche Reihe 31:185–199

Homberger DG, de Silva KN (2000) Functional microanatomy of the feather-bearing integument: implication for the evolution of birds and avian flight. Am Zool 40:553–574

Hu D, Hou L-H, Zhang L, Xu X (2009) A pre-*Archaeopteryx* troodontid theropod from China with long feathers on the metatarsus. Nature 461:640–643

Huxley TH (1868) On the animals which are most nearly intermediate between birds and reptiles. Ann Mag Nat Hist 2:66–75

Huxley TH (1870) Further evidence of the affinity between the dinosaurian reptiles and birds. Q J Geol Soc 26:12–31

Ji Q, Currie PJ, Norell MA, Ji S-A (1998) Two feathered dinosaurs from northeastern China. Nature 393:753–761

Li Q, Gao K, Vinther J, Shawkey MD, Clarke JA, D'Alba L, Meng Q, Briggs DEG, Prum RO (2010) Plumage color patterns of an extinct dinosaur. Science 327:1369–1372

Li Q, Gao K, Meng Q, Clarke JA, Shawkey MD, D'Alba L, Pei R, Ellison M, Norell MA, Vinther J (2012) Reconstruction of *Microraptor* and the evolution of iridescent plumage. Science 335:1215–1219

Livezey BC (2003) Evolution of flightlessness in rails (Gruiformes, Rallidae). Ornithol Monogr 53:1–654

Longrich N (2006) Structure and function of hindlimb feathers in *Archaeopteryx lithographica*. Paleobiology 32:417–431

Longrich NR, Vinther J, Meng Q, Li Q, Russell AP (2012) Origins and evolution of the avian wing: new evidence from *Archaeopteryx lithographica* and *Anchiornis huxleyi*. Curr Biol 22:1–6

Lucas AM, Stettenheim PR (1972) Avian anatomy: integument, vol 2. U.S. Government Printing Office, Washington, DC

Lüdicke M (1974) Radioaktive Markierungsversuche an Federn von *Casuarius casuarius*. J Ornithol 115:348–364

Manning PL, Edwards NP, Wogelius RA, Bergmann U, Barden HE, Larson PL, Schwarz-Wings D, Egerton VM, Sokaras D, Mori RA (2013) Synchrotron-based chemical imaging reveals plumage patterns in a 150 million year old early bird. J Anal At Spectrom 28:1024–1030

Martin LD, Lim J-D (2005) Soft body impression of the hand in *Archaeopteryx*. Curr Sci 89:1089–1090

Mayr G, Pohl B, Peters DS (2005) A well-preserved *Archaeopteryx* specimen with theropod features. Science 310:1483–1486

Mayr G, Pohl B, Hartmann S, Peters DS (2007) The tenth skeletal specimen of *Archaeopteryx*. Zool J Linnean Soc 149:97–116

McGowan C (1989) Feather structure in flightless birds and its bearing on the question of the origin of feathers. J Zool 218:537–547

Meseguer J, Chiappe LM, Sanz JL, Ortega F, Sanz-Andrés A, Pérez-Grande I, Franchini S (2012) Lift devices in the flight of *Archaeopteryx*. Spanish Journal of Palaeontology 27:125–130

Meyer HV (1862) *Archaeopteryx lithographica* aus dem lithographischen Schiefer von Solenhofen. Palaeontographica:53–56

Moyer AE, Zheng W, Johnson EA, Lamanna MC, Li D-Q, Lacovara KJ, Schweitzer MH (2014) Melanosomes or microbes: testing an alternative hypothesis for the origin of microbodies in fossil feathers. Sci Rep 4:4233

Nachtigall W, Kempf B (1971) Vergleichende Untersuchungen zur flugbiologischen Funktion des Daumenfittichs (Alula spuria) bei Vögeln. J Comp Physiol A Neuroethol Sens Neural Behav Physiol 71:326–341

Norberg UM (1985a) Evolution of vertebrate flight: an aerodynamic model for the transition from gliding to active flight. Am Nat 126:303–327

Norberg RA (1985b) Function of vane asymmetry and shaft curvature in bird flight feathers: inferences on flight ability of *Archaeopteryx*. In: Hecht MKO, Ostrom JH, Viohl G, Wellnhofer P (eds) The beginnings of birds: proceedings of the international Archaeopteryx conference. Freunde des Jura-Museums Eichstätt, Eichstätt, pp 303–318

Norberg UM (1990) Vertebrate flight. Springer, Berlin

Norberg RA (1995) Feather asymmetry in *Archaeopteryx*. Nature 374:221

Norell MA, Xu X (2005) Feathered dinosaurs. Annu Rev Earth Planet Sci 33:277–299

Norell MA, Clark JM, Weintraub R, Chiappe LM, Demberelyin D (1995) A nesting dinosaur. Nature 278:247–248

Norell MA, Ji Q, Gao K, Yuan C, Zhao Y, Wang L (2002) 'Modern' feathers on a non-avian dinosaur. Nature 416:36–37

Nudds RL, Dyke GJ (2010) Narrow primary feather rachises in *Confuciusornis* and *Archaeopteryx* suggest poor flight ability. Science 328:887–889

Olson S, Feduccia A (1979) Flight capability and the pectoral girdle of Archaeopteryx. Nature 278:247–248

Ostrom JH (1973) The ancestry of birds. Nature 242:136

Ostrom JH (1976) *Archaeopteryx* and the origin of birds. Biol J Linn Soc 8:91–182

Ostrom JH (1979) Bird flight: how did it begin? Am Sci 67:46–56

Owen R (1863) On the *Archaeopteryx* of Von Meyer, with a description of the fossil remains of a long-tailed species from the lithographic stone of Solnhofen. Philos Trans R Soc Lond 153:33–47

Paul GS (2002) Dinosaurs of the air: the evolution and loss of flight in dinosaurs and birds. The John Hopkins University Press, Baltimore, MD

Pei R, Li Q, Meng Q, Gao K-Q, Norell MA (2014) A new specimen of *Microraptor* (Theropoda: Dromaeosauridae) from the Lower Cretaceous of western Liaoning, China. Am Mus Novit 3821:1–28

Pei R, Li Q, Meng Q, Norell MA, Gao K (2017) New Specimens of Anchiornis huxleyi (Theropoda: Paraves) from the Middle-Late Jurassic of northeastern China. Bull Am Mus Nat Hist 411:1–66

Pianka ER, Vitt LJ (2003) Lizards: windows to the evolution of diversity. University of California Press, Oakland

Prum RO (2003) Dinosaurs take to the air. Nature 421:323–324

Rauhut OWM, Foth C, Tischlinger H (2018) The oldest *Archaeopteryx* (Theropoda: Avialiae): a new specimen from the Kimmeridgian/Tithonian boundary of Schamhaupten, Bavaria. PeerJ 6:e4191

Rauhut OWM, Tischlinger H, Foth C (2019) A non-archaeopterygid avialan theropod from the Late Jurassic of southern Germany. eLife 8:e43789

Rayner JMV (2001) On the origin and evolution of flapping flight aerodynamics in birds. In: Gauthier J, Gall LF (eds) New perspectives on the origin and early evolution of birds: proceedings of the international conference in honor of John H. Ostrom. Peabody Museum of Natural History, New Haven, CT, pp 364–381

Rietschel S (1985) Feathers and wings of Archaeopteryx and the question of her flight ability. In: Hecht MK, Ostrom JH, Viohl G, Wellnhofer P (eds) The beginnings of birds: proceedings of the international Archaeopteryx conference. Freunde des Jura-Museums Eichstätt, Eichstätt, pp 251–260

Sanz JL, Chiappe LM, Pérez-Moreno BP, Buscalioni AD, Moratalla JJ, Ortega F, Poyato-Ariza FJ (1996) An Early Cretaceous bird from Spain and its implications for the evolution of avian flight. Nature 382:442–445

Speakman JR, Thompson SC (1994) Flight capabilities of *Archaeopteryx*. Nature 370:514

Stavenga DG, Leertouwer HL, Marshall NJ, Osorio D (2011) Dramatic colour changes in a bird of paradise caused by uniquely structured breast feather barbules. Proc R Soc Lond B Biol Sci 278:2098–2104

Steiner H (1962) Befunde am dritten Exemplar des Urvogels *Archaeopteryx*. Vierteljahrsschrift der Naturforschenden Gesellschaft in Zürich 107:197–210

Stephan B (1985) Remarks on reconstruction of *Archaeopteryx* wing. In: Hecht MK, Ostrom JH, Viohl G, Wellnhofer P (eds) The beginnings of birds: proceedings of the international Archaeopteryx conference. Freunde des Jura-Museums Eichstätt, Eichstätt, pp 261–265

Stephan B (1987) Urvögel: Archaeopterygiformes. Ziemsen, Wittenberg

Sullivan C, Wang Y, Hone DW, Wang Y, Xu X, Zhang F (2014) The vertebrates of the Jurassic Daohugou Biota of northeastern China. J Vertebr Paleontol 34:243–280

Tischlinger H (2005) Neue Informationen zum Berliner Exemplar von *Archaeopteryx lithographica* H. v. Meyer 1861. Archaeopteryx 23:33–50

Tischlinger H (2009) Der achte *Archaeopteryx*–das Daitinger Exemplar. Archaeopteryx 27:1–20

Tischlinger H, Unwin D (2004) UV-Untersuchungen des Berliner Exemplars von *Archaeopteryx* lithographica H. v. Meyer 1861 und der isolierten *Archaeopteryx*-Feder. Archaeopteryx 22:17–50

Vinther J, Nicholls R, Lautenschlager S, Pittman M, Kaye TG, Rayfield E, Mayr G, Cuthill IC (2016) 3D camouflage in an ornithischian dinosaur. Curr Biol 26:2456–2462

Viohl G (1985) Geology of the Solnhofen lithographic limestone and the Habitat of *Archaeopteryx*. In: Hecht MKO, Ostrom JH, Viohl G, Wellnhofer P (eds) The beginnings of birds: proceedings of the international Archaeopteryx conference. Freunde des Jura-Museums Eichstätt, Eichstätt, pp 31–44

Voeten DFAE, Cubo J, de Margerie E, Röper M, Beyrand V, Bureš S, Tafforeau P, Sanchez S (2018) Wing bone geometry reveals active flight in *Archaeopteryx*. Nat Commun 9:923

Wang X, Nudds RL, Palmer C, Dyke GJ (2017a) Primary feather vane asymmetry should not be used to predict the flight capabilities of feathered fossils. Science Buelltin 62:1227–1228

Wang X, Pittman M, Zheng X, Kaye TG, Falk AR, Hartman SA, Xu X (2017b) Basal paravian functional anatomy illuminated by high-detail body outline. Nat Commun 8:14576

Wellnhofer P (2004) The plumage of *Archaeopteryx*: feathers of a dinosaur? In: Currie PJ, Koppelhus EB, Shugar MA, Wright JL (eds) Feathered dragons: studies on the transition from dinosaurs to birds. Indiana University Press, Bloomington, pp 282–300

Wellnhofer P (2009) Archaeopteryx: the icon of evolution. Verlag Dr. Friedrich Pfeil, München

Wogelius R, Manning P, Barden H, Edwards N, Webb S, Sellers W, Taylor K, Larson P, Dodson P, You H

(2011) Trace metals as biomarkers for eumelanin pigment in the fossil record. Science 333:1622–1626

Xing L, McKellar RC, Wang M, Bai M, O'Connor JK, Benton MJ, Zhang J, Wang Y, Tseng K, Lockley MG, Li G, Zhang W, Xu X (2016) Mummified precocial bird wings in mid-Cretaceous Burmese amber. Nat Commun 7:12089

Xu X, Li F (2016) A new microraptorine specimen (Theropoda: Dromaeosauridae) with a brief comment on the evolution of compound bones in theropods. Vertebrata PalAsiatica 54:269–285

Xu X, Norell MA (2004) A new troodontid dinosaur from China with an avian-like sleeping posture. Nature 431:838–841

Xu X, Zhou Z, Wang X, Kuang X, Zhang F, Du X (2003) Four winged dinosaurs from China. Nature 421:335–340

Xu X, Wang K, Zhang K, Ma Q, Xing L, Sullivan C, Hu D, Cheng S, Wang S (2012) A gigantic feathered dinosaur from the Lower Cretaceous of China. Nature 484:92–95

Xu X, Zhou Z, Sullivan C, Wang Y, Ren D (2016) An updated review of the Middle-Late Jurassic Yanliao Biota: Chronology, Taphonomy, Paleontology and Paleoecology. Acta Geologica Sinica (English Edition) 90:2229–2243

Xu X, Currie P, Pittman M, Xing L, Meng Q, Lü J, Hu D, Yu C (2017) Mosaic evolution in an asymmetrically feathered troodontid dinosaur with transitional features. Nat Commun 8:14972

Zhang F, Zhou Z (2004) Leg feathers in an Early Cretaceous bird. Nature 341:925

Zhang F, Zhou Z, Dyke G (2006) Feathers and 'feather-like' integumentary structures in Liaoning birds and dinosaurs. Geol J 41:395–404

Zheng X, Zhou Z, Wang X, Zhang F, Zhang X, Wang Y, Wei G, Wang S, Xu X (2013) Hind wings in basal birds and the evolution of leg feathers. Science 339:1309–1312

Zhou Z, Zhang F (2003a) *Jeholornis* compared to *Archaeopteryx*, with a new understanding of the earliest avian evolution. Naturwissenschaften 90:220–225

Zhou Z, Zhang F (2003b) Anatomy of the primitive bird *Sapeornis chaoyangensis* from the Early Cretaceous of Liaoning, China. Can J Earth Sci 40:731–747

9 The Plumage of Basal Birds

Jingmai O'Connor

Institutional Abbreviations
BMNHC, Beijing Museum of Natural History, Beijing, China; DIP-V, Dexu Institute of Palaeontology, Chaozhou, China; DNHM, Dalian Natural History Museum, Dalian, China; GSGM, Gansu Geological Museum, Lanzhou, Gansu Province, China; HPG, Hupoge Amber Museum, Tenchong, China; IVPP, Institute of Vertebrate Paleontology and Paleoanthropology, Chinese Academy of Sciences, Beijing, China; LH, Museo de las Ciencias de Castilla-La Mancha, Cuenca, Spain; MCCMLH, Museo de las Ciencias de Castilla-La Mancha, Cuenca, Spain; SDM, Shandong Museum, Jinan, China; STM, Shandong Tianyu Museum of Nature, Pingyi, Shandong, China; UFRJ-DG, Universidade Federal do Rio de Janeiro, Department of Geology collection, Rio de Janeiro, Brazil.

9.1 Introduction

The study of fossil birds begins with feathers. Described in 1861, the original holotype of *Archaeopteryx lithographica*, still the oldest and most primitive known avian taxon, consists of an isolated feather (von Meyer 1861). At this time, only birds were known to possess this unique integumentary structure, and thus this Late Jurassic feather was identified as avian. Skeletal specimens were later referred to *Archaeopteryx* based on the co-occurrence of preserved feathers arranged on the forelimbs forming wings (Dames 1884; Owen 1863). However, with the exception of *Archaeopteryx*, fossil feathers associated with skeletal remains are otherwise very rare in the Mesozoic fossil record and, until the discovery of feathered nonavian dinosaurs and birds in the rich Early Cretaceous Jehol Biota in the late 1990s (Chen et al. 1998; Hou et al. 1996; Ji et al. 1998), the plumage of early birds was largely unknown. Many specimens from the Lower Cretaceous volcanolacustrine deposits in northeastern China that produce the Jehol Biota preserve traces of soft tissues normally rare in the fossil record (Zhou et al. 2003). Thousands of specimens preserving remnants of feathers have been uncovered, most still waiting to be described. These discoveries provide an ever-increasing wealth of data on the plumage of early birds (Zhang et al. 2006). However, because most data comes primarily from a single Biota, albeit one that persisted for over 10 million years (Pan et al. 2013), evolutionary trends across greater periods of time are poorly understood, and it is difficult to make broad statements regarding clades not limited to the Jehol.

The Jehol avifauna accounts for approximately half of all known Mesozoic avian diversity, with

J. O'Connor (✉)
Key Laboratory of Vertebrate Evolution and Human Origins, Institute of Vertebrate Paleontology and Paleoanthropology of the Chinese Academy of Sciences, Beijing, China

CAS Center for Excellence in Life and Paleoenvironment, Beijing, China
e-mail: jingmai@ivpp.ac.cn

C. Foth, O. W. M. Rauhut (eds.), *The Evolution of Feathers*, Fascinating Life Sciences,
https://doi.org/10.1007/978-3-030-27223-4_9

representatives from every known Early Cretaceous clade (Zhou and Zhang 2006). It occurs in three formations: the 130.7 Ma *Protopteryx*-horizon of the Huajiying Formation, the 125 Ma Yixian Formation, and the 120 Ma Jiufotang Formation (the latter two strictly forming the Jehol Group) (Pan et al. 2013). Avian diversity and numbers increase through time, being highest in the Jiufotang Formation where long boney-tailed birds (*Jeholornis*) live alongside primitive pygostylians (Sapeornithiformes and Confuciusornithiformes) and a diversity of ornithothoracines (the clade formed by Enantiornithes and Ornithuromorpha) (Zhou and Zhang 2006) (Fig. 9.1). The second oldest avian-bearing deposits in the world, the *Protopteryx*-horizon of the Huajiying Formation, records the earliest members of the Enantiornithes (Wang et al. 2014c), the dominant clade of Cretaceous avians, and the Ornithuromorpha (Wang et al. 2015a), the lineage that includes Neornithes nested within, as well as basal confuciusornithiforms (Navalón et al. 2018; Zhang et al. 2008a; Zheng et al. 2017). Although birds are rare in this horizon (only seven taxa reported represented by a total of ten specimens), the remains of feathers preserved in these specimens are among the best known from China (Wang et al. 2014c; Liu et al. 2019).

Besides *Archaeopteryx*, nearly all skeletal specimens associated with feather remains known from outside the China are referable to the Enantiornithes with the exception of feathers associated with the holotype of *Ambiortus dementjevi* from Early Cretaceous deposits in central Mongolia (Kurochkin 1985) and potentially purported feather traces in a specimen of *Parahesperornis* from the Late Cretaceous of Kansas (Martin 1984; Williston 1896). Elsewhere in China, feathered specimens have also been collected from the Changma locality of the Aptian Xiagou Formation (e.g., enantiornithines *Feitianius paradisi* and GSGM-07-CM-001, and specimens of the ornithuromorph *Gansus yumenensis*) (O'Connor et al. 2012a, 2016a; You et al. 2006). In addition, the only reported Late Cretaceous bird, the holotype of the

Fig. 9.1 Reconstruction of the 120 Ma Jiufotang Avifauna by Michael Rothman

enantiornithine *Parvavis chuxiongensis* from the Jiangdihe Formation, preserves feather remains (Wang et al. 2014b). Outside China, several enantiornithines preserving feathers are known from the Early Cretaceous of Spain: the holotype of *Eoalulavis hoyasi* (Sanz et al. 1996), the pellet LH 11386 consisting of several juveniles (Sanz et al. 2001), and the isolated wing MCCMLH31444, all from the 125 Ma Las Hoyas Konservat-Lagerstätten (Navalón et al. 2015) and the holotype of *Noguerornis gonzalezi* from the Upper Berriasian–Lower Barremian "La Pedrera de Meiá" Limestones (Chiappe and Lacasa-Ruiz 2002; Lacasa-Ruiz 1988). The partial specimens representing the holotype materials of *Hoboltia ponomarenkoi* from Mongolia (Zelenkov and Averianov 2016) and *Enantiophoenix electrophyla* from Lebanon (Cau and Arduini 2008; Dalla Vecchia and Chiappe 2002) also preserve feather traces, but these are uninformative. An indeterminate juvenile (UFRJ-DG 031Av) from the Aptian Crato Formation in Brazil preserves excellent tail feathers (de Souza Carvalho et al. 2015), and most recently, associated skeletal and integumentary remains including feathers have been found preserved in mid-Cretaceous Cenomanian (~99 Ma) amber from Myanmar (Xing et al. 2016, 2017, 2018b). The latter specimens contribute important morphological information typically inaccessible from two-dimensional specimens. The cumulative knowledge distilled from these discoveries is summarized here.

9.2 Descriptions

9.2.1 *Archaeopteryx*

Impressions of feathers are preserved in several specimens of *Archaeopteryx lithographica* (Foth et al. 2014; Mayr et al. 2005). The plumage of this important fossil avian is thoroughly described in the literature, including a chapter in this volume; thus, this summary will be concise. The short, broad wings are surprisingly modern in morphology (Elzanowski 2002). The primary feathers are asymmetrical, a feature absent in the wings of nonavian dinosaurs (with the exception of the volant dromaeosaurid *Microraptor*). Although the leading-edge vane displays barb angles comparable to modern birds, those of the trailing edge vane are markedly smaller, also documented in basal pygostylians *Sapeornis* and *Confuciusornis* (Feo et al. 2015). There are 10–12 primaries (Foth et al. 2014) and an estimated 11 secondaries (Elzanowski 2002) which is within the range of living birds (most neognathus birds have 10 primaries with secondaries showing a greater range) (Gill 2007). The longest primary is less than twice the length of the manus. The dorsal coverts are half the length of the primaries, similar to living birds (Foth et al. 2014). The body coverts are symmetrical pennaceous feathers with thin, gently curved rachises. The crural feathers are roughly half the length of the tibia (Foth et al. 2014). Very short feathers, approximately one quarter the length of the crural feathers, are present on the proximal tarsometatarsus (Foth et al. 2014). The elongate boney tail bears symmetrical pennaceous feathers perpendicularly oriented from the bone surface along its entire length. Melanosomes sampled from the holotype, identified as a wing covert, suggest this feather was black (Carney et al. 2012).

9.2.2 *Jeholornis* and Kin (the Jeholornithiformes)

The Jeholornithiformes are the only definitive long-tailed birds currently recognized other than *Archaeopteryx* (O'Connor et al. 2011) (Fig. 9.2). Remnants of feathers are preserved in approximately a third of all known specimens (38 out of 95 reported specimens), but only the tail has been described (O'Connor et al. 2013). Notably, body feathers are only preserved in a single specimen (STM2-11) along the dorsal surface of the neck and nowhere else (O'Connor et al. 2013). The wings (incomplete but best preserved in STM2-37, STM2-8, STM2-23, and STM2-18) appear to be proportionately longer than in *Archaeopteryx* (longest primary twice the length of the hand) but are similarly broad (Fig. 9.2a). Two rows of dorsal coverts are clearly present over the

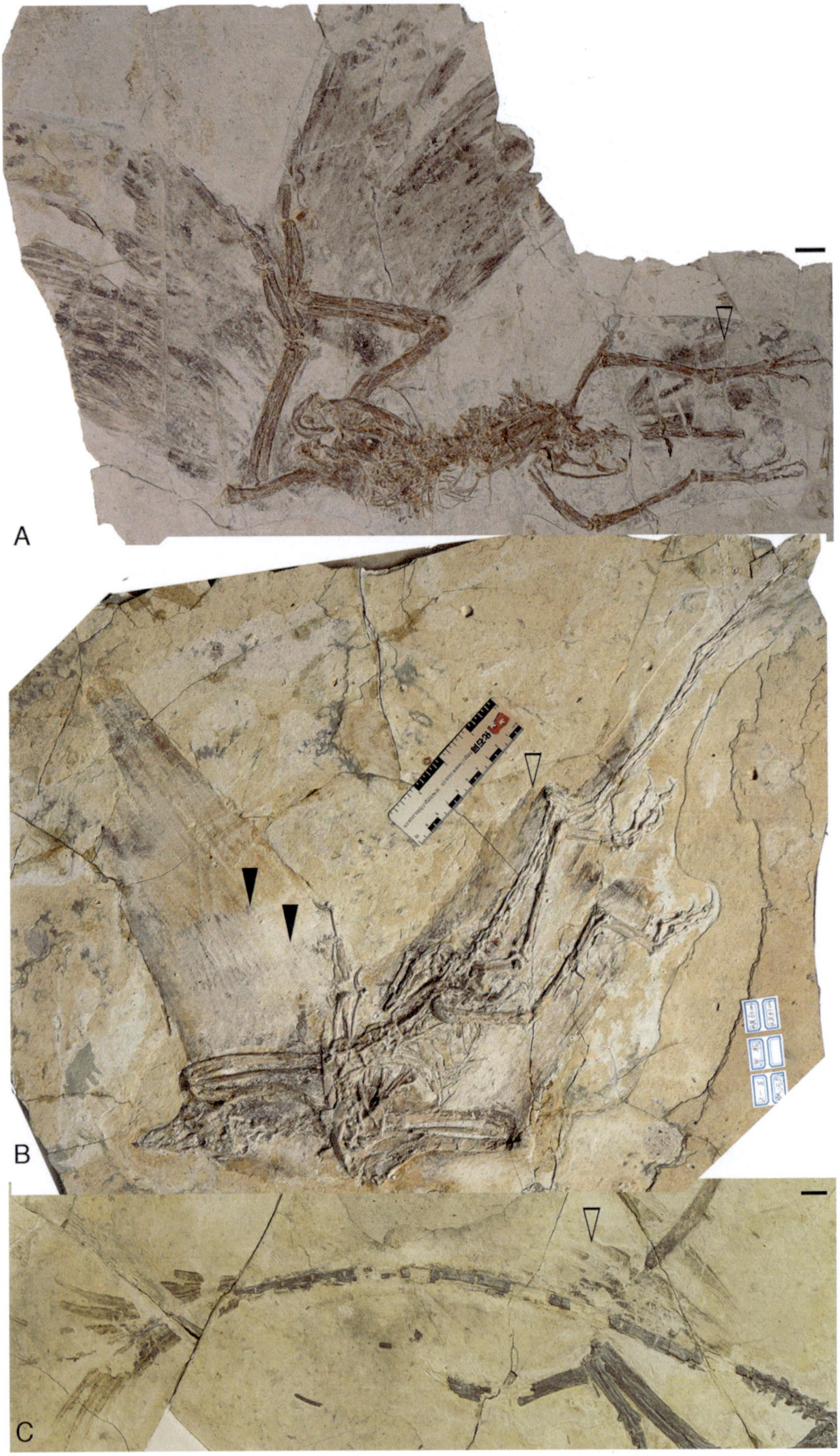

Fig. 9.2 Photographs of specimens of the long boney-tailed bird *Jeholornis*: (**a**) STM2-37 preserving wings and proximal tail feathers; (**b**) STM 2-8 preserving remiges and wing coverts and both tail tracts; (**c**) SDM 20090109.1 preserving proximal and distal tail feathers. Scale bars equal 2 cm. The two rows of secondary wing coverts are indicated by solid arrows; the proximal tail tract is indicated by an open arrow

secondaries in STM2-8 (Fig. 9.2b); the longer layer is half the length of the secondaries and the shorter layer is half that. The tail consists of two rectricial pterylae: a short proximal fan and a distally restricted "frond" (Fig. 9.2). The proximal pteryla is located over the short proximal caudal vertebrae and consists of six rectrices roughly 10 cm long with broad distal ends overlapping to form a fan-shaped surface (O'Connor et al. 2013). This surface is considered capable of generating lift to supplement the wings. The distal frond consists of five or six pairs of slightly shorter feathers perpendicularly oriented from the lateral surface of the distal vertebrae and a single feather attaching to the distal margin of the last caudal. Compared with the proximal tail feathers, these distal tail feathers are much more narrow, sharply tapered, and strongly curved so that their craniolateral surfaces are concave (Fig. 9.2c) (O'Connor et al. 2012b). The strong curvature and tapered morphology of these feathers suggests they were primarily ornamental. The tapered and narrow morphology of these feathers would also have decreased any incurred aerodynamic cost of this ornamentation (Møller and Hedenström 1999). However, the small cohesive surface formed by the proximal overlap of these distal tail feathers together with the overall elongated boney tail of *Jeholornis* (27 vertebrae compared to 22 in *Archaeopteryx*) suggests the distal frond also functioned as a stabilizer, generating pitching moments when the tail was raised or depressed (O'Connor et al. 2013). The feathers are considerably elongated past the point where they are aerodynamically functional, suggesting the distal rectricial pteryla was shaped by both natural and sexual selection (Fitzpatrick 1999). Compared with the "frond-like" morphology present in *Archaeopteryx*, the two rectricial tracts in *Jeholornis* are inferred to have both contributed less to the total body mass and incurred less drag, while at the same time generating approximately the same amount of lift (O'Connor et al. 2013). Due to the absence of a pygostyle, the complex caudal plumage of *Jeholornis* is unlike that of any living bird. However, the presence of a tail crafted through both natural and sexual selective forces parallels observations in neornithines (Thomas 1997).

9.2.3 *Confuciusornis* and Kin (Pygostylia: Confuciusornithiformes)

The basal pygostylian *Confuciusornis sanctus* is the most common species found in the Jehol Group, known from hundreds (if not thousands) of specimens mostly from the Yixian Formation (Chiappe et al. 1999; Hou et al. 1995). Roughly 300 preserve feather remains (Zheng et al. 2013b). Their narrow wings are proportionately much more elongate than in other early birds. The longest primaries are roughly three times the length of the hand, whereas they are only about twice the length of the hand in *Jeholornis*. The primaries are much longer than the secondaries forming a novel wing shape not observed in living birds (Falk et al. 2016). Descriptions of their rachis as thin and too weak to support powered flight are considered incorrect, based on measurements not representing the full rachis (Falk et al. 2016; Nudds and Dyke 2010; Zheng et al. 2010). Similar to rachis-dominated feathers, the rachis of the remiges preserves a faint, medially placed longitudinal stripe (STM 13-327). This may represent a ventral groove the same as that observed in the primaries of extant birds (Stettenheim 2000). However, the medial stripe appears to be visible in both ventral and dorsal view, which is somewhat inconsistent with this interpretation. The outermost primary is less than half the length of the second. The third through fifth outermost feathers are the longest (peaking in length in the fourth outermost primary), approximately 20–25% longer than the second primary, after which they decrease in length (IVPP V13156) (Fig. 9.3). The wing preserved in IVPP V13156 consists of 10 asymmetric and sharply tapered primaries (Falk et al. 2016). The secondaries are approximately half the length of the longest primary; they are symmetrical with rounded distal margins (IVPP V13156). Two layers of wing coverts are clearly preserved in

Fig. 9.3 Photographs of Confuciusornithiformes: (**a**) *Confuciusornis sanctus* IVPP V13156 showing the right wing in dorsal view and the proximal ends of the paired rachis dominated tail feathers; (**b**) *Eoconfuciusornis* sp. STM7-144 preserving dark spots on the tips of the greater coverts and secondaries—note the preservation of ovarian follicles and absence of elongate tail feathers, proximal ends of which are visible in (**a**). Scale bar equals 3 cm

Eoconfuciusornis STM7-144 (Zheng et al. 2017). The greater coverts are roughly half the length of the secondaries, and the lesser coverts are half the length of the greater coverts (Fig. 9.3). The distal ends of the secondaries and greater coverts in *Eoconfuciusornis* preserve dark spots that appear to be more heavily melanized, which both resists wear and serves as ornamentation (Fig. 9.3b) (Zheng et al. 2017). Some specimens of *Confuciusornis* also preserve direct evidence of their plumage patterns, preserving similar dark circular spots on the tips of some wing feathers. An alula is not preserved in any *Confuciusornis* (Chiappe et al. 1999) although *Eoconfuciusornis* STM7-144 preserves a single short pennaceous feather on the alular digit, probably representing a contour feather (Zheng et al. 2017).

Body feathers cover the dorsal and ventral neck and skull except the rostrum, which was presumably covered in a horny beak although this is very rarely preserved and even then only visible as a faint impression (Hou et al. 1999; Falk et al. 2019). Although only three specimens preserve traces of this feature, the impressions vary considerably in shape suggesting at least some different species were characterized by differences in beak shape (Falk et al. 2019). The dorsal coronal body feathers are longer than the ventral submalar feathers. In *Eoconfuciusornis* STM7-144 phaeomelanosomes were only recovered in the submalar tract suggesting a rufous throat patch (Zheng et al. 2017). However, samples within a single feather of *Confuciusornis* show both brown phaeomelanosomes and black eumelanosomes suggesting complex coloration (Zhang et al. 2010). Crural feathers decrease in length distally along the tibiotarsus and end proximal to the ankle (Fig. 9.3). They are reported to be pennaceous (Zheng et al. 2013b); although this interpretation is most likely correct, the elongate feathers assigned to the crus by Zheng et al. (2013b) and those on the proximal tarsometatarsus are reinterpreted here as remnants of the underlying wing feathers. Body feathers cover the dorsal and ventral surfaces of the body including the pygostyle (Fig. 9.3). Some specimens additionally preserve a pair of elongate rachis-dominated feathers (Chiappe et al. 1999), so named because the rachis is proportionately wide although this characteristic is more strongly developed in enantiornithines (Wang et al. 2014c) (Fig. 9.3). The feathers preserve a faint longitudinal stripe that extends nearly the entire length of the rachis (O'Connor et al. 2012a). This has been interpreted as a longitudinal groove, but it is now clear from three-dimensionally preserved RDFs found in amber that the entire rachis was extremely thin and concave and that the dark stripe represents a dorsal groove/ventral ridge in the cortex (Xing et al. 2018a; Carroll et al. 2019). The pennaceous vane in the paired tail feathers is distally restricted; referred to as racket plumes, this morphology is also present in numerous living birds (e.g., Momotidae, racket-tailed drongo). In *Confuciusornis* the distal "racket" is narrow with a gently tapered distal margin. Although described as symmetrical (Chiappe et al. 1999), the medial vade is typically preserved wider than the lateral vane (IVPP V13156). The vane proximally tapers and is continuous with a very narrow dark strip that lines the rachis for most of its length; this has been hypothesized to represent short barbs that were not differentiated through apoptosis (O'Connor et al. 2012a). Specimens preserving these elongate tail feathers are generally regarded as males (Hou et al. 1996). Interpretations regarding sexual dimorphism are supported by the absence of racket plumes in one specimen of *Eoconfuciusornis* preserving ovarian follicles (Zheng et al. 2017). Gender identification relying on purported medullary bone in a rectrix-less specimen of *Confuciusornis* is not supported here (Chinsamy et al. 2013; O'Connor et al. 2018).

9.2.4 *Sapeornis* and Kin (the Sapeornithiformes)

Sapeornis chaoyangensis is the largest pygostylian bird in the Early Cretaceous with proportionately elongate forelimbs (Zhou and Zhang 2002, 2003). Specimens, most of which come from the Jiufotang Formation, rarely preserve any remnants of their plumage (20 of 106 specimens) (Zheng et al. 2014b). The wings preserved in DNHM D3038, a young subadult specimen, consist of 11–13 elongate strongly asymmetrical primaries and approximately 11 secondaries (Gao et al. 2012). The outermost feather along the cutting edge of the wing is approximately half the length of the second, and the third-fourth outermost primaries are the longest after which the primaries reduce in length. The longest primary is approximately twice the length of the hand (Fig. 9.4a). The secondaries are shorter and subequal (Gao et al. 2012). An alula cannot be observed in this or any other known specimen. Aerodynamic analysis suggests *Sapeornis* utilized its broad wings for thermal soaring (Serrano and Chiappe 2017). Wing coverts are present, but details are poorly known, and they are described as small and thin (Gao et al. 2012). The tail is preserved in a single described specimen preserving what appears to be an elongate, graduated tail fan in lateral view (Wang et al. 2014c; Zheng et al. 2013b) (Fig. 9.4b). Two additional specimens confirm this interpretation and indicate the tail fan was formed by 8–10 medially striped rectrices (Wang and O'Connor 2017). This case is consistent with their proportionately short and plough-shaped pygostyle with a pygostyle lamina, features otherwise only present in ornithuromorphs and consistent with the presence of specialized rectricial musculature (Wang and O'Connor 2017). Although not preserved in any specimen, body feathers clearly would have been present. Two subadult specimens preserve feathers on the distal hindlimb projecting off of the caudal margin of the intertarsal joint (Zheng et al. 2013b). The feathers roughly extend the distal half of the tibiotarsus and the full length of the tarsometatarsus, becoming shorter distal on the hindlimb (Fig. 9.4b). They are curved, and their distal ends are splayed, suggesting these hindlimb feathers had ornamental qualities, whereas the preservation of these ankle feathers relative to other body feathers may suggest some aerodynamic qualities (O'Connor and Chang 2015). Most likely these feathers were shaped under both natural and sexual selection.

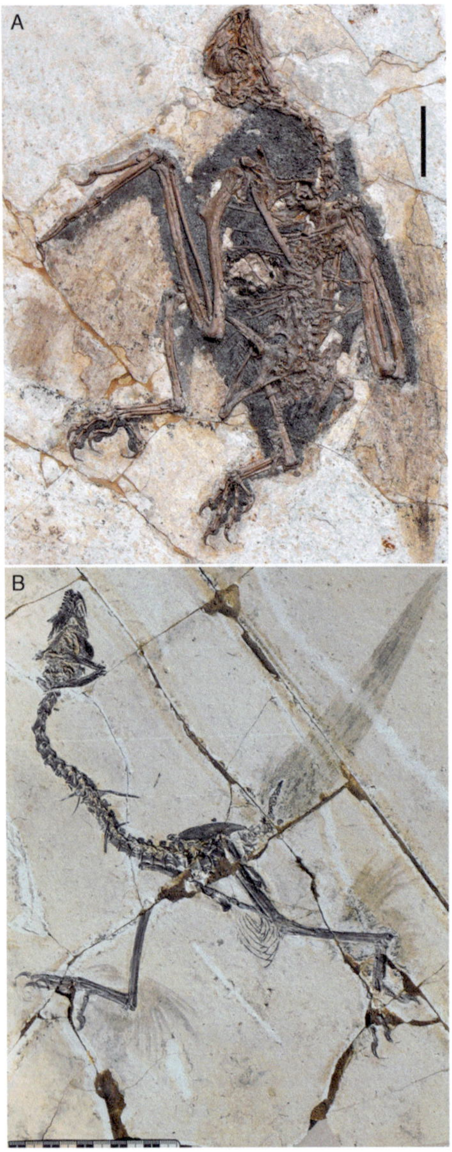

Fig. 9.4 Specimens of the basal pygostylian *Sapeornis chaoyangensis*: (**a**) 41HIII0405 preserving carbonized body tissue and impressions of the wings (scale bar equals 4 cm); (**b**) STM16-18 preserving tail and hindlimb feathers. Both are subadult specimens

9.2.5 The Diverse Enantiornithes (Aves: Ornithothoraces)

The enantiornithines have a global record that extends from the 130.7 Ma Huajiying Formation until the end Cretaceous and includes data from every continent with the exception of Antarctica (O'Connor et al. 2011). Their diversity and abundance is reflected by the available data regarding their plumage. Over a thousand specimens at the STM reportedly preserve some feather remains and most have yet to be studied (Zheng et al. 2013b). Remains of the remiges are preserved in the holotype specimens of *Chiappeavis magnapremaxillo* (O'Connor et al. 2016b), *Cruralispennia multidonta* (Wang et al. 2017), *Dapingfangornis sentisorhinus* (Li et al. 2006), *Eoenantiornis houi* (Zhou et al. 2005), *Eopengornis martini* (Wang et al. 2014c), *Fortunguavis xiaotaizicus* (Wang et al. 2014a), *Junornis houi* (Liu et al. 2017), *Grabauornis lingyuanensis* (Dalsätt et al. 2014), *Longirostravis hani* (Hou et al. 2004), *Noguerornis gonzalezi* (Lacasa-Ruiz 1988), *Orienantius ritteri* (Liu et al. 2019), *Parapengornis eurycaudatus* (Hu et al. 2015), *Protopteryx fengningensis* (Zhang and Zhou 2000), *Shanweiniao cooperorum* (O'Connor et al. 2009), and *Zhouornis hani* (Zhang et al. 2013), as well as in a bohaiornithid CUGB P1012 (Peteya et al. 2017), and two indeterminate specimens, DNHM D2884 (O'Connor 2009) and MCCMLH31444 (Navalón et al. 2015); three-dimensional specimens encased in amber include two partial wings (Xing et al. 2016), two partial skeletons (Xing et al. 2017, 2018b), a feathered foot (Xing et al. 2019a), and an isolated hindlimb with remiges—the holotype of *Elektorornis chenguangi* (Xing et al. 2019b). Most completely preserved in *Eopengornis martini* STM24-1 (Fig. 9.5a), the wings are broad, formed by asymmetrical rectrices with medially striped rachises (Wang et al. 2014c). The outermost primary is two-thirds the length of the next, which is two-thirds the length of the third outermost primary. These are followed by two longer primaries, after which the medial primaries are shorter. The longest primaries are twice as long as the hand. Unlike *Confuciusornis*, the secondaries are not significantly shorter than the primaries in *Eopengornis*, forming a broader wing with a larger surface area (Figs. 9.3a and 9.5a). The primaries display both leading and trailing edge vane with barb angles comparable to living birds (Feo et al. 2015). So far there is no strong evidence that the caudal margin of the wing was notched in any taxon. A distally incomplete but well preserved isolated partial wing from the Early Cretaceous of Spain preserves 8–9 primaries and 10–12 secondaries (Navalón et al. 2015). *Orienantius* BMNHC PH 1156 preserves the calami for 10 primary remiges visible in the post-patagium (Liu et al. 2019). Nine primaries form the wing in the more complete of the two amber specimens (Xing et al. 2016) but only eight are present in the hatchling HPG-15-1 (Xing et al. 2017). An alula is preserved in several specimens (e.g., *Eoalulavis hoyasi*, *Eoenantiornis houi*, *Protopteryx fengningensis*, and DIP-V-15100) and is considered plesiomorphic to Ornithothoraces and present throughout the clade (O'Connor 2009; Xing et al. 2016; Zhang and Zhou 2000; Zhou et al. 2005). As in some living birds, the alula is formed by two feathers that end approximately level with the distal end of the major digit (Xing et al. 2016) and a shorter proximal feather (Sanz et al. 1996). Wing coverts are more heavily obscured by overlap. Three rows are reported in DIP-V-15100 (Xing et al. 2016). The longest coverts appear to be approximately half the length of the remiges (Fig. 9.5a). The Burmese amber specimens further reveal that the contour feathers in at least some enantiornithines had proximodistally specialized barb morphologies, being plumaceous at the base as in living birds (Xing et al. 2016). Analysis of the wings preserved in *Junornis houi* suggests this taxon engaged in bounding flight, typical of extant arboreal passerines (Liu et al. 2017).

Nearly complete body feathers are present in some specimens, but preservation results in so much overlap that very little information is available regarding most pterylae. Body feathers are clearest along the neck, tail, and tibiotarsus, where there is less anatomical overlap (Fig. 9.5a). Generally, body feathers vary somewhat in length being longer on the dorsal surface of the body and around the pygostyle and shorter on the ventral surface of the body (a pattern generally observed in all Mesozoic birds). They are reported to be rachis-less with the exception of the crural feathers (O'Connor 2009; Zhang and Zhou 2004; Zhang et al. 2006) although this is likely a taphonomic artifact (see Sect. 9.3). Where preserved, these extend the full length of the tibiotarsus. Although crural feathers are short and weakly decrease in length distally in some taxa (e.g., *Eopengornis*; Fig. 9.5a) (Wang et al. 2014c), in others they are longer (40–60% the length of the tibiotarsus) and subequal the entire length of the tibiotarsus (e.g., *Longipteryx*, IVPP V13939, *Cathayornis* STM7-50) (Zhang and Zhou 2004; Zheng et al. 2013b). In some specimens, the crural feathers are preserved nearly perpendicular to the tibiotarsus, which has led some authors to infer an aerodynamic function for these feathers (Zheng et al. 2013b). However, body feathers on other parts of the body are similarly oriented (e.g., coronal and neck feathers), and this is interpreted as the result of compression during burial (Foth 2012), and not indicative of the presence of airfoils.

In one specimen (BMNHC-Ph1061B, Huajiying Formation), the hindlimb feathers extend onto the proximal two-thirds of the tarsometatarsus, where they continue to decrease in length distally (Chiappe and Meng 2016). An isolated foot trapped in amber (DIP V15105) from Burma preserves pennaceous feathers that extend onto the pedal digits themselves, present on digits III and IV and reaching the ungual phalanx on digit IV (Xing et al. 2019b). The pedal scales in DIP V15105, HPG-15-1 and Elektorornis chenguangi additionally have unusual projecting filaments; referred to as scutellate scale filaments, these structures are reportedly white, which may hinder their preservation and/or observation in most two-dimensional specimens (Xing et al. 2017, 2019a, 2019b). It has been suggested these scutellate scale filaments may have had a sensory function (Xing et al. 2019b).

Bizarre feathers, described as having a "proximally wire-like part with a short filamentous distal tip," preserved on the wing and hindlimb in *Cruralispennia* from the Huajiying Formation (Wang et al. 2017) may alternatively represent pin feathers (developing feathers still encased in the feather shaft) (O'Connor et al. 2020) or unusual taphonomic artifacts resultant from the

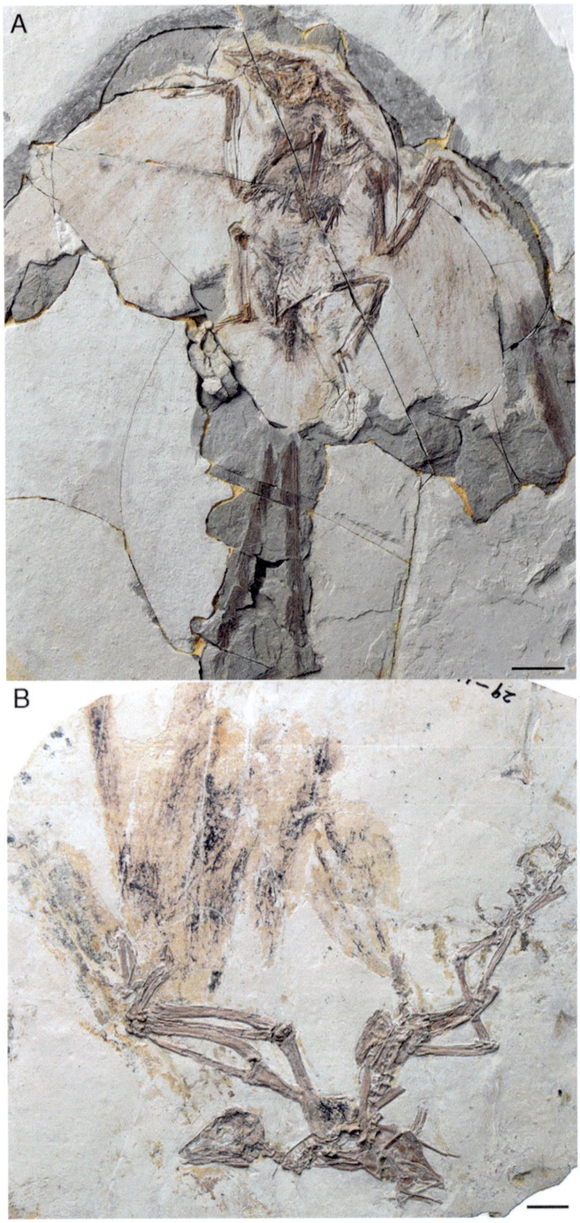

Fig. 9.5 Two pengornithid enantiornithines: (**a**) *Eopengornis martini* from the 130.7 Ma *Protopteryx*-horizon of the Huajiying Formation; (**b**) *Chiappeavis magnapremaxillo* from the 120 Ma Jiufotang Formation. The tail in the older taxon consists of a pair of elongate rachis-dominated streamers, whereas that of the younger taxon consists of a fan-shaped tail formed by eight rectrices. Scale bars equal 2 cm

lacustrine depositional environment (Kundrát 2004). Alternative interpretations are more consistent with their anatomical location in that ornamental feathers projecting craniolaterally from the wrist would presumably impede flight. Although their identification remains equivocal, the unusual feathers preserved in *Cruralispennia* strongly resemble pin feathers preserved in a juvenile enantiornithine STM34-1 (Zheng et al. 2012) (O'Connor et al. 2020). These feathers are preserved on several parts of the body. In STM34-1, the pin feathers are clearest in the caudal region and on the tibiotarsus but are also present on the forelimb, as in *Cruralispennia*, and on the dorsal surface of body. Notably, STM34-1 also appears to preserve some of the original patterning of the feathers; the distal portions of the remiges are preserved as a distinctly darker color suggesting the proximal and distal portions of the feathers were colored different in vivo.

Of the feather pterylae, the morphology of the rectrices is best documented and shows considerable diversity. Most commonly, as in *Confuciusornis*, a pair of rachis-dominated feathers is present, although compared with *Confuciusornis*, the rachis is proportionately even wider in enantiornithines (Wang et al. 2014c). Although for some taxa all that is known is a faint incomplete impression of the enlarged rachis (e.g., *Bohaiornis guoi* and *Protopteryx fengningensis*), enantiornithine rachis-dominated feathers show greater diversity than the confuciusornithiforms (which appear to share a single morphology): both streamers (fully pennaceous) and racket plumes (distally pennaceous) are documented (the fully pennaceous morphology is currently only documented in two species of the Pengornithidae, *Eopengornis martini* and *Parapengornis eurycaudatus*) (Hu et al. 2015; Wang et al. 2014c) (Fig. 9.5a). Within the racket-plume morphology, there is further diversity in the overall length of the feathers relative to the body, the number of pairs (one specimen reportedly preserves two, the holotype specimen of *Paraprotopteryx zhengi*, and *Shanweiniao cooperorum* is interpreted as having a similar morphology) (O'Connor et al. 2016b; Zheng et al. 2007) and the morphology of the distal "racket" (O'Connor et al. 2012a). In some taxa, the pennaceous portion is restricted to the distal most portion of the rachis (*Junornis*), rapidly expanding with a rounded distal margin (spoon-shaped as in *Dapingfangornis*), whereas the pennaceous portion expands gradually and extends for a greater proportion of the feather in others (e.g., GSGM-07-CM-001, an unnamed enantiornithine from the Xiagou Formation). Many specimens preserve an absence of rectrices (holotype specimens of *Avimaia schweitzeri*, *Cruralispennia multidonta*, *Fortunguavis xiaotaizicus*, *Eoenantiornis houi*, *Grabauornis lingyuanensis*, *Longipteryx chaoyangensis*, and *Zhouornis hani* and an indeterminate enantiornithine preserving follicles STM29-8), having only wispy body feathers preserved around the pygostyle. As in confuciusornithiforms, these specimens are regarded as females, similarly supported by the preservation of female reproductive tissue in one specimen (Zheng et al. 2013a) and the presence of an egg in *Avimaia* IVPP V25371 (Bailleul et al. 2019).

Outside China, rachis-dominated tail feathers are also preserved in "*Cratoavis*" UFRJ-DG 031 Av from Brazil (de Souza Carvalho et al. 2015) and numerous pairs of rachis-dominated feathers have been found in Burmese amber (Xing et al. 2018a; Carroll et al. 2019) indicating that this morphology was widespread and persisted in enantiornithines for over 30 million years. This specimen preserves the feathers primarily as a mold, confirming that the rachis is grooved along at least one of its surfaces, with the enlarged portion of the rachis on either side of the groove forming convex surfaces (de Souza Carvalho et al. 2015). Three-dimensional specimens preserved in amber indicate that the traces in UFRJ-DG 031 are dorsally exposed and flattened. The amber specimens reveal that the wide rachis observed in compression fossils in vivo was C shaped with a thickness between 3 and 10 microns (Xing et al. 2018a; Carroll et al. 2019). Unusual spots on the rachis suggest these ornamental tail feathers bore striking color patterns (de Souza Carvalho et al. 2015). These tail feathers have been documented in numerous young juvenile specimens (e.g., UFRJ-DG 031 Av, STM34-7, STM34-9, IVPP V15564) indicating that unlike living birds, enantiornithines developed sexually dimorphic ornaments at a very early ontogenetic age, before the onset of both sexual and skeletal maturity (Zheng et al. 2012). One specimen in amber (HPG-15-1) suggests that these feathers may even appear as early as the first postnatal molt (Xing et al. 2017).

Two specimens clearly preserve more than four elongate rectrices, the holotype specimens of *Feitianius paradisi* (O'Connor et al. 2016a) and *Chiappeavis magnapremaxillo* (O'Connor et al. 2016b). The latter is a pengornithid preserving a short, graded fan-shaped tail consisting of ten overlapping rectrices (Fig. 9.5b). This tail is considered to have aerodynamic qualities and be capable of generating lift although the presence of rectricial bulbs like those of neornithines is unlikely, being confirmed absent using UV light in several other enantiornithines (Liu et al. 2019; Chiappe and Meng 2016) and thus this tail may have been incapable of adjusting its width to suit aerodynamic demands during flight (O'Connor et al. 2016b). *Feitianius paradisi* GSGM-05-CM-004 preserves a tail consisting of at least three different rectricial morphologies, none of which are considered to have any strong aerodynamic properties (Fig. 9.6). Together they are inferred to form a sizeable tail display, the likes of which are only observed among living birds in males of polygamous species (O'Connor et al. 2016a). *Chiappeavis* and *Feitianius* represent the opposite extremes observed within the Enantiornithes, although notably ornamental morphologies dominate the spectrum. A recent study on the pygostyle has suggested that the dorsally excavated morphology typical of enantiornithines may have accommodated an enlarged caudal levator muscle, which it may have used to raise the tail in display similar to extant phasianids (Wang and O'Connor 2017) (Fig. 9.6c). *Cruralispennia* apparently had a small, "ornithuromorph-like" plough-shaped pygostyle but preserves only short caudal coverts (Wang et al. 2017). Additional specimens are required to clarify the pygostyle morphology in this unusual taxon.

A nearly complete juvenile trapped in amber (HPG-15-1) preserves the most complete plumage of any enantiornithine, documenting the early juvenile plumage present in at least one lineage of enantiornithines (Xing et al. 2017). This specimen reveals well-developed remiges (with banded color pattern) combined with sparse body feathers that consist of both pennaceous and filamentous morphotypes. The head and neck are covered with short, dark tufts consisting of 3–4 basally joined filaments which may represent newly erupting contour feathers. Body contour feathers appear to be proximally plumaceous and distally pennaceous. The femur and tibiotarsus have neoptile feathers consisting of basally joined plumaceous barbs with a short or absent rachis with additional isolated bristlelike filaments also present on the crus (Xing et al. 2017). The scutellae scales on the tarsometatarsus bear filamentous projections also present in some extant juvenile birds (Xing et al. 2017). The relatively sparse covering of body feathers is reminiscent of altricial hatchlings, whereas the mature remiges is consistent with the extreme precocial strategy inferred for the Enantiornithes (Zhou and Zhang 2004). Primitive filamentous morphotypes reminiscent of proto-feathers present in basal theropods may be lost during enantiornithine ontogeny or alternatively cannot be discerned in two-dimensionally preserved fossils due to overlap.

Thus far, only two specimens have been sampled for melanosomes: the holotype of *Cruralispennia* and an indeterminant bohaiornithid (Peteya et al. 2017; Wang et al. 2017). The latter analysis suggested the presence of iridescent plumage, as in the dromaeosaurid *Microraptor* (Li et al. 2012).

9.2.6 Earliest Members of the Ornithuromorpha (Aves: Ornithothoraces)

In the Jehol ornithuromorphs are generally less common than enantiornithines and thus comparatively less is known regarding their plumage. Remnants of remiges lacking anatomical detail are preserved in several specimens (the holotype specimens of *Hongshanornis longicresta*, *Jianchangornis multidonta*, *Juehuaornis zhangi*, *Longicrusavis houi*, *Piscivoravis lii*, *Schizooura lii*, *Tianyuornis cheni*, *Yixianornis grabaui*, and *Xinghaiornis lini* as well as referred specimens of *Yanornis* STM 9-15 and 9-49) (Wang et al. 2013,

Fig. 9.6 Enantiornithine from the Xiagou Formation, the holotype of *Feitianius paradisi* GSGM-05-CM-004: (**a**) photograph, (**b**) interpretative drawing, (**c**) reconstruction by M. Rothman. Scale bar equals 10 mm

2015b; Zheng et al. 2014a, c; Zhou et al. 2009, 2012, 2013; Zhou and Zhang 2001, 2005). Uninformative feather traces are also preserved in some specimens of *Gansus yumemensis* from the Xiagou Formation (Wang et al. 2015c; You et al. 2006) and the holotype of *Ambiortus dementjevi* from Mongolia (Kurochkin 1985). In no specimen is the wing complete and clearly preserved well enough to allow the number of remiges to be to be determined. The wing is best preserved in the two known specimens of the hongshanornithid *Archaeornithura meemanni*, the only ornithuromorph known from the *Protopteryx*-horizon (Wang et al. 2015a). The primaries are broad, strongly asymmetrical and bluntly tapered at their distal ends (Fig. 9.7). At least one short primary (possibly two) forms the leading edge of the wing; this outermost primary (or primaries) is narrower than the others and roughly two-fifths the length of the next primary, which is followed by two longer subequal feathers, after which the primaries decrease in length medial in the wing. The longest primary is approximately 2.5 times the length of the hand, proportionately longer than observed in other Early Cretaceous birds with the exception of *Confuciusornis*, whose manual digits are not reduced (Fig. 9.3). This morphology is confirmed, as least for hongshanornithids, in a well-preserved referred specimen of *Hongshanornis* (DNHM D2945/6) (Chiappe et al. 2014). The well-preserved primary feathers in *Archaeornithura* STM7-145 reveal the presence of a medial longitudinal stripe in the rachis, as in *Confuciusornis* (Wang et al. 2015a). Because the specimen is torn between two slabs with feather remains in both, it is unclear which surface bore the stripe, although this feature is more clearly visible in the main slab in which the skeleton is in dorsal view. A layer of coverts, approximately the length of the outermost primary, is present, but details of their morphology are obscured by overlap (Fig. 9.7). The morphology of these specimens indicates hongshanornithids had broad, long, and tapered wings. The alula (best preserved in specimens of *Archaeornithura*) is formed by three short asymmetrical feathers, within the range of modern birds (Lovette and Fitzpatrick 2004), ending proximal to the distal end of the major digit (proportionately shorter than in enantiornithines) (Wang et al. 2015a).

Body feathers are known on the head and neck (Fig. 9.7) but notably are not clearly documented on the tibiotarsus in any specimen except one, despite the fact that crural feathers are widespread in living members (Neornithes) and present in other groups of Early Cretaceous birds. Given that crural feathers are rarely preserved even in *Confuciusornis*, their absence in basal ornithuromorphs most likely reflects the comparatively smaller number of available specimens (Zheng et al. 2013b). Crural feathers are only preserved in a single exceptionally well-preserved specimen of *Yanornis* STM9-5, in which the feathers appear to be short and taper off above the ankle—the pes is clearly covered in scales (Zheng et al. 2013b). This is consistent with the semiaquatic ecology inferred for *Yanornis* and most basal ornithuromorphs—neornithines in aquatic and semiaquatic habitats commonly have proximally restricted and/or reduced crural feathers. The capital coverts in the holotype of *Hongshanornis longicresta* project dorsally, giving the species its specific name (Zhou and Zhang 2005); although incomplete preservation above the skull roof makes interpretations of a crest equivocal, the feathers are clearly visible tapering rostrally from mid-orbit, ending level with the rostral margin of the antorbital fenestra. A similar crest is not observed in any other hongshanornthid specimen (Chiappe et al. 2014).

Morphology of the rectrices and pygostyle itself suggest that rectricial bulbs were present in even basal members of the Ornithuromorpha (Clarke et al. 2006). All well-preserved rectricial morphologies display undeniable aerodynamic qualities. Most taxa preserve round rectricial fans (*Archaeornithura meemanae*, *Hongshanornis longicresta*, *Piscivoravis lii*, *Tianyuornis zhengi*, *Yanornis martini*, *Yixianornis grabaui*) (Chiappe et al. 2014; Clarke et al. 2006; Wang et al. 2015a; Zheng et al. 2014a, c; Zhou et al. 2013), although *Schizooura lii*

Fig. 9.7 *Archaeornithura meemannae* (STM7-145) the only ornithuromorph known from the 130.7 Ma *Protopteryx* horizon of the Huajiying Formation. This taxon possesses both a modern alula and a fan-shaped tail indicating advanced flight capabilities

preserves a forked tail, considered to be the optimal aerodynamic condition (Thomas 1993; Zhou et al. 2012). One specimen of *Archaeorhynchus spathula* preserves a pin-tail, a fan-shaped tail consisting of 12 medially striped feathers paired with two medial streamers of indeterminate distal morphology (Wang et al. 2018). No unequivocal primarily ornamental tail morphologies are documented (Zhou et al. 2014). However, all tail morphologies observed appear to be also somewhat shaped under the influence of sexual selection: rounded tail fans are considered to be ornamental relative to a square-shaped tail fans; elongate forked tails, like that in *Schizooura*, are less functional (and thus considered also to function as ornamentation) relative to short forked tails; and pin-tails readily serve both ornamental and aerodynamic functions, with the ornamental pins being narrow to reduce their aerodynamic cost (Fitzpatrick 1999; Møller and Hedenström

1999). In the absence of taxa highly specialized for aerodynamic performance at this grade in avian evolution, the tails of the earliest ornithuromorphs appear to be, as in other groups, shaped by both natural and sexual selection. Wispy body feathers, possibly poorly preserved tail coverts, are preserved with the pygostyle in some specimens (e.g., *Yanornis* STM9-15). Notably, there is no clear evidence than any specimen lacked elongate rectrices in vivo, as in some enantiornithine and confuciusornithiform specimens.

Melanosomes have only been sampled in two specimens, the holotype of *Changzuiornis ahgmi* (potentially a junior synonym of *Juehuaornis zhangi*) and a referred specimen of *Gansus yummenensis* (Barden et al. 2011; Huang et al. 2016). Samples from both specimens revealed eumelanosomes, which suggest the feathers were at least partially black.

Younger and more derived ornithuromorph specimens (those referable to Ornithurae) thus far do not preserve any remnants of their plumage. One hesperornithiform, a Late Cretaceous clade of diving ornithurines, purportedly preserves feathers on the foot (Williston 1896); although the impressions of scales are clear in the illustration, the reported feathers are not readily identifiable, and no description of these traces accompanied a later study, in which the specimen was recognized as distinct from *Hesperornis* and assigned to *Parahesperornis* (Martin 1984). Instead, the hespeornithiform toes are described as, "probably lobed and scutellate-reticulate" (Martin 1984: p. 147).

9.3 Discussion

The plumage of the earliest known birds from the Late Jurassic and Early Cretaceous is remarkably well known despite the enormous taphonomic filter presented by the fossilization process and over a hundred million years of geologic time. Among known specimens, there is recognizable inter- and intraclade diversity in the wings, tail, and hindlimb feathers (O'Connor et al. 2016a; Zheng et al. 2013b). Not unexpectedly, cumulative data still pales in comparison to the diversity of feather morphologies, plumages, molts, and patterns recognized among living birds. The numerous flight styles, ecologies, and reproductive strategies utilized by the over 10,000 species of living birds provide for a vast diversity of feather structures, plumages, and wing and tail shapes (Gill 2007). Although studies suggest that by the Late Cretaceous enantiornithines had evolved a diversity of flight styles comparable to neornithines (Dyke and Nudds 2008), specimens preserving feathers from the well-sampled Jehol Biota appear to occupy a more limited ecospace including a limited range of body sizes. Many basal taxa appear to be fairly generalist (e.g., *Archaeopteryx*, *Confuciusornis*), whereas enantiornithines show arboreal features and ornithuromorphs show features indicative of semiaquatic ecologies (O'Connor et al. 2011). Morphological specializations of the skeleton for flight and specialized environments do not reach the grade observed in Neornithes. Similarly, one would not expect to find in early birds the same degree of plumage variation observed in living taxa.

Basal birds further lacked many of the advanced biological characteristics that set living birds apart from other living reptiles, such as their rapid growth, high metabolic rates, and advanced pneumatic respiratory system, all paired with an extremely lightweight skeleton highly adapted for aerodynamic activity (Chinsamy et al. 1995; Gill 2007). Non-ornithuromorph birds lacked an enlarged, keeled sternum, notarium, well-developed synsacrum, and other compound bones that form the lightweight and rigid avian skeleton, and grew more slowly (Chinsamy et al. 1995; O'Connor et al. 2011). As in other aspects of their biology, several observable differences are present between the plumage of basal birds and that of neornithines. These include at least one feather morphotype that no longer exists, plumages consisting of feather combinations not observed in living birds, tail morphologies unknown in living birds (some due to the absence of a pygostyle in the most primitive birds), and ontogenetic differences in the appearance of sexually dimorphic ornaments. This information, continuously built upon by new discoveries, helps us to begin to understand the early evolution of the diverse and often highly specialized plumage observed in modern birds.

At least one feather morphotype, the rachis-dominated rectrix, is recognized as extinct and is so far found only in Early Cretaceous confuciusornithiforms and enantiornithines (O'Connor et al. 2012a; Zhang et al. 2006), and is documented in specimens from China, Myanmar, and Brazil (de Souza Carvalho et al. 2015; Carroll et al. 2019; Xing et al. 2018a). Both streamer and racket-plume morphologies are documented in the Enantiornithes, whereas only racket plumes are observed in the Confuciusornithiformes. The rachis is proportionately wider in enantiornithines compared with the Confuciusornithiformes, and these tail feathers are inferred to have evolved independently in these two clades (Wang et al. 2014c). Potentially, rachis-dominated rectrices may also be present in *Epidexipteryx* (Zhang et al. 2008b), a member of the Scansoriopterygidae—a group of volant nonavian maniraptorans of uncertain phylogenetic affinity (O'Connor and Sullivan 2014; Xu et al. 2015). Notably, the poorly preserved remains of four potentially rachis-dominated (or ribbonlike) rectrices is the only evidence for the presence of feathers with a rachis (presumably pennaceous) in this group (Xu et al. 2015; Wang et al. 2019). This may suggest that the tail ornaments in *Epidexipteryx* are not modified pennaceous feathers at all. Similarities between these tail feathers may turn out to be a product of the poor preservation of this feature in the only specimen of *Epidexipteryx*. Given their phylogenetic distribution, these tail ornaments are inferred to have evolved in parallel in scansoriopterygids and pygostylian birds (Wang et al. 2014c).

Recently our understanding of rachis dominated feathers has been transformed by the discovery of three-dimensional rachis dominated feathers preserved in 99 Ma amber from Myanmar. Based on lithic compression fossils, these feathers appeared to consist of a mediolaterally enlarged rachis that bears either a dorsal or ventral longitudinal groove (potentially both), considered to have functioned to strengthen these elongate feathers. The presence of a groove was inferred from the presence of a dark medial longitudinal stripe on the rachis and confirmed through the discovery of a three-dimensional cast of one surface of the tail feathers in the juvenile enantiornithine *Cratoavis*. They are interpreted as modified pennaceous feathers (O'Connor et al. 2012a; Prum and Brush 2002), supported by the discovery of fully pennaceous "streamer" forms and the presence of a medial longitudinal stripe in the remiges of confuciusornithiforms, unmodified rectrices of *Sapeornis*, and the remiges and unmodified rectrices of ornithothoracines (Wang et al. 2014c, 2015a). Rachis dominated feathers in amber have shown that the wide, flat morphology actually formed a C-shaped cross-section (Xing et al. 2018a). The dark medial stripe represents a ventral ridge and a dorsal groove and may have represented the so called rachial ridge (Xing et al. 2018a; Carroll et al. 2019). The rachis measures only 3–10 microns thick and completely lacks medullary pith. The absence of a medullary pith suggests that these feathers are not modified modern pennaceous feathers and likely formed through completely different developmental pathways (Carroll et al. 2019). It is unclear whether the medial stripe observed in the rachis of the flight feathers of some well-preserved basal birds (e.g., *Eopengornis, Archaeornithura*) is indicative of a similar pith-less structure, which would in turn indicate that the feathers of basal birds were far more primitive that previously thought, or if the medial stripe in these flight feathers results from a different morphology. The remiges in three-dimensional amber specimens do not show any medial stripe or groove (Xing et al. 2017).

A large number of bizarre feather morphologies have been described from feathered paravians (the group of maniraptorans that includes birds, dromaeosaurids, and troodontids) found in Jehol deposits (Wang et al. 2017; Xu and Guo 2009; Xu et al. 2010). Within Pennaraptora, the monophyletic clade of maniraptoran dinosaurs with pennaceous feathers (formed by Oviraptorosauria and Paraves), descriptions of the body feathers as rachis-less (unshafted) are widespread (O'Connor et al. 2012a; Zhang et al.

2006). However, the identification of these unusual feathers in Early Cretaceous Jehol birds is at odds with the presence of pennaceous body feathers in the basal Late Jurassic *Archaeopteryx* (Foth et al. 2014) and in some mid-Cretaceous Burmese enantiornithines (Xing et al. 2016, 2017). Feathers in these specimens are notably preserved as molds and in their entirety, respectively. In contrast, feather traces preserved in Jehol specimens typically consist of decay-resistant melanosomes, organelles that are often absent from the rachis (Zhang et al. 2010). The rachis is thus preserved as a gap between the melanosome-rich vanes, although often marred by a dark medial stripe of uncertain identity. Overlap and slight taphonomic distortion may well be obscuring the presence of a vane in the small body feathers of Jehol birds making them only appear non-pennaceous (Foth 2012). Water also obscures feather morphology; hence, even features in isolated feathers can be controversial (Kundrát 2004), as in the case of the isolated feathers preserved in the holotype of *Protopteryx fengningensis*, which have been interpreted both as shafted and unshafted (Foth 2012; Zhang et al. 2006). Therefore, the existence of rachis-less body feathers as an extinct morphotype cannot be confirmed until taphonomic factors are explored further (Saitta et al. 2018). Based on observations from *Archaeorhynchus* STM7-11, the absence of an observable rachis in the body feathers of many specimens may be due to the fact this structure is proportionately thin and delicate (Wang et al. 2018). If a rachis was indeed present in the body feathers of Jehol birds, which seems most likely, it appears in some feathers it would have been basally restricted with very long barbs, giving the feathers a downy appearance (Zhang and Zhou 2000). Such a morphology is reminiscent of neoptile feathers in living birds (Foth 2011; Kundrát 2004) that have also been observed in Burmese enantiornithines trapped in amber (Xing et al. 2017). It is possible the downy morphology observed in the body feathers of Jehol birds and nonavian dinosaurs may be an adaptation for a cooler climate, like that inferred for the paleoenvironment of the Jehol region based on isotope data (Amiot et al. 2011). The presence of proximally plumaceous–distally pennaceous body contour feathers in 99 Ma enantiornithines trapped in amber favors a taphonomic interpretation; however, given the bizarre combination of feather morphotypes revealed by these specimens, similarly unusual plumages in Jehol birds cannot be ruled out. Enantiornithines in amber indicate that primitive unshafted feather morphologies reminiscent of those present in more primitive theropods persisted into the Ornithothoraces at least during early ontogenetic stages (Xing et al. 2017).

Observations from super-precocial enantiornithine juveniles suggest this clade is characterized by a lesser degree of ontogenetic variation between molts compared with neornithines and even some nonavian dinosaurs (Xu et al. 2010). Like living super precocial birds (the Megapodidae) (Starck and Ricklefs 1998), the natal plumage in enantiornithines is characterized by fully developed remiges together with downy body feathers (Xing et al. 2016, 2017; Zhou and Zhang 2004). However, ornamental tail feathers are also documented in several juvenile specimens (de Souza Carvalho et al. 2015; Zheng et al. 2012). Thus, unlike the condition in living birds, in enantiornithines sexually dimorphic ornaments preceded the advent of sexual maturity, being present in the juvenile and possibly even natal plumage. This may suggest that the rachis-dominated racket plumes had a minimal detrimental effect on the overall flight performance in enantiornithines, as already suggested by their morphology (Wang et al. 2014c). This discovery also further supports the hypothesis that more complicated molting patterns are derived within Aves (Gill 2007).

Differences in plumage coloration certainly existed, but these have only just begun to be explored in Cretaceous birds. Mostly melanosome-based coloration is being studied—only six taxa have currently been sampled: two confuciusornithiforms, referred specimens of *Eoconfuciusornis* (Zheng et al. 2017) and *Confuciusornis* (Zhang et al. 2010); two enantiornithines, the holotype of *Cruralispennia* (Wang et al. 2017) and a bohaiornithid enantiornithine (Peteya et al. 2017); and two ornithuromorphs, the holotype of *Changzuiornis* (Huang et al. 2016) and isolated feathers referred to *Gansus* (Barden et al. 2011). Preliminary results mostly reveal eumelanosomes (Barden et al. 2011; Huang et al. 2016; Wang et al. 2017); so far, phaeomelanosomes have only been recovered in confuciusornithiforms (Zhang et al. 2010; Zheng et al. 2017). Variation in melanosome distributions within a single feather of *Confuciusornis* (Zhang et al. 2010) strongly suggests that high-density sampling is required to accurately reconstruct coloration through this method. However, this would be highly destructive and still yield only a partial understanding of coloration, limited to that formed by melanosomes. Only one specimen has been heavily sampled (although still not to the degree necessary to capture detailed patterns)—a female *Eoconfuciusornis*, revealing a dark plumage with a rufous throat patch (Zheng et al. 2017). The predominance of dark plumages recovered for early birds (including *Archaeopteryx*) (Carney et al. 2012) and the inability of melanosome morphology to distinguish visibly preserved color patterns (e.g., the spots in *Eoconfuciusornis*) (Zheng et al. 2017) both suggest the use of melanosomes to reconstruct plumage color is limited. Other pigments utilized by birds are proven to fossilize but have yet to be demonstrated in any Mesozoic bird (Thomas et al. 2014b; Vinther 2015). It is possible carotenoid pigmentation did not evolve in Aves until after the end-Cretaceous extinction (Thomas et al. 2014a).

Although individual basal lineages independently evolved advanced and sometimes complex plumages (e.g., the tail of *Jeholornis* and the hindlimb feathers in *Sapeornis*), among Cretaceous birds, there is a trend toward more advanced integumentary features in more derived clades. The flight apparatus in enantiornithines and ornithuromorphs is better suited for powered flight than that of more primitive clades, showing advanced aerodynamic features in both the pectoral girdle and plumage. Although asymmetrical feathers are an apomorphy of Aves, as increased data comes to light, detailed studies of feather anatomy indicate that vane asymmetry to the degree observed in modern birds did not evolve until the Ornithothoraces (Feo et al. 2015). The presence of an alula, the so-called bastard wing, also characterizes this latter clade. This feathery structure on the leading edge of the wing allows for greater angles of attack, mostly used to aid in take-off and landing and during slow flight (Wang et al. 2015a; Zhang and Zhou 2000). The Ornithuromorpha is characterized by the appearance of numerous advanced features that characterize living birds, but Early Cretaceous members still retained a few primitive traits, such as manual claws, gastralia, and distally contacting pubes (O'Connor et al. 2011). Similarly, their wings are proportionately longer than in most early birds, but wings and tail feathers show a limited diversity of shapes, and potentially they retained primitive "rachis-less" body feathers (Wang et al. 2015a) and lacked medullary pith. One major difference between the two ornithothoracine clades is the morphology of the pygostyle, which in basal ornithuromorphs appears nearly modern in structure, whereas enantiornithines retain a primitive morphology similar to that present in *Confuciusornis* and have similar tail feathers. In contrast, ornithuromorphs do not preserve primitive tail morphologies such as the absence of rectrices or rachis-dominated feathers like that observed in confuciusornithiforms and enantiornithines. Even the earliest-known member of the Ornithuromorpha (130.7 Ma *Archaeornithura meemanni*) possesses a fan-shaped tail (Wang et al. 2015a) and all taxa in which the tail is known preserve more than four feathers (most enantiornithines have four or less with the exception of *Chiappeavis* and *Feitianius*) arranged to form an overlapping surface capable of generating aerodynamic forces. A ploughshare-shaped pygostyle similar to that of ornithuromorphs evolved in parallel in the Sapeornithiformes and co-occurs with a long, fan-shaped tail strongly suggesting these two features coevolved (Wang and O'Connor 2017).

The Early Cretaceous Jehol avifauna reveals that early birds were diverse in their plumage (Fig. 9.1), particularly with regard to the tail, a conclusion supported by finds from other deposits (Fig. 9.5). Flight feathers are more commonly preserved than those that cover and protect the body (although in some cases only body feathers are preserved—the taphonomy of each specimen is unique and generally poorly understood). The feathers of the wing are often obscured by overlap, particularly when the forelimb is preserved folded. As a result, most feather data pertain to the tail. Among living volant birds, rectrices show more morphological variation than any other pteryla (Fitzpatrick 1999; Thomas 1997). Remiges are more strictly limited by the functional morphospace of volant behavior (plumage is unknown in all flightless Cretaceous birds), whereas the tail only serves to supplement the wing apparatus (Gill 2007) and was not crucial to flight in early pygostylians, as evidenced by the absence of elongate tail feathers in female confuciusornithiforms and at least some female enantiornithines. However, early members of the Ornithuromorpha do not share neornithine morphological diversity and all known Early Cretaceous members preserve tail morphologies that appear to be primarily aerodynamic in function (primarily, because as discussed tail morphologies are produced by complex interactions between natural and sexual selection). The fact that such morphologies nearly only occur in groups with a plough-share-shaped pygostyle (the Ornithuromorpha and Sapeornithiformes) supports the hypothesis that without the soft tissue and musculature to control the spread of the tail fan (e.g., the rectricial bulbs and m. bulbi rectricium), such a feature may not be greatly beneficial (O'Connor et al. 2016b). In contrast, most members of the Enantiornithes preserve tails that appear primarily ornamental in function. One taxon apparently possessed an extravagant tail plumage comparable to that observed in polygamous neornithines alive today (e.g., Phasianidae, Paradisaeidae) (O'Connor et al. 2016a). This adds to growing evidence that early in the evolution of feathers, sexual selection was a major driving force (O'Connor and Chang 2015). The robust enantiornithine pygostyle, which like Confuciusornithiformes lacks a dorsal lamina, may have been unsuited for supporting large rectricial bulb-like soft tissue, a hypothesis supported by the absence of significant soft tissue located lateral to the pygostyle as revealed by UV light (Chiappe and Meng 2016; Liu et al. 2019). However, it appears well suited for expanded caudal levator and depressor muscles, suggesting that these birds may have engaged in tail displays similar to extant phasianids (Wang and O'Connor 2017). The only known tail that can readily be considered aerodynamic is that preserved in the holotype of the pengornithid *Chiappeavis magnapremaxillo* (O'Connor et al. 2016b). This group is characterized by a pygostyle that is relatively shorter and less robust than that of other enantiornithines and in this regard is more similar to that of ornithuromorphs relative to other enantiornithines. Notably, the tail fan is shorter than in sympatric ornithuromorphs, likely due to the increase in muscular force required to control longer tail fans (Thomas and Balmford 1995). The expanded pygostyle present in phasianids, strongly reminiscent of the enantiornithine pygostyle (Wang and O'Connor 2017), still possesses rectricial bulbs, although they are just somewhat limited in dorsal expansion by the dorsal platform. The absence of a blade-like dorsal lamina and the presence of expanded dorsolateral and ventrolateral margins of the pygostyle in confuciusornithiforms and enantiornithines suggest strong levator and depressor musculature but do not rule out the presence of rectricial bulbs. Only the strong co-occurrence of a dorsal lamina and tail fans in *Sapeornis* and ornithuromorphs, the absence of taxa utilizing aerodynamic morphologies in the Confuciusornithiformes and Enantiornithes (with the exception of *Chiappeavis*), and the absence of significant soft tissue preserved surrounding the pygostyle in *Confuciusornis* (IVPP V13156) (Fig. 9.3a) and some enantiornithines (Chiappe and Meng 2016; Liu et al. 2019) suggest this feature was absent in the latter clades. Other members of the Pengornithidae preserve rachis-dominated streamers, indicating that intraclade plumage diversity was high, as in living birds (Gluckman 2014).

Hindlimb feathers in early birds are regarded with considerable significance due to the idea that avian flight evolved through a four-winged stage (Zheng et al. 2013b). This historical hypothesis

(Beebe 1915) received new life with the discovery of *Microraptor zhaoianus*, a dromaeosaurid dinosaur that clearly utilized hind-wings in some form of volant behavior (Xu et al. 2003; Pei et al. 2014). However, the presence of hindlimb feathers themselves is not unusual: crural feathers are present in all known living birds, although they are proximally restricted in some taxa, particularly those in aquatic and semiaquatic environments (e.g., Ardeidae, Scolopacidae). Metatarsal feathers are present in raptorial birds, and many owls additionally have feathered toes (Kelso and Kelso 1936). Hindlimb feathers extend onto the metatarsus in *Archaeopteryx* (Foth et al. 2014), *Sapeornis* (Zheng et al. 2013b), and some enantiornithines (Chiappe and Meng 2016). In *Archaeopteryx* and some enantiornithines, the symmetrical crural feathers are roughly half the length of the tibia, but below the ankle, these feathers are small and short. In *Sapeornis* hindlimb, feathers are only preserved around the ankle, although presumably the proximal portion of the hindlimb was also feathered. Potentially the longer crural feathers in *Archaeopteryx* and some enantiornithines and the ankle feathers in *Sapeornis* could somehow have assisted flight; however, this hypothesis requires greater study given that living birds both with and without elongate crural feathers engage in foot-braking during landing (Pennycuick 1968, 1971). Regardless, the condition in Aves differs considerably from the unique condition in *Microraptor* in which the metatarsal feathers are asymmetrical and the longest on the hindlimb, mimicking to the condition in winged forelimbs in which the feathers on the distalmost element (the hand) are the longest and asymmetrical (Xu et al. 2003). In comparison, the hindlimb feathers in *Sapeornis chaoyangensis* (Zheng et al. 2013b) are more proximally located (around the ankle rather than entirely on the tarsometatarsus), proportionately shorter, curved, and with a splayed morphology, which all together suggest a primarily ornamental function (O'Connor and Chang 2015). However, there is no doubt the cohesive surface formed by the proximal ends of the feathers would have had some effect on aerodynamics. The crural feathers in *Archaeopteryx* are very similar to those observed in some enantiornithines (Zhang and Zhou 2004), whereas in others (e.g., *Eopengornis*), the crural feathers are overall shorter and decrease in length distal on the leg, as in *Confuciusornis*. However, some enantiornithines had hindlimb feathering more extensive than *Archaeopteryx*, extending onto the toes, a condition that evolved in parallel to some neornithine lineages (e.g., *Strix*, *Lagopus*) (Xing et al. 2019b). Crural feathers are not preserved in any specimen of *Jeholornis* suggesting they were not likely endowed with strong aerodynamic qualities, and they are clearly short and taper distally in confuciusornithiforms. If distal hindlimb feathers were involved in the evolution of avian flight, their contribution ceased very early in avian evolution, with secondary reduction being the overall predominate trend. Given that asymmetrical distal hindlimb feathers are only present in *Microraptor zhaoianus*, it is likely that this dromaeosaurid independently evolved its volant abilities in parallel to those of early birds and uniquely utilized a tetrapteryx strategy (Foth et al. 2014), further highlighting the developmental plasticity of feathers.

Acknowledgements I would like to thank X-T. Zheng of the STM, H-L. Chang of the Henan Geological Museum, and C-L. Gao of the DNHM for access to specimens. I would also like to thank C. Sullivan and T. Stidham of the IVPP for useful discussions and M. Rothman for use of his art work.

References

Amiot R, Wang X, Zhou Z-H, Wang X-L, Buffetaut E, Lecuyer C, Ding Z-L, Fluteau F, Hibino T, Kusuhashi N, Mo J-Y, Suteethorn V, Wang Y-Q, Xu X, Zhang F-S (2011) Oxygen isotopes of East Asian dinosaurs reveal exceptionally cold Early Cretaceous climates. Proc Natl Acad Sci USA 108:5179–5183

Bailleul AM, O'Connor J, Zhang S-K, Li Z-H, Wang Q, Lamanna M, Zhu X-F, Zhou Z-H (2019) An Early Cretaceous enantiornithine (Aves) preserving an unlaid egg and probable medullary bone. Nat Commun 10 (1275):1–10

Barden HE, Wogelius RA, Li D-Q, Manning PL, Edwards NP, van Dongen BE (2011) Morphological and geochemical evidence of eumelanin preservation in the feathers of the Early Cretaceous bird, *Gansus yumenensis*. PLoS One 6:e25494

Beebe CW (1915) A tetrapteryx stage in the ancestry of birds. Zoologica 2:38–52

Carney RM, Vinther J, Shawkey MD, D'Alba L, Ackermann J (2012) New evidence on the colour and nature of the isolated *Archaeopteryx* feather. Nat Commun 3:1–8

Carroll NR, Chiappe LM, Bottjer DJ (2019) Mid-cretaceous amber inclusions reveal morphogenesis of extinct rachis-dominated feathers. Sci Rep 9 (18108):1–8

Cau A, Arduini P (2008) *Enantiophoenix electrophyla* gen. et sp. nov. (Aves, Enantiornithes) from the Upper Cretaceous (Cenomanian) of Lebanon and its phylogenetic relationships. Atti della Società Italiana di Scienze Naturali e del Museo Civico di Storia Naturale di Milano 149:293–324

Chen P-J, Dong Z, Zhen S (1998) An exceptionally well-preserved theropod dinosaur from the Yixian Formation of China. Nature 391:147–152

Chiappe LM, Lacasa-Ruiz A (2002) Noguerornis gonzalezi (Aves: Ornithothoraces) from the Early Cretaceous of Spain. In: Chiappe LM, Witmer LM (eds) Mesozoic birds: above the heads of dinosaurs. University of California Press, Berkeley, CA, pp 230–239

Chiappe LM, Meng Q-J (2016) Birds of stone. JHU Press, Pittsburgh

Chiappe LM, Ji S, Ji Q, Norell MA (1999) Anatomy and systematics of the Confuciusornithidae (Theropoda: Aves) from the Late Mesozoic of northeastern China. Bull Am Mus Nat Hist 242:1–89

Chiappe LM, Zhao B, O'Connor JK, Gao C-H, Wang X-R, Habib M, Marugan-Lobon J, Meng Q-J, Cheng X-D (2014) A new specimen of the early cretaceous bird *Hongshanornis longicresta*: insights into the aerodynamics and diet of a basal ornithuromorph. PeerJ 2:1–28

Chinsamy A, Chiappe LM, Dodson P (1995) Mesozoic avian bone microstructure: physiological implications. Paleobiology 21:561–574

Chinsamy A, Chiappe LM, Marugán-Lobón J, Gao C-H, Zhang F-J (2013) Gender identification of the Mesozoic bird *Confuciusornis sanctus*. Nat Commun 4:1–5

Clarke JA, Zhou Z, Zhang F (2006) Insight into the evolution of avian flight from a new clade of Early Cretaceous ornithurines from China and the morphology of *Yixianornis grabaui*. J Anat 208:287–308

Dalla Vecchia FM, Chiappe LM (2002) First avian skeleton from the Mesozoic of northern Gondwana. J Vertebr Paleontol 22:856–860

Dalsätt J, Ericson PGP, Zhou Z-H (2014) A new Enantiornithes (Aves) from the Early Cretaceous of China. Acta Geol Sin 88:1034–1040

Dames W (1884) Über *Archaeopteryx*. Palaeontologische Abhandlungen 3:119–196

de Souza Carvalho I, Novas FE, Agnolin FL, Isasi MP, Freitas FI, Andrade JA (2015) A Mesozoic bird from Gondwana preserving feathers. Nat Commun 6:1–5

Dyke GD, Nudds RL (2008) The fossil record and limb disparity of enantiornithines, the dominant flying birds of the Cretaceous. Lethaia 42:248–254

Elzanowski A (2002) Archaeopterygidae (Upper Jurassic of Germany). In: Chiappe LM, Witmer LM (eds) Mesozoic birds: above the heads of dinosaurs. University of California Press, Berkeley, CA, pp 129–159

Falk AR, Kaye TG, Zhou Z-H, Burnham DA (2016) Laser fluorescence illuminates the soft tissue and life habits of the Early Cretaceous bird *Confuciusornis*. PLoS One 11:e0167284

Falk AR, O'Connor J, Wang M, Zhou Z-H (2019) On the preservation of the beak in *Confuciusornis* (Aves: Pygostylia). Diversity 11(212):1–8

Feo TJ, Field DJ, Prum RO (2015) Barb geometry of asymmetrical feathers reveals a transitional morphology in the evolution of avian flight. Proc R Soc B Biol Sci 282:20142864

Fitzpatrick S (1999) Tail length in birds in relation to tail shape, general flight ecology and sexual selection. J Evol Biol 12:49–60

Foth C (2011) The morphology of neoptile feathers: ancestral state reconstruction and its phylogenetic implications. J Morphol 272:387–403

Foth C (2012) On the identification of feather structures in stem-line representatives of birds: evidence from fossils and actuopalaeontology. Paläontol Z 86:91–102

Foth C, Tischlinger H, Rauhut OWM (2014) New specimen of *Archaeopteryx* provides insights into the evolution of pennaceous feathers. Nature 511:79–82

Gao C-H, Chiappe LM, Zhang F-J, Pomeroy DL, Shen C-Z, Chinsamy A, Walsh MO (2012) A subadult specimen of the Early Cretaceous bird *Sapeornis chaoyangensis* and a taxonomic reassessment of sapeornithids. J Vertebr Paleontol 32:1103–1112

Gill FB (2007) Ornithology, 3rd edn. W.H. Freeman and Company, New York, p 758

Gluckman T-L (2014) Pathways to elaboration of sexual dimorphism in bird plumage patterns. Biol J Linn Soc 111:262–273

Hou L, Zhou Z-H, Martin LD, Feduccia A (1995) A beaked bird from the Jurassic of China. Nature 377:616–618

Hou L, Martin LD, Zhou Z, Feduccia A (1996) Early adaptive radiation of birds: evidence from fossils from northeastern China. Science 274: 1164–1167

Hou L, Martin LD, Zhonghe Z, Feduccia A, Zhang F (1999) A diapsid skull in a new species of the primitive bird *Confuciusornis*. Nature 399:679–682

Hou L, Chiappe LM, Zhang F, Chuong C-M (2004) New Early Cretaceous fossil from China documents a novel trophic specialization for Mesozoic birds. Naturwissenschaften 91:22–25

Hu H, O'Connor JK, Zhou Z-H (2015) A new species of Pengornithidae (Aves: Enantiornithes) from the Lower Cretaceous of China suggests a specialized scansorial habitat previously unknown in early birds. PLoS One 10:e0126791

Huang J-D, Wang X, Hu Y-C, Liu J, Peteya JA, Clarke JA (2016) A new ornithurine from the Early Cretaceous of China sheds light on the evolution of early ecological and cranial diversity in birds. PeerJ 4:e1765

Ji Q, Currie PJ, Norell MA, Ji S-A (1998) Two feathered dinosaurs from northeastern China. Nature 393:753–761

Kelso L, Kelso EH (1936) The relation of feathering of feet of American owls to humidity of environment and to life zones. Auk 53:51–56

Kundrát M (2004) When did theropods become feathered? Evidence for pre-*Archaeopteryx* feathery appendages. J Exp Zool 302B:1–10

Kurochkin EN (1985) A true carinate bird from Lower Cretaceous deposits in Mongolia and other evidence of Early Cretaceous birds in Asia. Cretac Res 6:271–278

Lacasa-Ruiz A (1988) An Early Cretaceous fossil bird from Montsec Mountain (Lleida, Spain). Terra Nova:45–46

Li L, Duan Y, Hu D, Wang L, Cheng S, Hou L (2006) New eoentantiornithid bird from the Early Cretaceous Jiufotang Formation of western Liaoning, China. Acta Geologica Sinica (English Edition) 80:38–41

Li Q-G, Gao K-Q, Meng Q-J, Clarke JA, Shawkey MD, D'Alba L, Pei R, Ellison M, Norell MA, Vinther J (2012) Reconstruction of *Microraptor* and the evolution of iridescent plumage. Science 335:1215–1219

Liu D, Chiappe LM, Serrano FJ, Habib M, Zhang Y-G, Meng Q-J (2017) Flight aerodynamics in enantiornithines: information from a new Chinese Early Cretaceous bird. PLoS One 12:e0184637

Liu D, Chiappe LM, Zhang Y-G, Serrano FJ, Meng Q-J (2019) Soft tissue preservation in two new enantiornithine specimens (Aves) from the Lower Cretaceous Huajiying Formation of Hebei Province, China. Cretac Res 95:191–207

Lovette IJ, Fitzpatrick JW (2004) The handbook of bird biology. Princeton University Press, Princeton, NJ

Martin LD (1984) A new hesperornithid and the relationships of the Mesozoic birds. Trans Kans Acad Sci 87:141–150

Mayr G, Pohl B, Peters DS (2005) A well-preserved *Archaeopteryx* specimen with theropod features. Science 310:1483–1486

Møller AP, Hedenström A (1999) Comparative evidence for costs of secondary sexual characters: adaptive vane emargination of ornamented feathers in birds. J Evol Biol 12:296–305

Navalón G, Marugán-Lobón J, Chiappe LM, Sanz JL, Buscalioni AD (2015) Soft-tissue and dermal arrangement in the wing of an Early Cretaceous bird: implications for the evolution of avian flight. Sci Rep 5:14864

Navalón G, Meng Q-J, Marugán-Lobón J, Zhang Y-G, Wang B-P, Xing H, Liu D, Chiappe LM (2018) Diversity and evolution of the Confuciusornithidae: evidence from a new 131-million-year-old specimen from the Huajiying Formation in NE China. J Asian Earth Sci 152:12–22

Nudds RL, Dyke GD (2010) Narrow primary feather rachises in *Confuciusornis* and *Archaeopteryx* suggest poor flight ability. Science 328:887–889

O'Connor JK (2009) A systematic review of Enantiornithes (Aves: Ornithothoraces). In: Geological sciences. Dissertation, University of Southern California, Los Angeles, pp 600

O'Connor JK, Chang H-L (2015) Hindlimb feathers in paravians: primarily 'wings' or ornaments? Biol Bull 42:1–6

O'Connor JK, Sullivan C (2014) Reinterpretation of the Early Cretaceous maniraptoran (Dinosauria: Theropoda) *Zhongornis haoae* as a scansoriopterygid-like non-avian, and morphological resemblances between scansoriopterygids and basal oviraptorosaurs. Vertebrata Palasiatica 52:3–30

O'Connor JK, Wang X-R, Chiappe LM, Gao C-H, Meng Q-J, Cheng X-D, Liu J-Y (2009) Phylogenetic support for a specialized clade of Cretaceous enantiornithine birds with information from a new species. J Vertebr Paleontol 29:188–204

O'Connor JK, Chiappe LM, Bell A (2011) Pre-modern birds: avian divergences in the Mesozoic. In: Dyke GD, Kaiser G (eds) Living dinosaurs: the evolutionary history of birds. Wiley, Hoboken, NJ, pp 39–114

O'Connor JK, Chiappe LM, Chuong C-M, Bottjer DJ, You H-L (2012a) Homology and potential cellular and molecular mechanisms for the development of unique feather morphologies in early birds. Geosciences 2:157–177

O'Connor JK, Sun C-K, Xu X, Wang X-L, Zhou Z-H (2012b) A new species of *Jeholornis* with complete caudal integument. Hist Biol 24:29–41

O'Connor JK, Wang X-L, Sullivan C, Zheng X-T, Tubaro PL, Zhang X-M, Zhou Z-H (2013) The unique caudal plumage of *Jeholornis* and complex tail evolution in early birds. Proc Natl Acad Sci USA 110:17404–17408

O'Connor JK, Li D-Q, Lamanna M, Wang M, Harris JD, Atterholt JA, You H-L (2016a) A new Early Cretaceous enantiornithine (Aves: Ornithothoraces) from northwestern China with elaborate tail ornamentation. J Vertebr Paleontol 36:e1054035

O'Connor JK, Wang X-L, Zheng X-T, Hu H, Zhang X-M, Zhou Z-H (2016b) An enantiornithine with a fan-shaped tail, and the evolution of the rectricial complex in early birds. Curr Biol 26:114–119

O'Connor J, Erickson GM, Norell MA, Bailleul AM, Hu H, Zhou Z-H (2018) Medullary bone in an Early Cretaceous enantiornithine (Aves) and discussion regarding its identification in fossils. Nat Commun 9:1–8

O'Connor J, Falk AR, Wang M, Zheng X-T (2020) First report of immature feathers in juvenile enantiornithines from the Early Cretaceous Jehol avifauna. Vert PalAs

Owen R (1863) On the *Archaeopteryx* of von Meyer, with a description of the fossil remains of a long-tailed species, from the lithographic stone of Solenhofen. Philos Trans R Soc Lond 153:33–47

Pan Y-H, Sha J-G, Zhou Z-H, Fürsich FT (2013) The Jehol Biota: definition and distribution of exceptionally preserved relicts of a continental Early Cretaceous ecosystem. Cretac Res 44:30–38

Pei R, Li Q-G, Meng Q-J, Gao K-Q, Norell MA (2014) A new specimen of *Microraptor* (Theropoda: Dromaeosauridae) from the Lower Cretaceous of western Liaoning, China. Am Mus Novit 3821:1–28

Pennycuick CJ (1968) A wind-tunnel study of gliding flight in the pigeon *Columba livia*. J Exp Biol 49:509–526

Pennycuick CJ (1971) Control of gliding angle in Rüppell's Griffon Vulture *Gyps rüppellii*. J Exp Biol 55:39–46

Peteya JA, Clarke JA, Li Q-G, Gao K-Q, Shawkey MD (2017) The plumage and colouration of an enantiornithine bird from the Early Cretaceous of China. Palaeontology 60:55–71

Prum RO, Brush AH (2002) The evolutionary origin and diversification of feathers. Q Rev Biol 77:261–295

Saitta ET, Clapham C, Vinther J (2018) Experimental subaqueous burial of a bird carcass and compaction of plumage. PalZ:1–6

Sanz JL, Chiappe LM, Pérez-Moreno BP, Buscalioni AD, Moratalla JJ, Ortega F, Poyato-Ariza FJ (1996) An Early Cretaceous bird from Spain and its implications for the evolution of avian flight. Nature 382:442–445

Sanz JL, Chiappe LM, Fernández-Jalvo Y, Ortega F, Sánchez-Chillon B, Poyato-Ariza FJ, Pérez-Moreno BP (2001) An Early Cretaceous pellet. Nature 409:998–999

Serrano FJ, Chiappe LM (2017) Aerodynamic modelling of a Cretaceous bird reveals thermal soaring capabilities during early avian evolution. J R Soc Interface 14:20170182

Starck JM, Ricklefs RE (1998) Patterns of development: the altricial-precocial spectrum. In: Starck JM, Ricklefs RE (eds) Avian growth and development. Oxford University Press, New York City, pp 3–30

Stettenheim PR (2000) The integumentary morphology of modern birds—an overview. Am Zool 40:461–477

Thomas ALR (1993) On the aerodynamics of birds' tails. Philos Trans R Soc Lond B Biol Sci 340:361–380

Thomas ALR (1997) On the tails of birds. Bioscience 47:215–225

Thomas ALR, Balmford A (1995) How natural selection shapes bird's tails. Am Nat 146:848–868

Thomas DB, McGraw KJ, Butler MW, Carrano MT, Madden O, James HF (2014a) Ancient origins and multiple appearances of carotenoid-pigmented feathers in birds. Proc R Soc B Biol Sci 281:1–9

Thomas DB, Nascimbene PC, Dove CJ, Grimaldi DA, James HF (2014b) Seeking carotenoid pigments in amber-preserved fossil feathers. Sci Rep 4:1–6

Vinther J (2015) A guide to the field of palaeo color. BioEssays 37:643–656

von Meyer H (1861) *Archaeopteryx litographica* (Vogel-Feder) und *Pterodactylus* von Solenhofen. Neues Jahrbuch für Mineralogie, Geognosie, Geologie, und Petrefakten-kunde: 678–679

Wang W, O'Connor JK (2017) Morphological coevolution of the pygostyle and tail feathers in Early Cretaceous birds. Vertebrata Palasiatica 55:289–314

Wang X-R, Chiappe LM, Teng F-F, Ji Q (2013) Xinghaiornis lini (Aves: Ornithothoraces) from the Early Cretaceous of Liaoning: an example of evolutionary mosaic in early birds. Acta Geologica Sinica English Edition 87:686–689

Wang M, O'Connor JK, Zhou Z-H (2014a) A robust enantiornithine bird from the Lower Cretaceous of China with scansorial adaptations. J Vertebr Paleontol 34:657–671

Wang M, Zhou Z-H, Xu G-H (2014b) The first enantiornithine bird from the Upper Cretaceous of China. J Vertebr Paleontol 34:135–145

Wang X-L, O'Connor JK, Zheng X-T, Wang M, Hu H, Zhou Z-H (2014c) Insights into the evolution of rachis dominated tail feathers from a new basal enantiornithine (Aves: Ornithothoraces). Biol J Linn Soc 113:805–819

Wang M, Zheng X-T, O'Connor JK, Lloyd GT, Wang X-L, Wang Y, Zhang X-M, Zhou Z-H (2015a) The oldest record of Ornithuromorpha reveals heterogeneous rates of morphological evolution among Early Cretaceous birds. Nat Commun 6:6987

Wang R-F, Wang Y, Hu D-Y (2015b) Discovery of a new ornithuromorph genus, *Juehuaornis* gen. nov. from Lower Cretaceous of western Liaoning, China. Global Geology 34:7–11

Wang Y-M, O'Connor JK, Li D-Q, You H-L (2015c) New information on postcranial skeleton of the Early Cretaceous *Gansus yumenensis* (Aves: Ornithuromorpha). Hist Biol 28:666–679

Wang M, O'Connor JK, Pan Y-H, Zhou Z-H (2017) A bizarre Early Cretaceous enantiornithine bird with unique crural feathers and an ornithuromorph plough-shaped pygostyle. Nat Commun 8:1–12

Wang X, O'Connor JK, Maina JN, Pan Y, Wang M, Wang Y, Zheng X, Zhou Z (2018) *Archaeorhynchus* preserving significant soft tissue including probable fossilized lungs. Proc Natl Acad Sci USA. Published online, 1–6. https://doi.org/10.1073/pnas.1805803115

Wang M, O'Connor J, Xu X, Zhou Z-H (2019) A new Jurassic scansoriopterygid and the loss of membranous wings in theropod dinosaurs. Nature 569(7755):256–259

Williston SW (1896) On the dermal covering of *Hesperornis*. Kansas University Quarterly 5:53–54

Xing L-D, McKellar RC, Wang M, Bai M, O'Connor JK, Benton MJ, Zhang J-P, Wang Y, Tseng K-W, Lockley M, Li G, Zhang W-W, Xu X (2016) Mummified precocial bird wings in mid-Cretaceous Burmese amber. Nat Commun 7:12089

Xing L-D, O'Connor JK, McKellar RC, Chiappe LM, Tseng K-W, Li G, Bai M (2017) A mid-Cretaceous enantiornithine (Aves) hatchling preserved in Burmese amber with unusual plumage. Gondwana Res 49:264–277

Xing L-D, Cockx P, McKellar RC, O'Connor J (2018a) Ornamental feathers in Cretaceous Burmese amber:

resolving the enigma of rachis-dominated feather structure. J Palaeogeogr 7(13):1–18

Xing L-D, O'Connor JK, McKellar RC, Chiappe LM, Bai M, Tseng K-W, Zhang J, Yang H-D, Fang J, Li G (2018b) A flattened enantiornithine in mid-Cretaceous Burmese amber: morphology and preservation. Sci Bull

Xing L-D, McKellar RC, O'Connor JK, Bai M, Tseng K-W, Chiappe LM (2019a) A fully feathered enantiornithine foot and wing fragment preserved in mid-Cretaceous Burmese amber. Sci Rep 9(927):1–9

Xing L-D, O'Connor J, Chiappe LM, McKellar RC, Carroll N, Hu H, Bai M, Lei F-M (2019b) A new enantiornithine with unusual pedal proportions found in amber. Curr Biol 29(14):2396–2401

Xu X, Guo Y (2009) The origin and early evolution of feathers: insights from recent paleontological and neontological data. Vertebrata Palasiatica 47:311–329

Xu X, Zhou Z, Wang X, Kuang X, Du X (2003) Four-winged dinosaurs from China. Nature 421:335–340

Xu X, Zheng X-T, You H-L (2010) Exceptional dinosaur fossils show ontogenetic development of early feathers. Nature 464:1339–1341

Xu X, Zheng X-T, Sullivan C, Zhang F-C, O'Connor JK, Wang X-L (2015) A bizarre Jurassic maniraptoran theropod with preserved evidence of membranous wings. Nature 521:70–73

You H-L, Lamanna MC, Harris JD, Chiappe LM, O'Connor J, Ji S-A, Lü J-C, Yuan C-X, Li D-Q, Zhang X, Lacovara KJ, Dodson P, Ji Q (2006) A nearly modern amphibious bird from the Early Cretaceous of northwestern China. Science 312:1640–1643

Zelenkov NZ, Averianov AO (2016) A historical specimen of enantiornithine bird from the Early Cretaceous of Mongolia representing a new taxon with a specialized neck morphology. J Syst Palaeontol 14:319–338

Zhang F, Zhou Z (2000) A primitive enantiornithine bird and the origin of feathers. Science 290:1955–1960

Zhang F, Zhou Z (2004) Leg feathers in an Early Cretaceous bird. Nature 431:925

Zhang F, Zhou Z, Dyke GJ (2006) Feathers and 'feather-like' integumentary structures in Liaoning birds and dinosaurs. Geol J 41:395–404

Zhang F, Zhou Z, Benton MJ (2008a) A primitive confuciusornithid bird from China and its implications for early avian flight. Sci China Ser D Earth Sci 51:625–639

Zhang F-C, Zhou Z-H, Xu X, Wang X-L, Sullivan C (2008b) A bizarre Jurassic maniraptoran from China with elongate ribbon-like feathers. Nature 455:1105–1108

Zhang F-C, Kearns SL, Orr PJ, Benton MJ, Zhou Z-H, Johnson D, Xu X, Wang X-L (2010) Fossilized melanosomes and the colour of Cretaceous dinosaurs and birds. Nature 463:1075–1078

Zhang Z-H, Chiappe LM, Han G, Chinsamy A (2013) A large bird from the Early Cretaceous of China: new information on the skull of enantiornithines. J Vertebr Paleontol 33:1176–1189

Zheng X, Zhang Z, Hou L (2007) A new enantiornitine bird with four long rectrices from the Early Cretaceous of northern Hebei, China. Acta Geologica Sinica (English Edition) 81:703–708

Zheng X-T, Xu X, Zhou Z-H, Miao D, Zhang F-C (2010) Comment on "Narrow primary feather rachises in *Confuciusornis* and *Archaeopteryx* suggest poor flight ability". Science 330:320

Zheng X-T, Wang X-L, O'Connor JK, Zhou Z-H (2012) Insight into the early evolution of the avian sternum from juvenile enantiornithines. Nat Commun 3:1–8

Zheng X-T, O'Connor JK, Huchzermeyer FW, Wang X-L, Wang Y, Wang M, Zhou Z-H (2013a) Preservation of ovarian follicles reveals early evolution of avian reproductive behaviour. Nature 495:507–511

Zheng X-T, Zhou Z-H, Wang X-L, Zhang F-C, Zhang X-M, Wang Y, Wei G-J, Wang S, Xu X (2013b) Hind wings in basal birds and the evolution of leg feathers. Science 339:1309–1312

Zheng X-T, O'Connor JK, Huchzermeyer FW, Wang X-L, Wang Y, Zhang X-M, Zhou Z-H (2014a) New specimens of *Yanornis* indicate a piscivorous diet and modern alimentary canal. PLoS One 9:e95036

Zheng X-T, O'Connor JK, Wang X-L, Wang M, Zhang X-M, Zhou Z-H (2014b) On the absence of sternal elements in *Anchiornis* (Paraves) and *Sapeornis* (Aves) and the complex early evolution of the avian sternum. Proc Natl Acad Sci USA 111:13900–13905

Zheng X-T, O'Connor JK, Wang X-L, Zhang X-M, Wang Y (2014c) New information on Hongshanornithidae (Aves: Ornithuromorpha) from a new subadult specimen. Vertebrata Palasiatica 52:217–232

Zheng X-T, O'Connor JK, Wang X-L, Pan Y-H, Wang Y, Wang M, Zhou Z-H (2017) Exceptional preservation of soft tissue in a new specimen of *Eoconfuciusornis* and its biological implications. Natl Sci Rev 4:441–452

Zhou Z, Zhang F (2001) Two new ornithurine birds from the Early Cretaceous of western Liaoning, China. Chin Sci Bull 46:1258–1264

Zhou Z, Zhang F (2002) Largest bird from the Early Cretaceous and its implications for the earliest avian ecological diversification. Naturwissenschaften 89:34–38

Zhou Z, Zhang F (2003) Anatomy of the primitive bird *Sapeornis chaoyangensis* from the Early Cretaceous of Liaoning, China. Can J Earth Sci 40:731–747

Zhou Z, Zhang F (2004) A precocial avian embryo from the Lower Cretaceous of China. Science 306:653

Zhou Z, Zhang F (2005) Discovery of an ornithurine bird and its implication for Early Cretaceous avian radiation. Proc Natl Acad Sci USA 102:18998–19002

Zhou Z-H, Zhang F-C (2006) Mesozoic birds of China—a synoptic review. Vertebrata Palasiatica 44:74–98

Zhou Z, Barrett PM, Hilton J (2003) An exceptionally preserved Lower Cretaceous ecosystem. Nature 421:807–814

Zhou Z, Chiappe LM, Zhang F (2005) Anatomy of the Early Cretaceous bird *Eoenantiornis buhleri* (Aves: Enantiornithes) from China. Can J Earth Sci 42:1331–1338

Zhou Z-H, Zhang F-C, Li Z-H (2009) A new basal orithurine (*Jianchangornis microdonta* gen. et sp. nov.) from the Lower Cretaceous of China. Vertebrata Palasiatica 47:299–310

Zhou S, Zhou Z-H, O'Connor JK (2012) A new toothless ornithurine bird (*Schizooura lii* gen. et sp. nov.) from the Lower Cretaceous of China. Vertebrata Palasiatica 50:9–24

Zhou S, Zhou Z-H, O'Connor JK (2013) A new piscivorous ornithuromorph from the Jehol Biota. Hist Biol 26:608–618

Zhou S, O'Connor JK, Wang M (2014) A new species from an ornithuromorph dominated locality of the Jehol Group. Chin Sci Bull 59:5366–5378

10 A Morphological Review of the Enigmatic Elongated Tail Feathers of Stem Birds

Christian Foth

10.1 Introduction

The tail plumage of Mesozoic Pennaraptora is characterized by a huge shape diversity (Foth et al. 2014; Wang et al. 2014), which is influenced by the length and morphology of the caudal series (e.g., Felice 2014; Rashid et al. 2014), the distribution of contour feathers along the tail (e.g., O'Connor et al. 2013; Foth et al. 2014; Wang et al. 2014), and the morphology of the tail feathers (=rectrices) (e.g., O'Connor et al. 2012; Wang et al. 2014). The tail plumage of Confuciusornithidae and many species of Enantiornithes is of special interest, as it frequently contains one or two pairs of conspicuously elongated, distally vaned tail feathers, herein called rhachis-dominated racket plumes (Fig. 10.1a, c), which are attached to the distal end of the pygostyle (e.g., Chiappe et al. 1999; Zheng et al. 2007; Zhang et al. 2008a, b; O'Connor et al. 2012; Carvalho et al. 2015a). Similar tail feathers were also described for the enigmatic scansoriopterygid *Epidexipteryx hui* (Zhang et al. 2008a, b) and an early juvenile individual of the oviraptorosaur *Similicaudipteryx yixianensis* (Xu et al. 2010b), but their likeness to rhachis-dominated racket plumes is not fully accepted. Apart from the unclear phylogenetic position of Scansoriopterygidae within Maniraptora (see Xu et al. 2010a, 2015; Agnolín and Novas 2013; O'Connor and Sullivan 2014), the distal portions of the tail feathers of *Epidexipteryx* are not preserved. Thus, it is not clear at the moment if they represent rhachis-dominated racket plumes or an own distinct feather type. The presence of rhachis dominated racket plumes in *Similicaudipteryx* as was questioned by various authors (Prum 2010; Foth 2012; O'Connor et al. 2012), and the structure can alternatively be interpreted as pin feathers, that are developing pennaceous feathers, which are still covered by the feather sheath. Thus, both species will not be included into the actual comparison.

In analogy to the elongated rectrices of modern birds (Andersson 1982; Bleiweiss 1987; Peters and Peters 2009), the elongated tail feathers of Confuciusornithidae and Enantiornithes probably had an ornamental function (Peters and Peters 2009; O'Connor et al. 2012), which in some cases may have been related to sexual dimorphism (Zheng et al. 2017). The actual morphology of rhachis-dominated racket plumes, however, seems to be quite different from those of modern examples. As a result, there is no true consensus regarding their morphology, and various interpretations have been published in the past (e.g., Xu and Guo 2009; Prum 2010; Foth 2012; O'Connor et al. 2012; Carvalho et al. 2015b), which often relied on differences in the quality of preservation. In the current chapter, these different morphological interpretations are reviewed and compared with each other. After extracting the main organization, the single individual morphological components of these enigmatic tail feathers are compared with different modern feather types.

C. Foth (✉)
Department of Geosciences, University of Fribourg, Fribourg, Switzerland

C. Foth, O. W. M. Rauhut (eds.), *The Evolution of Feathers*, Fascinating Life Sciences,
https://doi.org/10.1007/978-3-030-27223-4_10

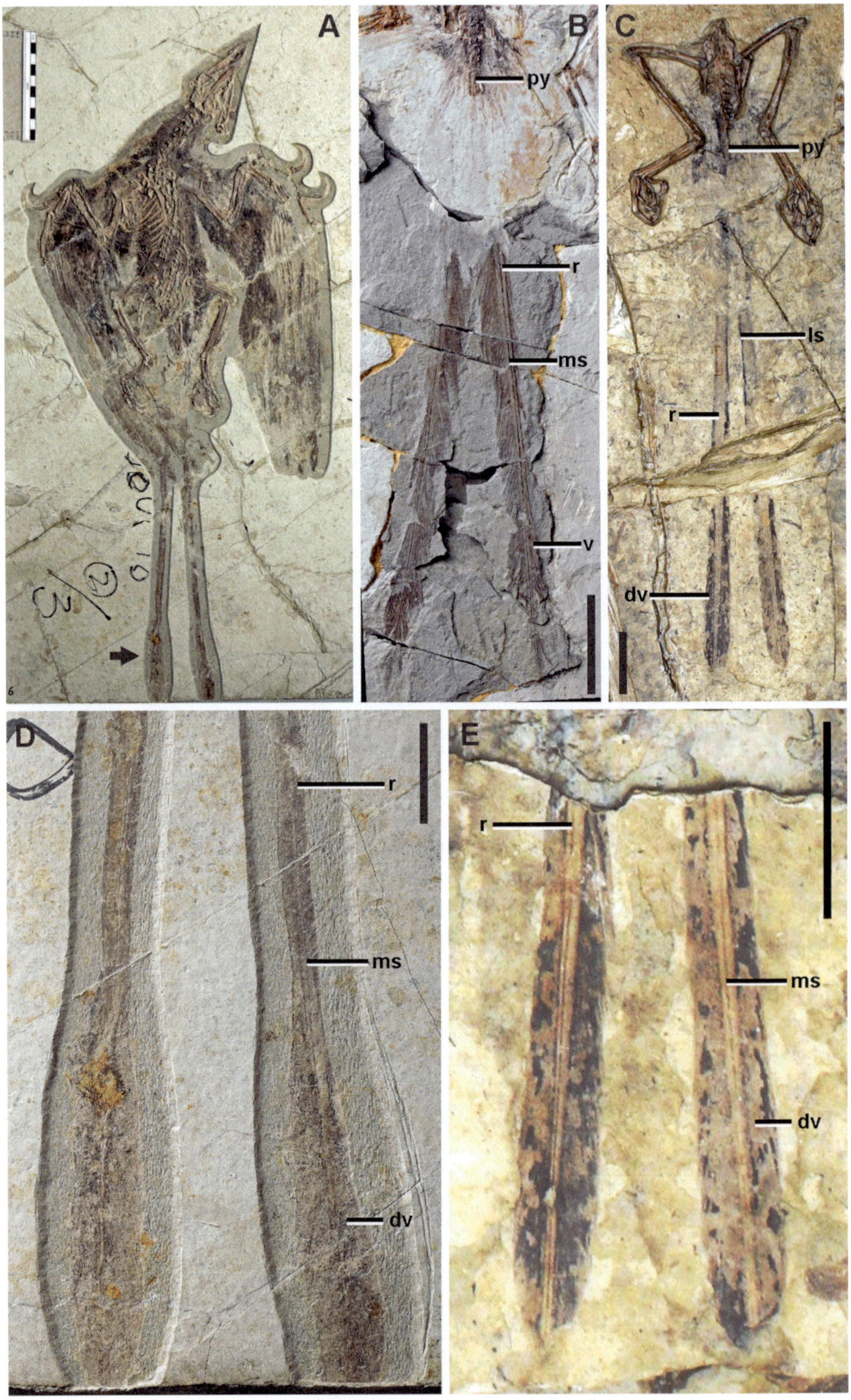

Fig. 10.1 Examples of elongated rhachis-dominated feathers in stem birds. (**a**) *Confuciusornis sanctus* (IVPP V13156) with rhachis-dominated racket plumes indicated by the arrow. (**b**) *Eopengornis martini* (STM24-1) with

By extending this comparison beyond the morphological spectrum of tail feathers, it is possible to track down analog structures, helping to understand the actual morphology of this extinct feather type.

10.2 Institutional Abbreviations

GSGM Gansu Geological Museum, Lanzhou, China; *IVPP* Institute of Vertebrate Paleontology and Paleoanthropology, Beijing, China; *NHMF* Natural History Museum Fribourg, Switzerland; *STM* Shandong Tianyu Museum of Natural History, Linyi, China

10.3 Previous Morphological Interpretations

In the past, the enigmatic, elongated rectrices of Confuciusornithidae and Enantiornithes were addressed with varying terms, including elongate ribbonlike tail feathers (ETFs) (Zhang et al. 2008a, b), proximally ribbonlike pennaceous feathers (PRPFs) (Xu et al. 2010b), rhachis-dominated tail feathers (O'Connor et al. 2012), or rhachis-dominated racket plumes (Wang et al. 2014). Despite these different terms, the rectrices of the taxa in question possess a characteristic morphology, consisting of a broad, elongated central element with a dark, median stripe. The proximal portion of the central element is naked, exhibits dark lateral margins, and shows no sign of branching, while the distal quarter is vaned, being pennaceous (Fig. 10.1c–e). As is typical for fossilized plumages from the Jehol beds, the elongated tail feathers are usually preserved as carbonized traces, which has been shown to result from the preservation of melanosomes, showing the original pigmentation of the fossilized feathers, in several taxa (e.g., Vinther et al. 2008; Li et al. 2010; Zhang et al. 2010).

Originally, this feather type was described as scalelike (Zhang and Zhou 2000), which was classified as an ancestral unbranched feather type. However, this interpretation was based on an incomplete specimen, where the distal portions of the feathers were not preserved (O'Connor et al. 2012). On the basis of complete feathers, three different morphological interpretations were published so far as follows:

1. Originally, the dark, median stripe of the central element was interpreted as a thin rhachis with two undivided, sheetlike vanes or laminae emerging on either side. In the distal portion, the pennaceous barbs were thought to extend outwards from the sheetlike vane and to not be directly connected with the rhachis (Zhang et al. 2006, 2008a, b; Xu and Guo 2009; Xu et al. 2010b; Fig. 10.2a).
2. Later, Xu et al. (2010c) and O'Connor et al. (2012) argued that the whole central element represents an extremely long, broad rhachis, which ends in a distal, pennaceous portion (see also Prum 2010). The calamus, which is not preserved due to the probable lack of melanosomes in this region (see Benton et al. 2008; Vinther et al. 2008), is restricted to the most proximal portion of the feather, while the dark stripe in the middle of the rhachis was interpreted to be a preservational artifact resulting from a ventral furrow (Fig. 10.2b). Additionally, O'Connor et al. (2012) interpreted the dark lateral margins of the proximal ribbonlike portion of the central elements (Fig. 10.1c) to be possible remains of narrow, undifferentiated vanes (see above).
3. Based on the branching pattern of the rectrices, Foth (2012) argued that the rhachis is only a broad and short element, restricted to the

Fig. 10.1 (continued) rhachis-dominated rectrices showing the common pennaceous morphology. (**c**) Enantiornithes indet. (GSGM-07-CM-001) with rhachis-dominated racket plumes. (**d**) Details of the distal end of the rhachis-dominated racket plumes of *Confuciusornis sanctus*. (**e**) Details of the distal end of the rhachis-dominated racket plumes of GSGM-07-CM-001. *dv* distal vane, *ls* lateral stripe, *ms* median stripe, *r* rhachis, *v* vane. Scale bars in (**b**)–(**e**) is 2 cm

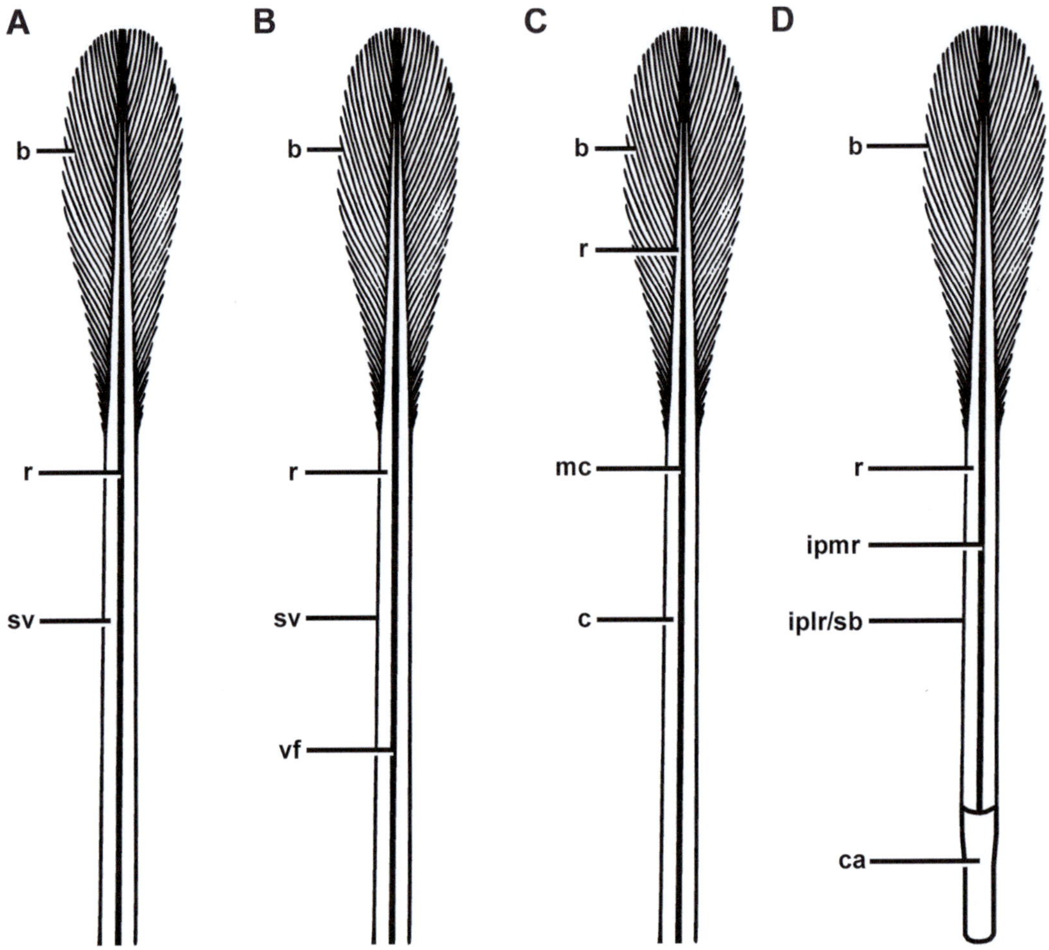

Fig. 10.2 Different interpretations of the morphology of rhachis-dominated racket plumes in stem birds. (**a**) Morphology after Zhang et al. (2006, 2008a, b), Xu and Guo (2009), and Xu et al. (2010b). (**b**) Morphology after Xu et al. (2010c) and O'Connor et al. (2012). (**c**) Morphology after Foth (2012). (**d**) Current interpretation based on the comparison with various modern feather types, including penguin body feathers. *b* barbs of the distal vane, *ca* calamus, *ipmr* internal pigmented median ridge, *iplr* internal pigmented lateral ridge, *mc* medullary cavity, *r* rhachis, *sb* short barbs, *sv* sheetlike vanes, *vf* ventral furrow. Illustration of elongated tail feathers modified after Xu et al. (2010a)

pennaceous portion, while the proximal ribbonlike structure in fact represents a prolonged, broadened calamus (Fig. 10.2c). The dark median line of the central element was interpreted as a pigmentation of the internal pith inside the rhachis and calamus or as remains of a ventral furrow (see O'Connor et al. 2012).

All three interpretations are problematic for various reasons: The presence of sheetlike vanes, running along the whole central element, as proposed by Zhang et al. (2006, 2008a, b), Xu and Guo (2009), and Xu et al. (2010b), or being restricted to the lateral margin of the proximal ribbonlike portion (see O'Connor et al. 2012), is incorrect from a semantic point of view, as feather vanes consist of a series of parallel arranged barbs (Lucas and Stettenheim 1972) and thus, by definition, cannot be undifferentiated. Along those lines, the Enantiornithes *Cratoavis cearensis* from the Early Cretaceous Crato Formation of Brazil (see below), whose elongated rectrices are preserved as impression and not carbonized traces (Carvalho et al. 2015a, b), indicates that, at least in this bird, the whole central element is strap-like (see Prum 2010; Xu et al. 2010c; Foth 2012; O'Connor et al. 2012), showing no morphological signs of undifferentiated vanes in the proximal portion.

Based on the hierarchical organization of feathers in terms of morphology and development (Lucas and Stettenheim 1972; Prum and Dyck 2003), the whole central structure in the pennaceous portion has to be classified as the rhachis (see Prum 2010; Foth 2012; O'Connor et al. 2012). However, contrary to Foth (2012), the central element of the rectrices shows no sign of interruption in the form of a superior umbilicus between the distal pennaceous and proximal ribbonlike portion, which would mark the transition from rhachis to calamus. In fact, the median stripe runs without interruption along the entire central element, and the dark lateral margins are continuous with the barbs of the pennaceous portion (O'Connor et al. 2012). Thus, the whole central element can, in fact, be interpreted as one single strap-like structure, that is, an elongated, dorsoventrally flattened rhachis, as previously interpreted by Prum (2010), Xu et al. (2010c), and O'Connor et al. (2012).

The median stripe itself is usually preserved as a narrow carbonized trace, but not as an impression (O'Connor et al. 2012). According to the interpretation of Xu et al. (2010a) and O'Connor et al. (2012), the stripe would be exposed ventrally. As demonstrated by several studies, carbonized traces are often the result of melanosome preservation, which show the original pigmentation pattern of the fossilized feather (e.g., Vinther et al. 2008; Li et al. 2010; Zhang et al. 2010). Because the integument is usually preserved as a film, this type of preservation provides no direct evidence as to which side of the feather is exposed. By contrast, as the lateral parts of the strap-like rhachides, which surround the median stripe, are not pigmented (see O'Connor et al. 2012), the dark median stripe could be a potential "eye-catcher" for other members of the species. Thus, assuming an ornamental function, it is more plausible that the median stripe was located dorsally or part of the internal pith (see below) and is actually not homologous with the ventral furrow of the rhachis.

In this context, Carvalho et al. (2015b) described the presence of a thin midline furrow along the broadened rhachis of the racket plumes of *Cratoavis*, which was interpreted as dorsal groove, a structure unknown for recent bird feathers. The authors apparently presumed that the rhachis-dominated racket plumes of *Cratoavis* are preserved in dorsal view without giving any explanation other than that the proximal caudal vertebrae and pygostyle are preserved in that view as well. On the basis of the Berlin specimen of *Archaeopteryx lithographica*, this equation of skeletal and integumental orientation is taphonomically not always valid, as in this particular specimen the skeleton is visible in dorsolateral view (Wellnhofer 2009), while the wings clearly show the ventral aspect (Wellnhofer 2009; Longrich et al. 2012). Such preservational artifacts result when the fossil is unevenly split between the two plates. Thus, without providing further evidence that the rectrices of *Cratoavis* are actually preserved in dorsal view, it is alternatively possible that the longitudinal furrow actually represents the ventral furrow of the rhachis (see Lucas and Stettenheim 1972).

10.4 Morphological Comparison with Modern Feather Types

Hereinafter, the hierarchical organization and the morphology of single structures common in rhachis-dominated racket plumes are compared with similar-looking structures in modern feather types. As rhachis-dominated racket plumes are extinct as morphotype, this comparison is restricted to single feather structures, while it is simultaneously extended to feather types from other body regions, which often fulfill a very different biological role. In consequence, functional aspects cannot be transmitted to morphological structure one to one.

10.5 Modern Feather Types with a Distally Branching Portion

Distally restricted vanes are known in racket plumes (Bleiweiss 1987; Fig. 10.3a–d) and filoplumes (Lucas and Stettenheim 1972; Fig. 10.3e). Modern racket plumes represent a type of display feathers, which occurs in the head and tail regions of various recent birds (e.g., *Prioniturus discurus*, *Ocreatus underwoodii*, *Loddigesia mirabilis*, *Tanysiptera carolinae*, *Parotia carolae*). Their distal portion consists of a thin rhachis with distinct pennaceous vanes, which merge proximally into a thin, "naked" wire section. This wire section, however, is not truly naked, but consists of narrow vanes of densely packed, rudimentary barbs, running along both sides of the rhachis. The most proximal portion of the racket plumes, however, can be fully vaned again, showing the typical pennaceous morphology (Bleiweiss 1987). Despite overall similarities, the rhachis-dominated racket plumes of the stem birds discussed above seem to show no indication for the presence of short barbs in the proximal portion of the rhachis (see *Cratoavis*), at least under normal light (see below; Fig. 10.3a–d). This situation is also evident in various rhachis-dominated feathers found in the Upper Cretaceous Burmese amber (Xing et al. 2018). Instead, the feathers often possess a dark lateral stripe on each side of the rhachis (O'Connor et al. 2012, see below).

By contrast, fully grown filoplumes, which fulfill a biological role as sensory organs, possess a small number of distal barbs, which are fused into a thin rhachis. The rhachis itself is elongated, showing a long naked portion, before it anastomoses ventrally into a short calamus (Lucas and Stettenheim 1972; Fig. 10.3e). Consequently, the gross organization of filoplumes (i.e., the portion of vanes and rhachis) resembles to some degree the morphology of the rhachis-dominated racket plumes as interpreted by Prum (2010), Xu et al. (2010c) and O'Connor et al. (2012). However, filoplumes are much smaller in size, possess a very short open (not pennaceous) vane, a thin rhachis, and, in contrast to most other feather types (except of bristles and semibristles), are associated with the nervous system located within the follicle (Lucas and Stettenheim 1972).

10.6 Modern Feather Types with Broadened Rhachis

In most feathers, the rhachis is a four-sided element and not conspicuously broadened and flattened (Lucas and Stettenheim 1972). However, display feathers of several bird species show a distal expansion. In the scale-feathered malkoha (*Phaenicophaeus cumingi*) and curl-crested aracari (*Pteroglossus beauharnaesii*), the distal expansions are caused by the lateral fusion of several barbs (Brush 1965, 1967). In contrast, in the rail species *Rallus aquaticus*, *Rallus elegans*, and *Rallus longirostris* and the cedar waxwing (*Bombycilla cedrorum*), a similar morphology results from the broadening of the terminal barb, which forms the tip of the rhachis (Brush 1967). The display feathers of the African openbill (*Anastomus lamelligerus*) show a mixture of both morphologies, as the most terminal barb is elongated and broadened, while additional, distally located barbs are fused to the terminal barb proximally (Vigneron et al. 2006). However, in all of these examples, the lateral expansion of the rhachis is restricted to the distal tip of the feather. More proximally, the rhachis thins to the common pennaceous condition.

The only example of modern feathers possessing a broadened, flattened rhachis over their entire length is known from penguins (Wohlauer 1901; Chandler 1916; Rutschke 1965). The rhachis of the body feathers, for instance, emerges from a short, cylindrical calamus. Proximally the rhachis is oval in cross section, but continuously expands laterally, while flattening dorsoventrally, before tapering at the feather tip (Chandler 1916; Rutschke 1965; Fig. 10.4). Feathers from the belly region possess a thin and shallow ventral furrow in the proximal half of the rhachis (Fig. 10.4b), while such a structure is absent in the back feathers. In contrast, prominent ventral furrows giving the rhachis an open C-shaped cross section are present in the remiges

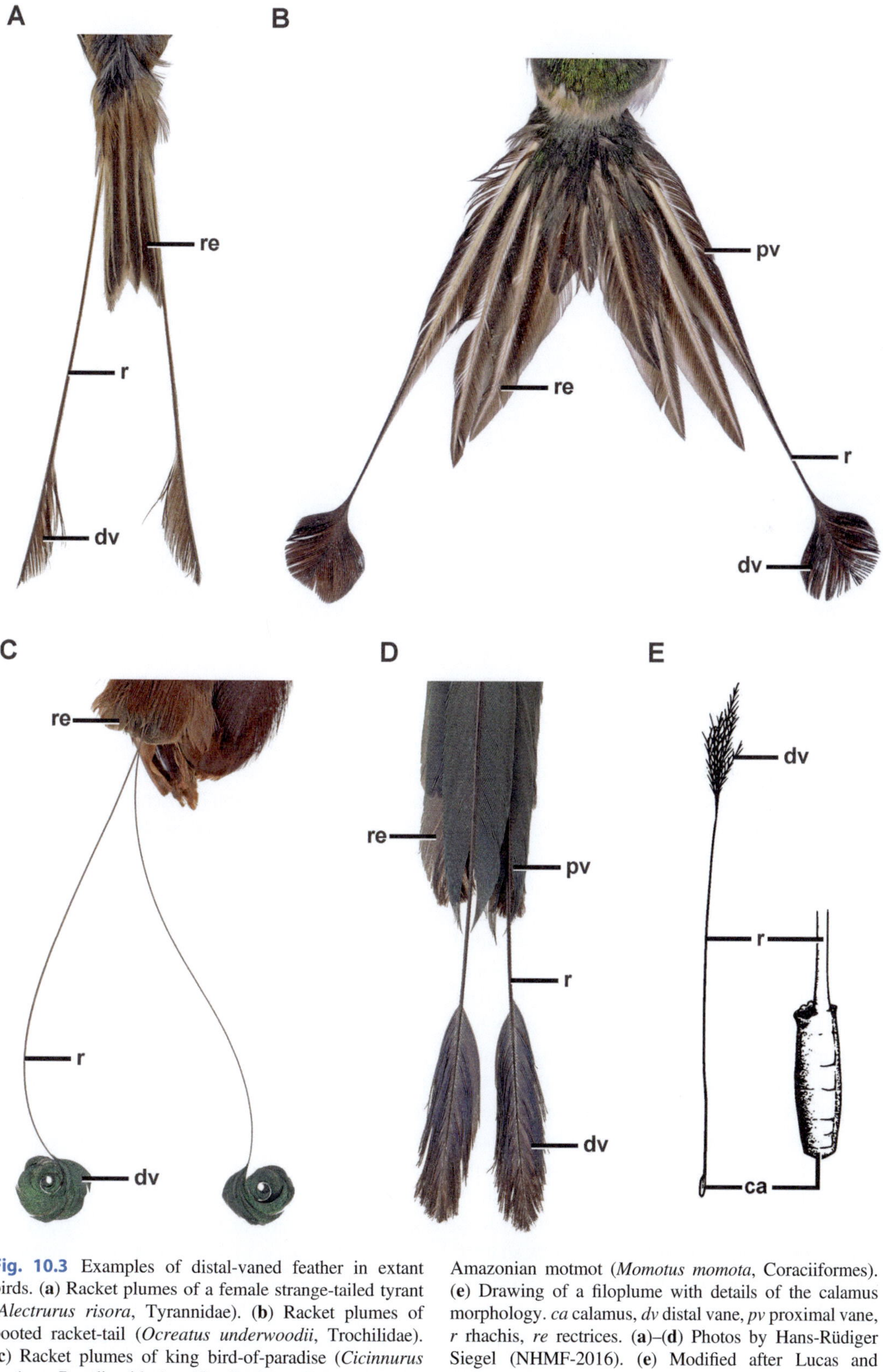

Fig. 10.3 Examples of distal-vaned feather in extant birds. (**a**) Racket plumes of a female strange-tailed tyrant (*Alectrurus risora*, Tyrannidae). (**b**) Racket plumes of booted racket-tail (*Ocreatus underwoodii*, Trochilidae). (**c**) Racket plumes of king bird-of-paradise (*Cicinnurus regius*, Paradisaeidae). (**d**) Racket plumes of the Amazonian motmot (*Momotus momota*, Coraciiformes). (**e**) Drawing of a filoplume with details of the calamus morphology. *ca* calamus, *dv* distal vane, *pv* proximal vane, *r* rhachis, *re* rectrices. (**a**)–(**d**) Photos by Hans-Rüdiger Siegel (NHMF-2016). (**e**) Modified after Lucas and Stettenheim (1972)

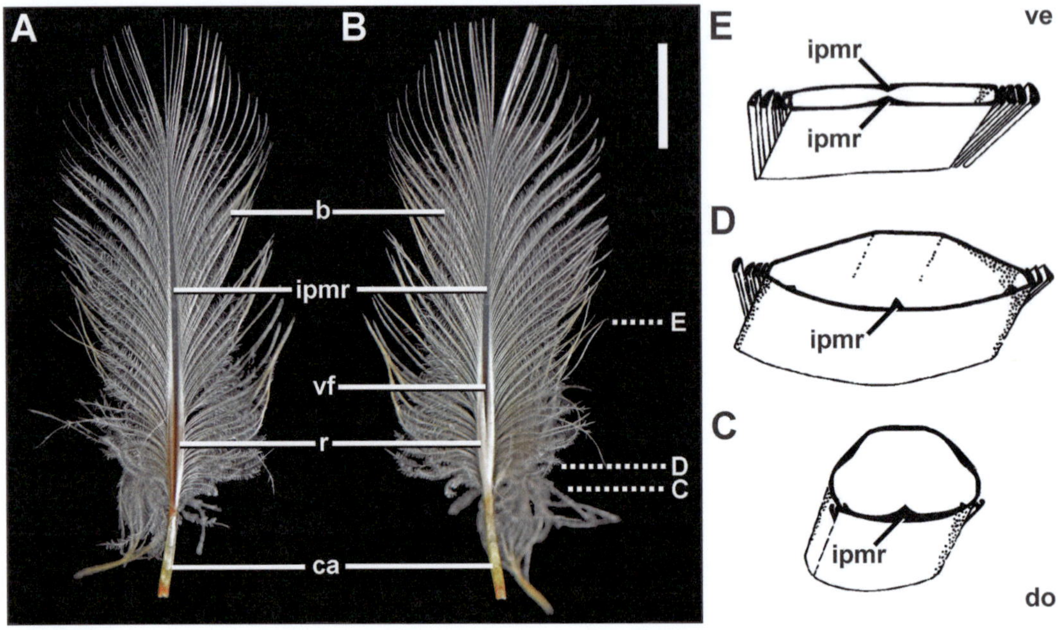

Fig. 10.4 Morphology of a pennaceous body feather of the emperor penguin (*Aptenodytes forsteri*). (**a**) Dorsal view. (**b**) Ventral view. (**c–e**) Drawings of the cross section of rhachis of a pennaceous body feather of *Aptenodytes forsteri* from different portions. *do* dorsal, *ve* ventral, *b* pennaceous barbs, *ca* calamus, *ipmr* internal pigmented median ridge, *r* rhachis, *vf* ventral furrow. Scale bars in (**a**)–(**b**) is 1 cm. (**c**)–(**e**) modified after Rutschke (1965)

and rectrices of penguins (Rutschke 1965). Thus, although fully vaned and shorter in relative length, the rhachis morphology of penguin body feathers resembles the observations of rhachis-dominated racket plumes and the fully pennaceous rhachis-dominated rectrices of *Eopengornis martini* and *Parapengornis eurycaudatus* (see discussion). Recent discoveries of rhachis-dominated feathers from Upper Cretaceous Burmese amber seem to contradict this comparison, by showing a central (rhachidal) ridge surrounded by two undifferentiated laminae, which lack an internal pith, but having a ventrally opened C-shaped cross section (Xing et al. 2018). However, as the central element of these feathers measures less than 1 mm in a diameter, this particular morphology could result from miniaturization, showing a broadened rhachis without a pith. In fact, many modern feather types with delicate barbs (e.g., small down feathers, many neoptile down feathers) or rhachides (e.g., filoplumes, small bristle feathers) also lack an internal pith (Lucas and Stettenheim 1972; Foth 2011). Thus, due to significant size differences, the rhachis morphology of the Upper Cretaceous Burmese amber does not necessarily correspond to the larger tail streamers found in the birds from the Jehol group. However, as stated above, the rhachides of remiges and rectrices in penguins also have a C-shaped in cross section (Rutschke 1965), resembling to a certain degree the condition found in the Burmese feathers.

10.7 Rhachis Pigmentation in Modern Feathers

In analogy to the general preservation of feathers as dark carbonized traces, the dark median stripe found in rhachis-dominated racket plumes is most likely based on preservation of melanosomes (see Vinther et al. 2008; Li et al. 2010; Zhang et al. 2010) and thus indicates a color pattern along the rhachis. In modern feathers, very complex color patterns can be present, but are usually exposed on the dorsal surface of the vanes of pennaceous feathers (Prum and Williamson 2002). The rhachis itself is often monochromatic, sometimes shaded, but not complexly pigmented. Here, pigments can be concentrated in the pith or in the cortex of the rhachis (Rutschke 1965; Brush 1967). For instance, in the scale-feathered malkoha, curl-crested aracari, different rail species, and the cedar waxwing (see above), high concentrations of melanin are present in the rhachidal pith (Brush and Allen 1963; Brush 1967). This kind of pigmentation results in a plane, dark, monochromatic appearance of the rhachis, which is, however, different from the situation found in the fossil examples. In contrast, the whitish, broadened rhachis of penguin feathers possess a thin, dark median stripe, which is usually expressed on the dorsal side of the feather (Wohlauer 1901; Rutschke 1965; Fig. 10.4a). This structure results from the presence of a high concentration of melanin pigments, which are located within a longitudinal, internal ridge that runs along the dorsal side of the cortex, while the rest of the cortex is unpigmented (Rutschke 1965; Fig. 10.4a, c–e). In some penguin feathers, a similar, median stripe is additionally present on the ventral side, which fuses with the dorsal ridge in the distal portion of the rhachis (Rutschke 1965; Fig. 10.4b, e). Also slightly different in morphology, the rhachis-dominated feathers from the Upper Cretaceous Burmese amber show a median ridge along the rhachis that is strongly pigmented (Xing et al. 2018).

In this context, the dark, lateral margins, originally described as undifferentiated vanes (O'Connor et al. 2012), could be the result from similar, highly pigmented, internal cortical ridges, running along the lateral side of the rhachis. Alternatively, the dark, lateral stripes could be also a preservational artifact caused by the conservation of highly pigmented, very short, but densely packed barbs, which cannot be detected with the help of normal light microscopy techniques. In this case, the proximal portion of the elongated rectrices would not be truly naked, but similar to the wire structures found in racket plumes (Bleiweiss 1987, see above). Here, the usage of laser-stimulated fluorescence (LSF) may be able to help to clarify the morphology of these structures in the future, as this autofluorescence method was successfully employed to visualize the remains of tiny barbules in fossil feathers, which were hardly detectable under white and polarized light conditions (Kaye et al. 2015). However, the rhachis-dominated feathers from *Cratoavis* and the Upper Cretaceous Burmese amber (Carvalho et al. 2015a; Xing et al. 2018) indicate that the proximal portion of the rhachis was actually naked.

10.8 Discussion

Within Pygostylia, rhachis-dominated racket plumes evolved at least two times independently within the stem line of birds, in Confuciusornithidae, and Enantiornithes (Foth et al. 2014; Wang et al. 2014). The presence of a pair of elongated, fully pennaceous, but rhachis-dominated, rectrices in the two Enantiornithes *Eopengornis* and *Parapengornis* (Fig. 10.1b) and the discovery of an enantiornithine bird with a rectricial fan have led to the conclusion that rhachis-dominated racket plumes were highly modified pennaceous feathers (O'Connor et al. 2012, 2016; Wang et al. 2014; Hu et al. 2015). This is further supported by the occurrence of delicate median stripes in the rhachis of wing feathers (=remiges) of some stem birds like *Confuciusornis* spp. (Confuciusornithidae) and *Eopengornis* (Enantiornithes) (Wang et al. 2014, 2015). With this review, I attempt to clarify a number of problematic aspects regarding the morphology of these feathers that have been published in the last years. In particular, the gross organization of these feathers resembles either that of filoplumes, containing a distally branched portion fused into a long, naked rhachis. The comparison with the wire section of modern racket plumes might be inadequate, as it possesses a series of short, densely arranged barbs running along the elongated "naked" portion (Fig. 10.3). Nevertheless, the distal portion itself was fully pennaceous as in modern racket plumes. The long, central element probably represents a single, elongated, strap-like rhachis, which most likely merges proximally into a short, cylindrically shaped calamus, thereby resembling the condition of the fully pennaceous rhachis-dominated rectrices of *Eopengornis martini* and *Parapengornis* and potentially that of modern penguin body feathers (Fig. 10.4). However, this particular morphology could be modified to a more laminar shape (Xing et al. 2018) due to miniaturization, resulting in a reduction of the internal rhachidal pith (Lucas and Stettenheim 1972). In further analogy to penguin feathers, the dark median stripe running along the broadened rhachis might represent a strongly pigmented internal cortical ridge (Fig. 10.4). [The situation for the median stripe in the wing feathers of some stem birds (see above) is not evaluated here due to the unexplored situation in terms of the presence of this particular character in modern bird wing remiges.] If one assumes an ornamental function, this pigmented ridge would probably have been located on the dorsal side of the rhachis, although a (additional) ventral expression, as in some penguin feathers, cannot be ruled out. Taking the variety of pigmentation patterns of modern feathers into account and the fact that these ornamental feathers originated at least two times independently, the occurrence of the pigmented ridge on the dorsal or ventral side could be variable and differ between taxa. In the strong miniaturized feathers from the Upper Cretaceous Burmese amber, the median ridge is even externally recognizable from both dorsal and ventral side (Xing et al. 2018), which could be caused by the reduction of the internal pith, leading to the extreme lamination of the rest of the rhachis. In analogy to modern pennaceous feathers, the longitudinal groove found in the rectrices in *Cratoavis* most likely represents the ventral furrow of the rhachis and not a dorsal groove as originally interpreted. The dark lateral margins in the proximal ribbonlike portion (Fig. 10.1c) could result from either pigmented internal lateral cortical ridges or very short, densely packed pigmented barbs running along the rhachis. As the rhachis-dominated feathers of *Cratoavis* and the Upper Cretaceous Burmese amber seem to have smooth lateral margins, the second alternative seems to be less likely, at least for Enantiornithes. And, once again, given that this feather type evolved two times independently (see above), it cannot be ruled out that the dark lateral margins evolved differently among Confuciusornithidae and Enantiornithes. To test this, the morphology of the lateral margins has to be investigated in more detail in the future using autofluorescence methods (see Kaye et al. 2015).

Despite these uncertainties, all proposed structures can be verified with an analog example found in modern feather types. This in turn implies that this very specialized fossil feather type falls into the morphological, and therefore developmental (including the genetic control), spectrum of modern feathers. Previously, O'Connor et al. (2012) proposed a hypothetical molecular developmental model, where rhachis enlargement is caused by changes in the BMP (bone morphogenetic protein), Noggin, and Shh (sonic hedgehog) activity (see also Yu et al. 2002). Due to the great similarities with the rhachis morphology of penguin feathers, this model can now be tested directly by studying feather morphogenesis in this group of birds.

While the broad rhachides of penguin feathers represent one of the many morphological adaptions of the plumage to the semiaquatic lifestyle (Rutschke 1965), the enigmatic, rhachis-dominated racket plumes of Confuciusornithidae and Enantiornithes had probably an ornamental function (Peters and Peters 2009; O'Connor et al. 2012; Foth et al. 2014) similar to the distally expanded or elongated feather examples mentioned above (see Brush 1965, 1967; Bleiweiss 1987; Vigneron et al. 2006). In extant birds, the expanded portions are usually highlighted by color patterns created by pigments or nanostructural organization to the cortex and pith. For instance, the internal organization of parallel layers in the cortex of the body feathers of the Africa openbill (Vigneron et al. 2006) creates thin-film interferences due to refraction and reflection along the surfaces of each single layer, resulting in a gleaming color pattern. Thus, the presence of a broadened rhachis in the feathers discussed herein may be a strong indicator for a complex, gleaming color pattern with delicate dark highlights resulting from the median and lateral stripes. In addition to these ornamental functions, it was also hypothesized that the long rectrices had an aerodynamic function (Zhang et al. 2006). Vane asymmetry in the fully pennaceous rectrices of *Eopengornis* indicates that aerodynamics was an important biological role in the precursor of rhachis-dominated racket plumes (Wang et al. 2014). However, as the short pennaceous tip of the latter possesses a symmetric shape and cannot produce much lift, an evolutionary shift toward a stronger ornamental function was hypothesized (Wang et al. 2014).

10.9 Conclusions

The enigmatic, elongated tail feathers of Confuciusornithidae and Enantiornithes are here interpreted as highly modified pennaceous feathers that originated independently from each other during evolution. A review of previous morphological interpretations and taphonomic preservation of this feather type and a careful comparison with modern feather morphologies shows that these feathers are very similar to the body feathers of penguins in terms of rhachis morphology and pigmentation pattern, while the gross organization resembles that of filoplumes. Assuming a similar cortical structure to the rhachis of the African openbill, the rectrices of these stem birds can be inferred to probably have been iridescent, supporting a possible ornamental function. As all morphological structures can be verified with an analog example, found in modern feather types, this fossil feather type falls into the morphological spectrum of modern feathers. This in turn indicates that both the morphogenesis (including the genetic control) could be potentially studied with the help of the modern analogues presented herein.

Acknowledgements I thank Jingmai O'Connor and Gao Wei (both Institute of Vertebrate Paleontology and Paleoanthropology) for photos of the elongated tail feathers of basal birds. I further thank Emanuel Gerber and Hans-Rüdiger Siegel (both Natural History Museum Fribourg, NHMF-2016) for photos of the elongated tail feathers of recent birds. Norma Schmitz and Ragnar Kinzelbach (both Zoological collection of the University of Rostock) are acknowledged for feather samples of *Aptenodytes forsteri* and Walter Joyce (University of Fribourg) for proofreading the manuscript. Jingmai O'Connor is further thanked for her critical comments on the previous version of the chapter. The study was supported by the Swiss National Science Foundation under grant PZ00P2_174040.

References

Agnolín FL, Novas FE (2013) Avian ancestors. A review of the phylogenetic relationships of the theropods Unenlagiidae, Microraptoria, *Anchiornis* and Scansoriopterygidae. Springer, Dordrecht

Andersson M (1982) Female choice selects for extreme tail length in a widowbird. Nature 299:818–820

Benton MJ, Zhou Z, Orr PJ, Zhang F, Kearns SL (2008) The remarkable fossils from the early cretaceous Jehol biota of China and how they have changed our knowledge of Mesozoic life. Proc Geol Assoc 119:209–228

Bleiweiss R (1987) Development and evolution of avian racket plumes: fine structure and serial homology of the wire. J Morphol 194:23–39

Brush AH (1965) The structure and pigmentation of the feather tips of the scaled cuckoo (*Lepidogrammus cumigi*). Auk 82:155–160

Brush AH (1967) Additional observations on the structure of unusual feather tips. Wilson Bull 79:322–327

Brush AH, Allen K (1963) Astaxanthin in the Cedar Waxwing. Science 142:47–48

Carvalho IS, Novas FE, Agnolín FL, Isasi MP, Freitas FI, Andrade JA (2015a) A new genus and species of enantiornithine bird from the early cretaceous of Brazil. Brazilian J Geol 45:161–171

Carvalho IS, Novas FE, Agnolín FL, Isasi MP, Freitas FI, Andrade JA (2015b) A Mesozoic bird from Gondwana preserving feathers. Nat Commun 6:7141

Chandler AC (1916) A study of the structure of feathers, with reference to their taxonomic significance. Univ Calif Publ Zool 13:243–446

Chiappe LM, Ji S, Ji Q, Norell MA (1999) Anatomy and systematics of the Confuciusornithidae (Theropoda: Aves) from the late Mesozoic of northeastern China. Bull Am Mus Nat Hist 242:1–89.17

Felice RN (2014) Coevolution of caudal skeleton and tail feathers in birds. J Morphol 275:1431–1440

Foth C (2011) The morphology of neoptile feathers: ancestral state reconstruction and its phylogenetic implications. J Morphol 272:387–403

Foth C (2012) On the identification of feather structures in stem-line representatives of birds: evidence from fossils and actuopalaeontology. Paläontol Z 86:91–102

Foth C, Tischlinger H, Rauhut OWM (2014) New specimen of *Archaeopteryx* provides insights into the evolution of pennaceous feathers. Nature 511:79–82

Hu H, O'Connor JK, Zhou Z (2015) A new species of Pengornithidae (Aves: Enantiornithes) from the lower cretaceous of China suggests a specialized scansorial habitat previously unknown in early birds. PLoS One 10:e0126791

Kaye TG, Falk AR, Pittman M, Sereno PC, Martin LD, Burnham DA, Gong E, Xu X, Wang Y (2015) Laser-stimulated fluorescence in paleontology. PLoS One 10: e0125923

Li Q, Gao K, Vinther J, Shawkey MD, Clarke JA, D'Alba L, Meng Q, Briggs DEG, Prum RO (2010)

Plumage color patterns of an extinct dinosaur. Science 327:1369–1372
Longrich NR, Vinther J, Meng Q, Li Q, Russell AP (2012) Primitive wing feather arrangement in *Archaeopteryx lithographica* and *Anchiornis huxleyi*. Curr Biol 22:1–6
Lucas AM, Stettenheim PR (1972) Avian anatomy. Integument, part I and II. U.S. Government Printing Office, Washington, DC
O'Connor JK, Sullivan C (2014) Reinterpretation of the early cretaceous maniraptoran (Dinosauria: Theropoda) *Zhongornis* haoae as a scansoriopterygid-like non-avian, and morphological resemblances between scansoriopterygids and basal oviraptorosaurs. Vertebrata PalAsiatica 52:3–30
O'Connor JK, Chiappe LM, Chuong C, Bottjer DJ, You H (2012) Homology and potential cellular and molecular mechanisms for the development of unique feather morphologies in early birds. Geosciences 2:157–177
O'Connor JK, Wang X, Sullivan C, Zheng X, Tubaro P, Zhang X, Zhou Z (2013) Unique caudal plumage of *Jeholornis* and complex tail evolution in early birds. Proc Natl Acad Sci 110:17404–17408
O'Connor JK, Wang X, Zheng X, H H, Zhang X, Zhou Z (2016) An enantiornithine with a fan-shaped tail, and the evolution of the rectricial complex in early birds. Curr Biol 26:114–119
Peters WS, Peters DS (2009) Life history, sexual dimorphism and "ornamental" feathers in the mesozoic bird *Confuciusornis sanctus*. Biol Lett 5:817–820
Prum RO (2010) Moulting tail feathers in a juvenile oviraptorosaur. Nature 468:E1
Prum RO, Dyck J (2003) A hierarchical model of plumage: morphology, development, and evolution. J Exp Zool (MOL DEV EVOL) 298B:73–90
Prum RO, Williamson S (2002) Reaction-diffusion models of within-feather pigmentation patterning. Proc R Soc Lond B 269:781–792
Rashid DJ, Chapman SC, Larsson HCE, Organ CL, A-G Bebin CS, Merzdorf RB, Horner JR (2014) From dinosaurs to birds: a tail of evolution. EvoDevo 5:1–20
Rutschke E (1965) Beiträge zur Morphologie der Pinguinfeder. Z Morphol Okol Tiere 55:835–858
Vigneron JP, Lousse V, Colomer J-F, Rassart M, Louette M (2006) Complex optical structure in the ribbon-like feathers of the African open-bill stork. Proc SPIE 6320:632014
Vinther J, Briggs DEG, Prum RO, Saranathan V (2008) The colour of fossil feathers. Biol Lett 4:522–525
Wang X, O'Connor JK, Zheng X, Wang M, Hu H, Zhou Z (2014) Insights into the evolution of rhachis dominated tail feathers from a new basal enantiornithine (Aves: Ornithothoraces). Biol J Linn Soc 113:805–819
Wang M, Zheng X, O'Connor JK, Lloyd GT, Wang X, Wang Y, Zhang X, Zhou Z (2015) The oldest record of Ornithuromorpha from the early cretaceous of China. Nat Commun 6:6987
Wellnhofer P (2009) *Archaeopteryx*: the icon of evolution. Verlag Dr. Friedrich Pfeil, München
Wohlauer E (1901) Die Entwicklung des Embryonalgefieders von *Eudyptes chrysocome*. Z Morphol Anthropol 4:149–178
Xing L, Cockx P, McKellar RC, O'Connor JK (2018) Ornamental feathers in cretaceous Burmese amber: resolving the enigma of rachis-dominated feather structure. J Palaeogeogr 7:13
Xu X, Guo Y (2009) The origin and early evolution of feathers: insights from recent paleontological and neontological data. Vertebrata PalAsiatica 47:311–329
Xu X, Ma Q, Hu D (2010a) Pre-*Archaeopteryx* coelurosaurian dinosaurs and their implications for understanding avian origins. Chin Sci Bull 55:3971–3977
Xu X, Zheng X, You H (2010b) Exceptional dinosaur fossils show ontogenetic development of early feathers. Nature 464:1338–1341
Xu X, Zheng X, You H (2010c) Reply: moulting tail feathers in a juvenile oviraptorosaur. Nature 468:E2
Xu X, Zheng X, Sullivan C, Wang W, Xing L, Wang Y, Zhang X, O'Connor JK, Zhang F, Pan Y (2015) A bizarre Jurassic maniraptoran theropod with preserved evidence of membranous wings. Nature 521:70–73
Yu M, P W, Widelitz RB, Chuong C (2002) The morphogenesis of feathers. Nature 420:308–312
Zhang F, Zhou Z (2000) A primitive enantiornithine bird and the origin of feathers. Science 290:1955–1959
Zhang F, Zhou Z, Dyke GJ (2006) Feathers and "feather-like" integumentary structures in Liaoning birds and dinosaurs. Geol J 41:395–404
Zhang F, Zhou Z, Benton MJ (2008a) A primitive confuciusornithid bird from China and its implications for early avian flight. Sci China Ser D Earth Sci 51:625–639
Zhang F, Zhou Z, Xu X, Wang X, Sullivan C (2008b) A bizarre Jurassic maniraptoran from China with elongate ribbon-like feathers. Nature 455:1105–1108
Zhang F, Kearns SL, Orr PJ, Benton MJ, Zhou Z, Johnson D, Xu X, Wang X (2010) Fossilized melanosomes and the colour of cretaceous dinosaurs and birds. Nature 463:1075–1078
Zheng X, Zhang Z, Hou L (2007) A new enantiornitine bird with four long rectrices from the early cretaceous of northern Hebei, China. Acta Geol Sin 81:703–708
Zheng X, O'Connor JK, Wang X, Pan Y, Wang Y, Wang M, Zhou Z (2017) Exceptional preservation of soft tissue in a new specimen of *Eoconfuciusornis* and its biological implications. Natl Sci Rev 4: 441–452

Palaeocolour: A History and State of the Art

11

Fiann Smithwick and Jakob Vinther

Alongside being the most diverse and widespread of all tetrapod clades, birds show the most remarkable range of colours in any group of terrestrial vertebrates (Baker and Parker 1979; Hill and McGraw 2006a). From the cryptic browns and greys of dunnocks and sparrows to the glorious rich palette of colours in the birds of paradise, parrots and hummingbirds, almost every colour imaginable to the human observer appears to be present in extant birds (Hill and McGraw 2006a, b). Thanks to features such as tetrachromacy and ultraviolet (UV) vision, many birds can also see colours invisible to di- or trichromatic mammals (Vorobyev et al. 1998). Colour plays a key role in ecology. Be it for camouflage from predators or prey, sexual display, species recognition, or as warning signalling (aposematism), colour has doubtlessly been a key driver in the evolution of bird plumage (Baker and Parker 1979; Hill and McGraw 2006a; Vinther 2015a). The range of colours exhibited by birds can likely be explained, at least in part, by the importance of visual cues in avian signalling due to their excellent tetrachromatic visual capabilities (Vorobyev et al. 1998; Koschowitz et al. 2014). These are only rivalled in vertebrates by the equally colourful teleost fishes (Cuthill 2006). Bird colouration has fascinated naturalists and scientists for centuries and helped to galvanise the theories of evolution by both natural and sexual selection (Baker and Parker 1979; Darwin 1859, 1871).

The dazzling array of colours seen in birds has traditionally been attributed to two mechanisms of colour production: the utilisation of pigments, biopolymers that differentially absorb and reflect specific wavelengths of light, and nanostructural arrays within feathers (McGraw et al. 2005; McGraw 2006a, b, c; Prum 2006). Structural colours are produced in two primary ways in bird plumage. Iridescence is angle-dependent refraction, most often associated with pigment layers and keratin interacting with incident light to modulate it resulting in the reflection of specific wave lengths of light (McGraw 2006b; Prum 2006; Igic et al. 2016). Non-iridescent structural colour is produced by a complex network of quasi-ordered air bubbles inside the keratin matrix (Babarović et al. 2019). This serves to scatter certain light waves, while an underlying melanin layer usually serves to absorb the remaining unscattered light (Prum 2006).

The hues, saturation and brightness of pigmentary colours are also controlled in part by their arrangement within the keratin matrix. Thus, a likely continuum in colour production involving both structural and pigmentary mechanisms, rather than a strictly dichotomous division of pigmentary and structural production, exists (McGraw et al. 2005; McGraw 2006b; Prum 2006; Galván and Solano 2016). Novel nuances of bird colour production are still being

F. Smithwick · J. Vinther (✉)
School of Earth Sciences, University of Bristol, Bristol, UK
e-mail: Jakob.Vinther@bristol.ac.uk

C. Foth, O. W. M. Rauhut (eds.), *The Evolution of Feathers*, Fascinating Life Sciences,
https://doi.org/10.1007/978-3-030-27223-4_11

recognised, and it is likely that we are far from fully appreciating the variations that exist in how certain hues are produced in extant bird plumages (Igic et al. 2016).

Pigmentary colour production involves the use of multiple different pigment types in birds. These include melanins, carotenoids, porphyrins, flavins, psittacofulvins, pterins and purines among others (McGraw and Nogare 2004; McGraw 2006a, b, c; Galván and Solano 2016). The most ubiquitous and likely ancient of these are the melanins (McGraw 2006b; Galván and Solano 2016; D'Alba and Shawkey 2018). Melanins are synthesised within the animal, whereas numerous other pigment classes including carotenoids, porphyrins and pterins are obtained directly from the diet. These diet-derived pigments may however be altered in their molecular structure after ingestion (McGraw 2006a, c). These less common pigment classes usually confer different hues to melanins, such as the reds, oranges and yellows of carotenoids and psittacofulvins. They are presumed to often be honest indicators of quality due to the need to obtain them through the diet (Olson and Owens 1998; McGraw 2006a; LaFountain et al. 2015).

Melanin is thought to be highly conserved throughout vertebrates, which synthesise the pigment within organelles called melanosomes (McGraw 2006b; Vinther 2015a; Clements et al. 2016). Melanosomes are found in multiple tissue types in vertebrates, including the eyes, internal organs and integument (Vinther 2015a; Clements et al. 2016; McNamara et al. 2018). It is those of the integument that are involved in the key role of colour production, although they serve multiple other functions alongside this including thermoregulation, as antimicrobial barriers and protection against UV radiation (McGraw 2006b; Margalida et al. 2008; Vinther 2015a; Galván and Solano 2016). In birds, most melanin involved in colour production is found in the feathers, where it is deposited by keratinocytes after being synthesised in melanocytes (McGraw 2006b; Galván and Solano 2016).

Two distinct types of melanin are known in vertebrates, including birds (McGraw 2006b; Galván and Solano 2016). Eumelanin, which is the most common of the two, imparts black, dark grey and dark brown hues (Fig. 11.1a–d) (McGraw 2006b; Vinther 2015a). Phaeomelanin produces lighter brown, yellow and reddish hues (Fig. 11.1e, f) (McGraw 2006b; Vinther 2015a). The two melanin types have distinct chemistries and are synthesised through different pathways (see below) (McGraw 2006b; Galván and Solano 2016). The physical structures of the two melanosome types are also distinct. Eumelanosomes exhibit a 'sausage-shaped' oblong morphology, while phaeomelanosomes show a spherical to ovoid morphology (Fig. 11.1; Vinther 2015a). The evolutionary origins of melanin are also

Fig. 11.1 Example of melanised colours in birds and the associated melanosome morphologies. (**a**) Black—*Buceros rhinoceros*. (**b**) Ellipsoidal eumelanosomes extracted from a black feather of *B. rhinoceros*. (**c**) Grey—*Semnornis ramphastinus*. (**d**) Large ellipsoidal eumelanosomes extracted from a grey feather of *S. ramphastinus*. (**e**) Rufous—*Alcedo atthis*. (**f**) Spherical phaeomelanosomes extracted from a rufous feather of *A. atthis*. (**g**) Glossy iridescent—*Collocalia isonota*. (**h**) Long, thin melanosomes extracted from *C. isonota*. (**i**) Bright iridescent—*Amazilia tobaci*. (**j**) Flattened plate-like melanosomes extracted from a green feather of *A. tobaci*. Scale bars represent 1 μm. Modern bird images from Pixabay, Flickr and Wikipedia: Perry Quan (**a**), Luis Fernando Serna Agudelo (**c**), Obsidian Soul (**g**) and Dominic Sherony (**i**)

currently unclear. Due to its ubiquity throughout vertebrates, the occurrence in fossils as far back as at least the Carboniferous and the presence of eumelanin in invertebrate clades as well as fungi, the pigment and its synthesis pathway likely have ancient origins (Glass et al. 2012; Clements et al. 2016; Galván and Solano 2016; D'Alba and Shawkey 2018). Cambrian vertebrate fossils, like *Haikouichthys* and *Metaspriggina* (Shu et al. 2003, Morris and Caron 2014), preserve their eyes and liver as an organic stain, which has been shown to be due to preserved melanin in younger occurrences. It has been suggested that due to melanin's ability to protect cells from UV radiation, owing to its broadband absorbance spectrum, it may have been essential for the evolution of life on Earth (Galván and Solano 2016). This may have been particularly the case when microorganisms began to inhabit environments exposed to the harmful effects of the sun. While these concepts relate to the darker melanin form, eumelanin, the origin and original functional role of phaeomelanin is much less clear (Galván and Solano 2016). Although the origins of phaeomelanin are uncertain, its importance in colouring the vertebrate integument and, in particular, that of birds and mammals is undoubted.

Much work has been performed in recent decades to better characterise the way in which melanosomes contribute to colour production in bird feathers (McGraw et al. 2005; McGraw 2006b, 2008; Galván and Solano 2016; Igic et al. 2016). A basic and longstanding assumption in bird colouration and, in particular, in the portrayal of palaeocolour has been that melanin-based colouration boils down to a dichotomy: black melanised colours produced by eumelanin and reddish to brown melanised colours produced by phaeomelanin (McGraw et al. 2005; McGraw 2006b). To date, most studies that have looked in detail at the melanin composition of extant bird feathers have concluded that both melanin types are present in almost every instance of melanised colouration. It is likely that the relative concentrations of each dictate the hue produced (McGraw et al. 2005; McGraw 2006b; Galván and Wakamatsu 2016).

Alongside work looking at the way melanins colour bird feathers, the discovery that melanin can preserve in the fossil record has helped achieve what was once thought an impossible endeavour: reconstructing colour in extinct animals (Vinther 2015a). This new field, known as palaeocolour, is helping us to uncover features of ancient animals hitherto thought intractable. It is also helping us understand new aspects of ancient ecologies like how colours reflect evolutionary changes in the dynamics between predators and their prey as well as intraspecific display (Vinther 2015a; Vinther et al. 2016; Brown et al. 2017; Smithwick et al. 2017a). It is obvious that the importance of colour in organismal evolution was as important in the past as it is in the present. This has likely been the case for as long as vision has existed—extending back to the Cambrian Explosion. This is particularly true for birds and their theropodan ancestors. Birds are important ecosystem engineers in most extant terrestrial and indeed aquatic environments. Dinosaurs would have been similarly important, also filling several niches comparable to modern terrestrial mammals. This invites several questions about how comparable the dynamics among predators and prey was in the Mesozoic to extant analogues and also in what respects it was different (Brown et al. 2017).

11.1 Overturning the Paradigm: From Bacteria to Coloured Dinosaurs

Soft tissues, that is, those that are not mineralised in life, are generally rare in the fossil record (Allison and Briggs 1993; Parry et al. 2017). When found however, they can offer unique additional insights (Allison and Briggs 1993; Briggs and Kear 1993; Wilby and Briggs 1997; Parry et al. 2017). Integumentary preservation in vertebrates is known from several exceptional fossil localities and includes scales, skin, hair and, importantly for the understanding of avian evolution, feathers (Davis and Briggs 1995; Vinther 2015a). The first fossil feather was found in the lithographic limestone of Solnhofen, in around 1860 (von Meyer 1861a, b, 1862; Griffiths 1996; Carney et al. 2012). This well-preserved and very detailed feather is preserved as a dark organic film in a buff coloured limestone and was used to erect a new taxon: *Archaeopteryx lithographica* (Carney et al. 2012; Fig. 11.2a). The discovery of this single isolated feather was followed just a year later by that of a remarkably complete animal from which it was assumed the isolated feather came (Fig. 11.2b) (von Meyer 1861a, b, 1862; Griffiths 1996). *Archaeopteryx* displayed a mixture of bird-like and reptile-like features. Importantly, this included a full covering of pennaceous feathers that helped to spark the now near universally accepted idea that birds are directly descended from, and for all intents and purposes are living representatives of, dinosaurs (Ostrom 1974; Carney et al. 2012; Foth et al. 2014; Brusatte et al. 2015). The feathers of the first complete *Archaeopteryx* specimen, the London specimen, as well as the later Berlin specimen (Fig. 11.2b) were preserved as inorganic impressions just like all subsequent finds.

In the 160 years since the initial iconic discovery of *Archaeopteryx*, numerous fossil localities have been uncovered globally where feathers have been found to preserve (Vinther 2015a). Among the most important of these include the Early Eocene formations of Messel (near Darmstadt, Germany), Fur (Denmark) and Green River (Wyoming, USA), exceptional Early Cretaceous deposits such as the Jehol Biota of Liaoning Province (China) and the Santana Formation (Brazil) and the Mid-Cretaceous Burmese amber deposits of Myanmar (Davis and Briggs 1995; Kellner 2002; Chang 2003; Vinther et al. 2008; Prado et al. 2016; Xing et al. 2016).

Despite fossil feathers being well known for over 150 years, in-depth questioning of the nature

Fig. 11.2 The iconic Late Jurassic paravian theropod *Archaeopteryx* from Solnhofen, Germany. (**a**) The first fossil feather to be discovered in 1860—an organically preserved isolated wing feather assigned to *Archaeopteryx* and recently shown to contain abundant melanosomes (Carney et al. 2012). (**b**) The 'Berlin specimen' (HMN 1880) showing the complete animal with feathered wings and tail displayed. Scale bars represent 5 mm in (**a**) and 50 mm in (**b**). Reproduced and modified with permission from Nature Publishing

of their preservation did not start until electron microscopic investigations became a commonplace in geology and palaeontology in the late twentieth century (Wuttke 1983; Davis and Briggs 1995; Vinther 2015a). The ultrastructure of fossil feathers was first examined alongside fossil hair and frog integument from the extraordinary Messel Formation using electron microscopy in the early 1980s (Wuttke 1983; Vinther 2015a). All integumentary structures were shown to be preserved due to the presence of abundant tiny sausage-shaped microbodies, around a micron each in length that were only found within the dark patches of the integumentary structures. In feathers, they were aligned together along the axis of the barbs and barbules (shown in our Fig. 11.3 of an undescribed isolated Messel feather). It was argued that these microbodies represented bacterial colonies and bacterial glycocalyx, which were preserving an outline of the features on which they grew (Wuttke 1983). Several mechanisms through which feathers could become preserved were subsequently postulated by Davis and Briggs (1995). These included preservation as carbonised mats, bacterial autolithification and imprintation, all of which showed similar microstructures attributed to preserved bacteria. This paradigm, that feathers (and other integumentary structures) were preserved due to the presence of the bacteria that were involved in their decay, pervaded for another decade until a novel realisation came about through an investigation of exceptional preservation in an invertebrate.

Coleoid cephalopods have been known to preserve soft tissue for around 170 years (Owen 1863). An unusual feature of their preservation often observed is that the ink sacs are usually three-dimensionally preserved due to preservation of the ink melanin pigment inside the organ. The rest of the soft tissues (including the muscular mantle) are most often preserved two-dimensionally compressed (Fig. 11.4a) (Vinther 2015a). It has long been recognised that fossil coleoid ink reveals an infrared spectroscopy chemical signature consistent with

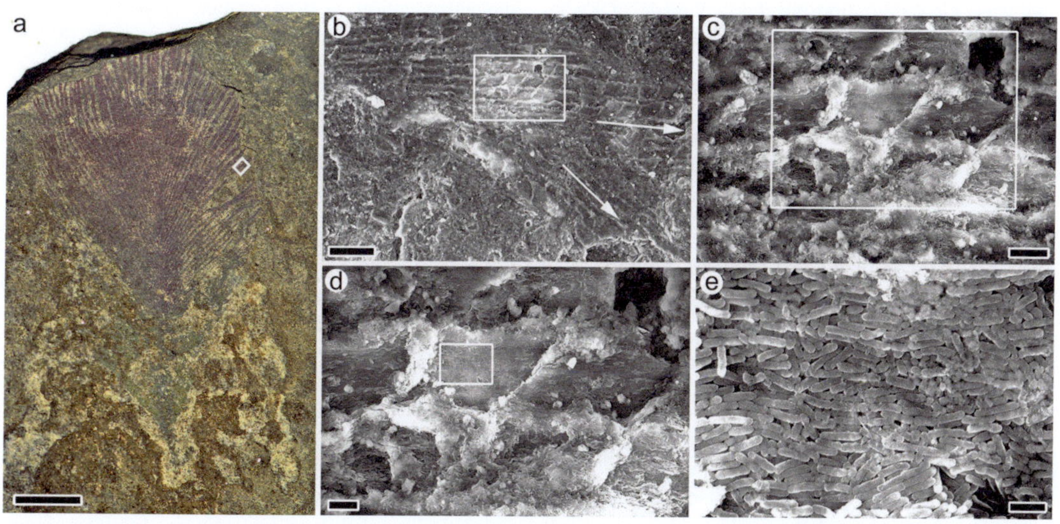

Fig. 11.3 The preservation and arrangement of melanosomes (previously identified as lithified bacteria) in an exceptional fossil feather from the Eocene Messel Formation, Germany. White boxes indicate the next SEM image in the series. (**a**) An isolated feather from an unknown bird (SMF-ME 3937) showing exceptional preservation of barbs and barbules. (**b**) SEM image of a barb showing preservation of individual barbules arranged as in a modern pennaceous feather. White arrows show the orientation of the barbules. (**c**) SEM image of the barbules showing apparent cell-like structure with each cell measuring around 10 μm in length. (**d**) Close-up SEM image of four individual cell-like structures. (**e**) SEM image of the melanosomes that explain the preservation of the barb and barbule structures, with no other original feather features (including keratin) present. The melanosomes are in their original arrangement sitting on the matrix, and it is clear that all keratin has degraded away. Scale bars represent 5 mm in (**a**), 50 μm in (**b**), 10 μm in (**c**), 5 μm in (**d**) and 1 μm in (**e**)

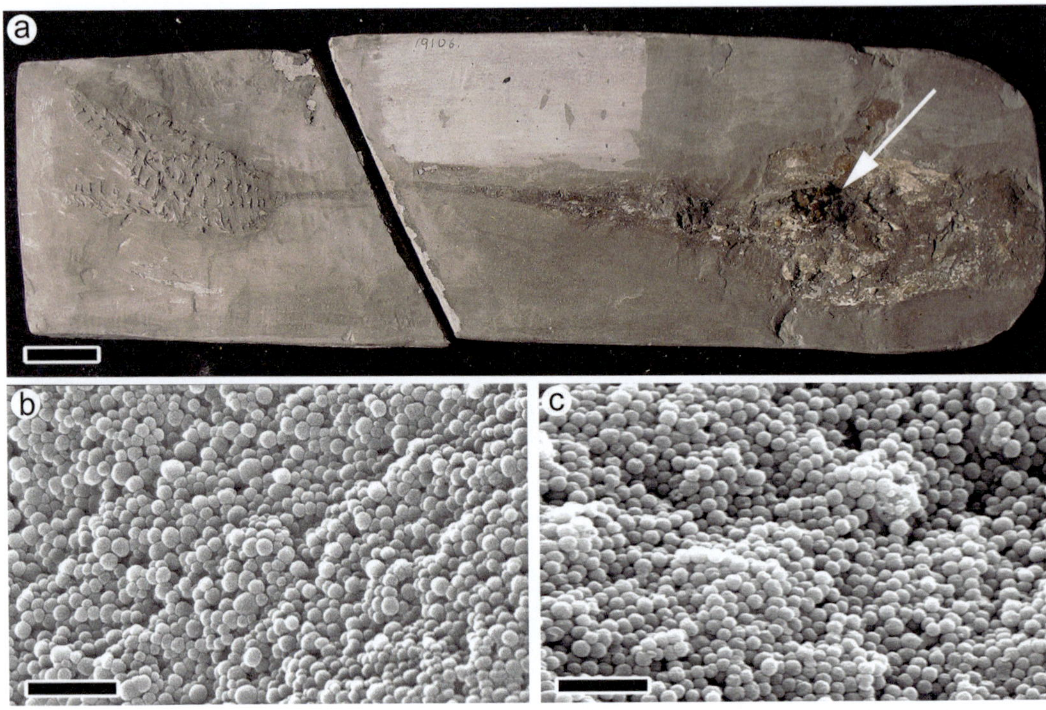

Fig. 11.4 Preserved melanin granules in fossil coleoid cephalopod ink demonstrating the remarkable recalcitrancy of the pigment. (**a**) A complete coleoid fossil of the genus *Geoteuthis* from the Early Jurassic of Lyme Regis, UK (YPM 19106), showing compressed soft tissue preservation apart from the 3D black ink sac (white arrow), uncompressed due to its constituent melanin. (**b**) SEM image of modern coleoid cephalopod (*Sepia*) ink showing the morphology of the melanin granules. (**c**) SEM image of the preserved ink from YPM 19106 showing identical melanin granule morphology to the extant *Sepia* ink. Scale bars represent 2 cm in (**a**) and 1 μm in (**b**, **c**)

melanin (Beyermann and Hasenmaier 1973). In the late 2000s, researchers examined the microstructure of Jurassic coleoid ink under an electron microscope. The preserved ink was composed of tightly packed spherical granules (Fig. 11.4c) (Vinther 2015a) identical in size and shape to modern cephalopod ink granules, which are composed of pure eumelanin (Doguzhaeva et al. 2004). The fossil ink therefore appears to be little changed over 196 million years (Fig. 11.4b) (Vinther 2015a). That the fossilised cephalopod ink sacs often showed three-dimensional preservation, despite compression of the rest of the animal, highlighted a key feature of melanin, its high recalcitrancy and resistance to degradation (Vinther 2015a, b). Subsequent work found that the precise chemical nature of the fossil ink was similar to that of modern melanin from the cuttlefish *Sepia officinalis* (Glass et al. 2012, 2013).

As modern bird feathers have been long known to be predominantly coloured by melanin, fossil feathers were an obvious next step in looking for melanin in the geological record. In 2008, Vinther et al. analysed a fossil feather from the Early Cretaceous Crato Formation of Brazil which showed distinct dark and light transverse bands (Fig. 11.5a). Within the dark bands of the feather, they found abundant oblong structures around 1–2 μm in length aligned along the barbs and barbules, identical to those previously identified as lithified bacteria in feathers from Messel (Fig. 11.5b; Vinther et al. 2008). In contrast, the light bands showed no microstructures other than the matrix itself (Fig. 11.5c). Vinther et al. (2008) noted that the oblong structures were identical to eumelanosomes found in modern bird feathers (Fig. 11.1). Due to the newly appreciated preservation potential of melanin together with

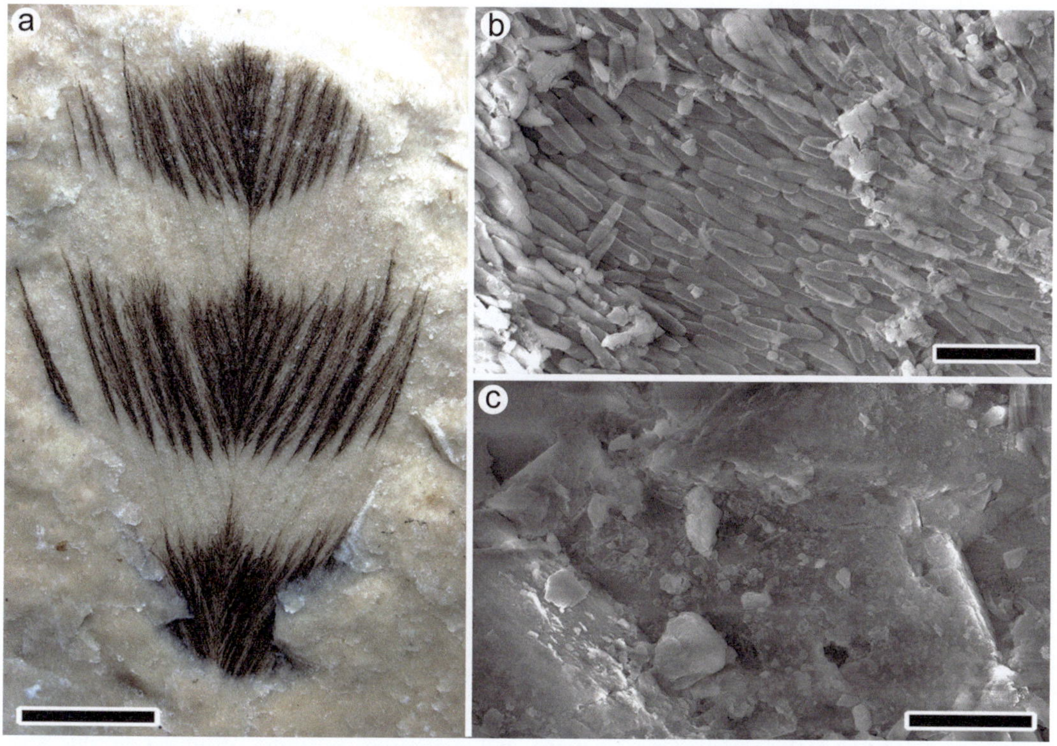

Fig. 11.5 A striped fossil feather from the Early Cretaceous Crato Formation, Brazil, the first to be unequivocally shown to preserve due to the presence of melanin (Vinther et al. 2008). (**a**) The isolated feather (LEIUG 115562) showing distinct dark and light banding. (**b**) SEM image of a sample taken from a dark band showing abundant aligned melanosomes. (**c**) SEM image of a sample taken from a light band showing no melanosomes, only the rock matrix. This shows that the striping is due to preservation of original colour patterning and the presence and absence of melanin in the dark and light bands, respectively. Scale bar represents 3 mm in (**a**), 2 μm in (**b**) and 5 μm in (**c**)

their unique arrangement and dimensions, it was considered more parsimonious that these micrometre-sized structures were in fact exceptionally preserved melanosomes rather than lithified bacteria. Furthermore, they were still arranged roughly as they would have been in the feather in life, although all remnants of the keratin matrix in which they sat had been lost. They suggested therefore that the feather was showing its original colour pattern of pigmented and unpigmented banding, i.e. the presence and absence of melanin. Based on their shape, it was surmised that the feather would have been striped black and white since it was known that spherical/oblate melanosomes contain reddish-brown phaeomelanin (Vinther et al. 2008). The bacterial explanation for this preservation was harder to justify, given that bacteria are associated with all parts of an animal during decay and thus would be highly unlikely to form distinct bands within individual structures or precisely mirror the morphology and arrangement of feather melanosomes (Vinther 2015a). Energy-dispersive X-ray spectroscopy (EDX) of the dark bands backed this up, showing high concentrations of carbon, whereas no carbon was present in the light bands. Alongside this isolated feather, a complete bird skull from the Fur Formation was also analysed in the study, which showed the same oblong-shaped structures that were aligned along the barbs and barbules in the preserved plumage forming a halo around the skull (Vinther et al. 2008). Furthermore, a dark patch within the eye orbit of the bird showed both oblong and subspherical microstructures which resembled melanosomes found within the retinas of modern birds (Vinther et al. 2008; Vinther 2015b).

This study overturned the orthodoxy that lithified bacteria were the chief explanation for integumentary preservation in vertebrate fossils and proposed that, instead, exceptionally preserved melanosomes were a more likely scenario (Vinther 2015a). The palaeocolour literature has subsequently adopted the term melanosome for their occurrence in fossils; however, some have advocated the use of the term melanin granules instead (Galván and Solano 2016). The term melanosome refers to the living organelle within a cell, while palaeocolour most often deals with melanin deposited into inert keratin. However, the morphology of the two is identical as the melanin granules are formed through dense packing of melanin inside the organelle's membrane, which is then lost as they are transported to the keratin. For consistency with the literature, we refer to both the living organelle and the fossil occurrences as melanosomes. More recently, the bacterial hypothesis has been revisited, with suggestions that bacteria fossilise easily and are indistinguishable from melanosomes (Moyer et al. 2014; Lindgren et al. 2015; Schweitzer et al. 2015). However, these criticisms fall short in their consideration of their relative preservation potential, the actual dimensions that bacteria exhibit and the unique fashion in which melanosomes localise themselves in fossil tissues (Vinther 2015a, b). Bacteria are certainly key to inducing conditions that make soft tissue fossilisation possible and promoting many of the processes that lead to exceptional preservation (Briggs and Kear 1993; Parry et al. 2017 and references therein). As yet however, they have not been conclusively demonstrated to be organically preserved in any fossil vertebrate. Furthermore, the bacterial model does not explain why only tissues that are known to contain melanin preserve as organic stains and not other tissues that would be likely to host thriving microbiomes, such as muscles and the intestinal tract.

11.2 Mechanism of Melanin Preservation

The ubiquity and varied functions of modern melanins have made them a focus of scientific study since the nineteenth century (Wolfenden 1884; Chittenden and Albro 1899; Galván and Solano 2016). Despite such a long history of research, certain aspects of the structure and synthesis of melanin are still incompletely understood in living animals (Galván and Solano 2016). Melanins are complex biopolymers thought to be formed through subtly different pathways depending on the exact melanin type. In modern birds, melanins are synthesised in melanocytes through a process known as melanogenesis (Galván and Solano 2016). The current state of knowledge of the full process is described in detail elsewhere (Galván and Solano 2016), but in brief melanogenesis involves a number of steps, which differ between eumelanin and phaeomelanin. The initial step common to both melanin types is the oxidation of L-tyrosine to L-dopaquinone by oxygen, catalysed by the enzyme tyrosinase (Galván and Solano 2016). After this step, the chemical composition of the solution appears to determine whether eumelanogenesis or phaeomelanogenesis follows through different polymerisation reactions resulting in the final large and complexly cross-linked pigment molecules (Galván and Solano 2016). The chemical structures of eumelanin and phaeomelanin differ in key aspects that have significant implication for fossil melanin. Eumelanin comprises cross-linked 5,6 dihydroxyindole (DHI) and 5,6 dihydroxiindole-2-carboxylic acid (DHICA) polymers. Phaeomelanin incorporates monomers of benzothiazines and benzothiazoles rather than DHI and DHICA. The presence of these distinctive sulphur compounds distinguishes phaeomelanin and also provides key signatures that can be looked for in the fossil record (Glass et al. 2012; Brown et al. 2017). The complex cross-linking of melanin may help to explain its remarkable recalcitrance and ability to survive diagenetic processes.

Despite numerous studies into the preservation of melanin in fossils, its precise taphonomy and preservation are at present still to be fully understood. This is in part due to the lack of complete clarity on the molecular structure of modern melanin, but also due to the relative novelty of the field. Although a small number of references to melanin being inferred to be present in fossils were made throughout the twentieth century (e.g., Voigt 1936, 1988; Whitear 1956; Beyermann and Hasenmaier 1973; Gottfried 1989; Mapes and Davis 1996), it has only been the past decade that has seen any serious advances made in unravelling the intricacies of the pigment's preservation in deep time (Vinther 2015a). Since the initial confirmation that melanin survives in fossils, it has become apparent that the pigment may be relatively common in the fossil record where soft tissues are present (Glass et al. 2012; Vinther 2015a, b; Clements et al. 2016). While it is generally accepted that melanin can survive taphonomic and diagenetic processes, work to better elucidate the mechanisms by which the pigment can become preserved has thus far been comparatively limited, and the subject is still in its infancy. A small number of important studies over the past decade have however attempted to further our understanding of the nature of melanin preservation and have provided key information.

Identifying chemical signatures for melanin and its diagenetic products has been attempted using a number of different techniques (e.g. Vinther et al. 2009; Barden et al. 2011; Wogelius et al. 2011; Glass et al. 2012, 2013; Lindgren et al. 2012; Colleary et al. 2015; Pan et al. 2016; Gren et al. 2017). Maturation experiments have also been performed to investigate how the effects of heat and pressure (both key factors in diagenesis) affect the morphology and chemical structure of melanin and melanosomes (McNamara et al. 2013; Colleary et al. 2015; Saitta et al. 2017).

While preliminary work in the 1970s studied fossil melanin using infrared spectrometry in order to understand melanin preservation (Beyermann and Hasenmaier 1973), only more recently after the discovery of widespread melanin preservation have scientists attempted to understand the chemical makeup of fossil melanins. Some studies were able to identify chelating metals associated with melanin using synchrotron X-ray fluorescence (XRF) (Wogelius et al. 2011). These could be mapped across a whole specimen. Similar elemental mapping confirmed the carbonaceous nature of melanosomes, using scanning electron microscopy (SEM) and EDX (Vinther et al. 2008). While melanin is unique for its ability to chelate an assortment of metals (Liu and Simon 2005), the use of these techniques in identifying fossil melanins is limited because other organic fossils will bind metals in life and after death (Vinther 2015a). Furthermore, the elemental mapping does not identify areas in which melanosomes have dissolved due to oxidation, but can be identified as impressions, giving a false appearance of colour pattern (Vinther 2015a).

Studies into the chemical nature of melanin in fossils using techniques other than whole-specimen elemental surface mapping have provided more insight into the nature of its preservation. Barden et al. (2011) performed Fourier transform infrared spectroscopy (FTIR), electron paramagnetic resonance (EPR) and pyrolysis gas chromatography mass spectrometry (py-GCMS) alongside SEM and EDS on isolated fossil feathers from the Early Cretaceous Xiagou Formation of China attributed to the amphibious bird *Gansus yumenensis*. Imaging demonstrated the presence of melanosomes, and the infrared spectra and chromatograms of the organics were very similar to a modern eumelanin standard. FTIR allows for the identification and quantification of functional groups, such as hydroxyls, ketones and carboxyls. Hydroxyls and carboxyls tend to be lost during maturation and hence allow for understanding the alterations of melanins. While the py-GCMS did not serve to identify melanin markers in this study, it importantly showed no significant contribution of bacterial hopanoid biomarkers, which helped rule out the possibility of a bacterial origin of the microbodies.

In 2012, Glass et al. performed what is still one of the most comprehensive chemical studies into fossil melanin preservation to date. The study focussed on one of the least controversial and likely purest sources of eumelanin in the fossil record: that of coleoid cephalopod ink sacs. Using a suite of analytical techniques, including py-GCMS, alkaline hydrogen peroxide oxidation, FTIR and X-ray photoelectron spectroscopy (XPS), Glass et al. determined that the fossil ink of coleoids from the Middle Jurassic of Christian Malford (Wiltshire, UK) and Early Jurassic of

Lyme Regis (Dorset, UK) was unequivocally preserved as eumelanin and its breakdown products. In fact, around 10–15% of the preserved ink was still composed of intact eumelanin with carboxyl and hydroxyl groups intact. This further demonstrated the incredible recalcitrance of the pigment and its ability to survive relatively unchanged for over 190 million years. Numerous melanin-derived products were found in the Jurassic ink as well as multiple diagenetic components using py-GCMS. Some had been secondarily sulphurised, that is, they had reacted with sulphur to form various thiophenes (Glass et al. 2012; Vinther 2015a).

A follow-up study by Glass et al. (2013) examined ink from coleoids found at the late Early Jurassic locality of Holzmaden (Germany). Although these were of an age intermediate to the two English coleoids studied previously, the melanin content was found to be just 1% that of those specimens and just 0.1% that of a modern eumelanin extract. The only major difference between the fossils was the burial and thus diagenetic history of the formations. The Posidonia Shale of Holzmaden was buried deeper than the other two Jurassic localities. This meant it had entered the oil window (the point at which insoluble organic matter, kerogen, thermally matures into oil, which is dependent on burial depth-controlled temperature). The level of organic maturation was therefore significantly higher, resulting in the alteration and enhanced breakdown of the original melanin. Despite this, numerous pyrolysate compounds were found in the Holzmaden specimens that were similar to the other less-matured fossils. This showed that py-GCMS is capable of detecting melanin-derived breakdown products even in heavily matured specimens (Glass et al. 2013; Vinther 2015a). Py-GCMS is a destructive technique however (although modern GCMS machines can analyse down to 50 μg), meaning that samples are destroyed in the analysis. This destructive technique may therefore prove unsuitable for particularly rare specimens, as is often the case with exceptionally preserved fossils.

A recent development for exploring fossil melanins is the relatively nondestructive time-of-flight secondary ion mass spectrometry (ToF-SIMS). This sensitive technique involves firing a pulsed ion beam at small areas of the surface of samples and determining the mass of secondary ions removed from the outermost surface (Vickerman and Briggs 2001). Both positively and negatively charged ions can be detected but have to be collected separately in successive runs. These secondary ions can give information about the elemental and molecular makeup of the sample surfaces (Vickerman and Briggs 2001; Lindgren et al. 2012; Colleary et al. 2015). For melanins, the technique is best suited to characterise low-molecular-weight secondary ions—fragments of the larger and poorly understood molecule. Spectra can be generated and compared to those of other samples with known compositions (Colleary et al. 2015). Importantly, each individual peak, composed of a single or a couple of secondary ions, is not diagnostic of melanin, but the relative intensity creates a spectral fingerprint that can be compared between samples.

Lindgren et al. (2012) used ToF-SIMS to probe the preserved eye of a fish from the early Eocene Fur Formation and compared the resulting spectra to the surrounding sediment, other body regions and, importantly, a modern melanin standard. The results from this showed that the mass spectra derived from the eye were very similar to that of the modern melanin standard and dissimilar to that of the surrounding sediment (Lindgren et al. 2012). ToF-SIMS has its limitations however, as it only provides fragments of the molecular makeup from a sample surface. The fragmentation of the molecules in situ leads to complex arrays of secondary ion fragments that could be derived from several sources in a heterogenous sample. Furthermore, the mass of the secondary ions is obtained from the time-of-flight from the sample to the detector, which gives poor mass resolution, further conflating the ability to characterise distinct ions of similar relative mass.

In spite of its limitations, ToF-SIMS has now become something of a standard in analysing melanin in the fossil record. It is currently considered to be one of the most suitable and easily applicable techniques to probe the molecular makeup of fossil samples without excessive damage to the specimen (Colleary et al. 2015; Clements et al. 2016; Gabbott et al. 2016; Gren et al. 2017). The use of multivariate statistics, such as principal component analysis (PCA), also allows more objective comparison of the appearance of the ToF-SIMS spectra. Incorporating PCA provides both the ability to distinguish melanin from negative non-melanin samples in an objective fashion (Fig. 11.6c) and the possibility of identifying alterations to melanin that occurred during maturation (Colleary et al. 2015).

11.2.1 Maturation Experiments

It has become apparent that melanin, while recalcitrant, does not survive molecularly intact. Instead, it must undergo alterations similar to other fossil organic materials (Eglinton and Logan 1991). Hence, researchers have begun to experimentally understand the nature of alterations to melanins during fossilisation, driven by maturation under elevated heat and pressure (McNamara et al. 2013; Colleary et al. 2015). Artificial maturation experiments have been carried out on both feathers and pure melanin samples under experimental conditions to elucidate how the pigment may alter during diagenesis (McNamara et al. 2013; Colleary et al. 2015; Saitta et al. 2017). As the geologic processes of organic alteration are contingent on time and the levels of temperature and pressure they have experienced, experimentalists resort to using more elevated P/T conditions in a shorter time window to speed the process up.

It has been observed that impressions of fossil melanosomes are often larger than the actual preserved ones (Clarke et al. 2010; Carney et al. 2012). Maturation experiments have demonstrated that this difference can be explained by dehydrating alterations taking place during maturation causing condensation reactions (Eglinton and Logan 1991) and hence shrinkage without affecting the preservation and overall morphology of the melanosome organelle (McNamara et al. 2013). Melanosome geometry has been shown to change between 7.6 and 20% depending on the conditions to which they were exposed (lower temperature regimes result in less shrinkage). Although the potential shrinkage of melanosomes had been highlighted previously (Clarke et al. 2010; Carney et al. 2012), this study (McNamara et al. 2013) highlighted how the effect of thermal maturation can induce the phenomenon. As the temperature controlled the degree of shrinkage, burial history and, in particular, the level of thermal maturation should therefore ideally be taken into account when melanosomes are analysed for palaeocolour reconstructions by their morphology. However, as the shrinkage appears to be isometric (i.e. the aspect ratio of melanosomes remains the same), it is observed to have a negligible effect. This is because it has been shown that aspect ratio is one of the most important variables in statistical comparisons for colour prediction by melanosome morphology (see below). Shrinkage can in theory be mitigated by scaling up fossil melanosome measurements by 10–20% in statistical analyses. Additionally, measuring both mouldic impressions and actual 3D preserved melanosomes (if present within a single sample) could allow the potential degree of shrinkage to be examined.

While melanosomes may shrink during diagenesis, the presence and absence of the pigment should not change; therefore, original colour patterns are likely to be visible even in highly matured deposits provided that soft tissue preservation occurs (Vinther et al. 2008; Vitek et al. 2013; Colleary et al. 2015; Vinther 2015a). This is highlighted by deposits such as those of the Jehol Biota, which are thought to have undergone a deeper burial history than many other feather-bearing sites, yet still show remarkable integumentary structure preservation including original colour patterning (Vinther 2015a, Vinther et al. 2016, Smithwick et al. 2017a, b). Younger deposits that have undergone less harsh burial conditions, such as Messel, can also preserve colour patterns in exquisite detail, suggesting minimal alteration (Fig. 11.7).

The most comprehensive maturation study looked at the effects of elevated temperatures

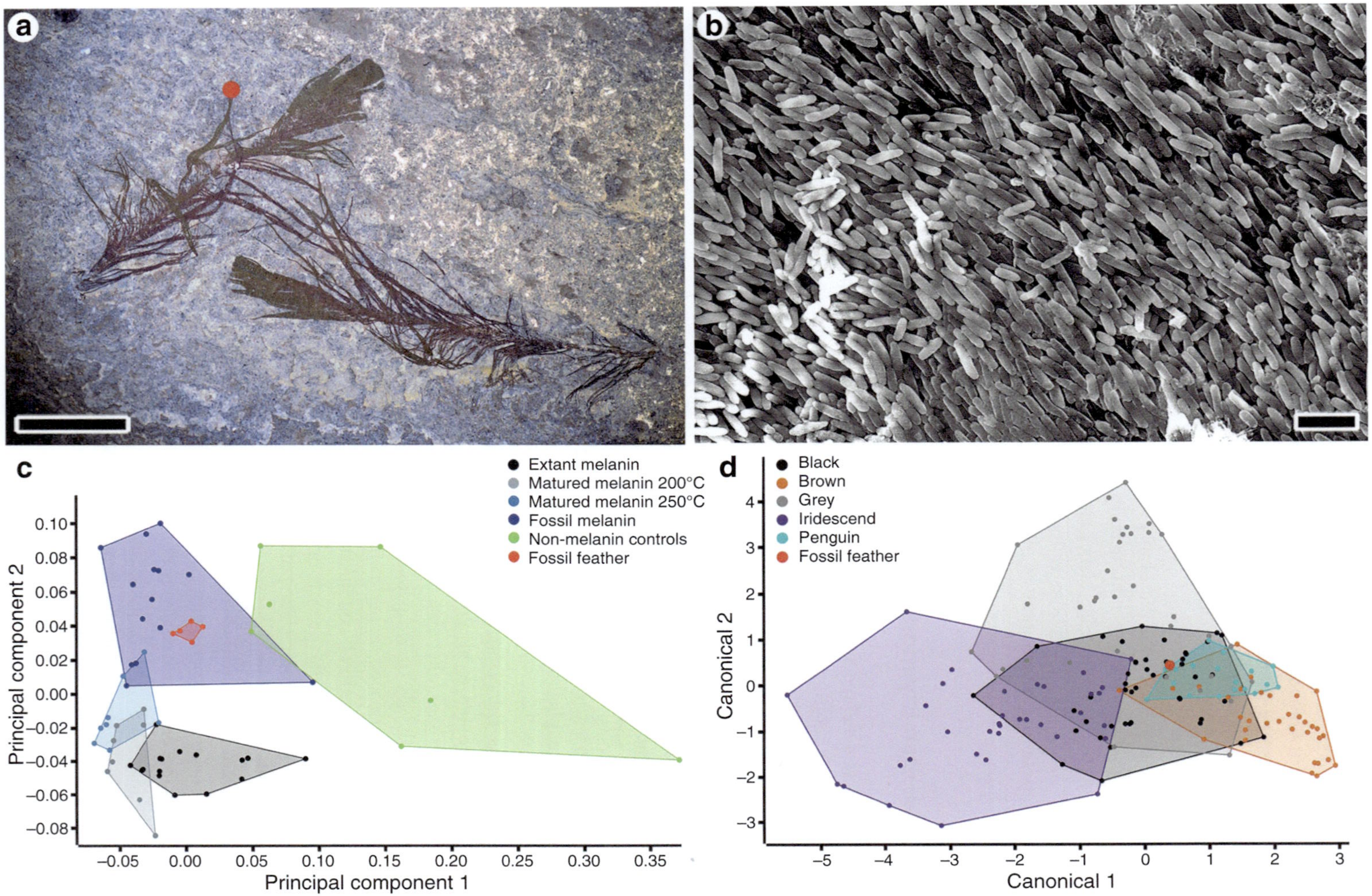

Fig. 11.6 Melanin preservation in Eocene feathers and palaeocolour inferences possible from minimal sampling. (**a**) A pair of feathers (SMF-ME 3855) from an unknown bird from the Eocene Messel Formation, Germany. The red dot indicates the point at which a sample was removed for analysis. (**b**) SEM image of melanosomes that mark

and pressures on melanins in conjunction with ToF-SIMS (Colleary et al. 2015). In addition to effects on the morphology of melanosomes, Colleary et al. (2015) also investigated how the chemical structure of melanin was altered. They analysed a range of fossil taxa, including bird feathers, mammal hair, fish eyes, amphibian skin and coleoid cephalopod ink. This provided for a broad sample straddling almost 300 million years of the fossil record, bracketing bilaterian metazoans and including melanosomes of varying morphology and hence likely different eu- and phaeomelanin composition. Maturation experiments were also run on extracts of pure melanin from modern bird feathers. As with previous experiments, samples were subjected to elevated temperatures and pressures (200 and 250 °C and 250 bar) for 24 h sealed in gold tubes. To compare the more than 55 ToF-SIMS spectra, PCA was used. Most significantly, the ToF-SIMS spectra could show that fresh and fossil melanins were distinct from negative samples and that they in turn showed subtle differences with matured melanins being intermediate in spectral composition (reproduced in our Fig. 11.6c). Furthermore, the spectra showed differences correlating the morphology of fossil melanosomes, suggesting that ToF-SIMS is also able to characterise different compositions of eu- and phaeomelanin. These results also show that the fossil samples did not cluster according to lithology, age or locality, showing that the framework mainly allows for characterising differences in original melanin chemistry. Some differences in how melanin chemistry spread in extant and fossil samples in PCA need further scrutiny and are the focus of current research (Colleary et al. 2015).

11.2.2 Melanin and Sulphurisation

In the comprehensive chemical characterisation of cephalopod ink by Glass et al. (2012), it was shown that the eumelanin had reacted with sulphur to form a host of thiophenes among the pyrolysates in the py-GCMS spectrum. Sulphurisation is a well-known phenomenon in organic geochemistry (Sinninghe Damsté et al. 1989) and is noted to likely be an important pathway for preserving both bone marrow and muscle tissue in a particularly sulphuric Miocene fossil deposit (McNamara et al. 2006, 2010). Since phaeomelanin is composed of monomers of benzothiazines and benzothiazoles, which contain sulphur, concerns had been raised about whether secondary sulphurisation could lead to unwanted conflation between original pheomelanin and secondarily sulphurised eumelanin (McNamara et al. 2016a). However, thiophenes and thiazines/thiazoles are molecularly distinct due to the presence of nitrogen in the latter. Hence, the secondary ions chosen by Lindgren et al. (2014) to characterise pheomelanin in ToF-SIMS analyses (of which many combine C, N and S) would be distinct from the thiophenes (which would not contain N). It is therefore observed that the sulphurised fossil coleoid ink analysed by Colleary et al. (2015) is not conflated with both modern and fossil melanins that contain benzothiazoles (Brown et al. 2017) or modern and fossil melanosomes that possess the distinct small and ovoid morphology characteristic of phaeomelanin-rich compositions (Colleary et al. 2015).

Fig. 11.6 (continued) out the barbules of the fossil feather. (**c**) Time-of-flight secondary ion mass spectrometry (ToF-SIMS) principal component analysis (PCA) plot showing the distribution of extant, matured and fossil melanin samples based on the first two PC axes derived from 54 mass peaks known to be associated with melanin (adapted using data from Colleary et al. 2015). The isolated fossil feather plots in the centre of the range of other fossil melanin samples, separate from the non-melanin controls, confirming a likely melanin affinity. (**d**) Canonical discriminant analysis results showing the relationship between melanosome morphology and colours produced. Modern feather melanosomes with known associated colours were categorised by colour and used to predict the likely colour of the fossil feather, shown here to fall within the range of black and 'penguin' feather melanosomes. Data adapted from Li et al. (2012). Scale bars represent 1 cm in (**a**) and 2 μm in (**b**)

Fig. 11.7 Examples of preserved original colour patterns in fossil birds from the Eocene Messel Formation, Germany. (**a**) A near-complete specimen of the genus *Hassiavis* (SMF-ME 9047a) missing the head but showing exceptional feather preservation including dark pigmented wings and a finely banded tail. (**b**) A complete specimen of the species *Messelirrisor grandis* (HLMD-Be 178) showing a striking thickly banded tail. These colour patterns are retained due to the presence of melanosomes. Scale bars represent 10 mm

11.3 Preservation of Non-melanin Feather Components

A feature that has been frequently noted in studies on fossil feathers is the absence of any feather ultrastructure apart from the preserved melanosomes (Vinther et al. 2008, 2009; Carney et al. 2012; Li et al. 2012; Field et al. 2013; Vitek et al. 2013; Vinther 2015a). Most SEM analyses of fossil feathers to date have shown that when feathers are preserved, melanosomes mark out the structure of the barbs and barbules, but lie freely in the matrix with little morphological evidence of the other key component of feathers, keratin, present (highlighted in Fig. 11.3, also clear in Figs. 11.5, 11.6 and 11.8) (Vinther et al. 2008, 2009; Carney et al. 2012; Li et al. 2012; Field et al. 2013; Vitek et al. 2013; Vinther 2015a, b). Maturation experiments have shown that keratin, like other proteins, does not survive in diagenetic environments and becomes fluid and water soluble readily when temperatures and pressures are elevated (Saitta et al. 2017), leaving only melanosomes (Colleary et al. 2015). McNamara et al. reported on experiments in which feather keratin appeared to survive (McNamara et al. 2013), but it turned out that there was a mix up in the presentation of their experimental protocol and only experiments that had been performed under brief intervals (1 h vs. 24 h) were reported (McNamara et al. 2017, recent correction), which is not a standard protocol.

Some have however claimed that keratin is preserved organically in fossil feathers and that molecular signatures remain (Schweitzer et al. 1999; Moyer et al. 2016; Pan et al. 2016). A recent study of the Early Cretaceous bird *Eoconfuciusornis zhengi,* from the Jehol Biota of China, looked at the possibility of keratin being preserved using immunohistochemistry techniques alongside SEM imaging, transition electron microscopy (TEM) and ChemiSTEM techniques (Pan et al. 2016). Antibody antigen binding suggested that original components of the feather beta keratin were preserved and

Fig. 11.8 Three-dimensional preservation in a fossil feather from the Eocene Messel Formation, Germany, shown to be originally iridescent (from Vitek et al. 2013). (**a**) Overview of the pennaceous portion of an isolated feather (SMF Me 3850) from an unknown bird. (**b**) Close-up of an individual barb showing preserved barbules and a silver sheen. (**c**) SEM image of a cross section of a barb and associated barbules cut using a focused ion beam (FIB). The barb is outlined in red, and the individual barbules are marked in blue. The structures can be seen to be three-dimensionally preserved, thanks to the presence of melanosomes. All keratin has degraded away, but the melanosomes sit in their original arrangement on the rock matrix and show distinct similarity in their arrangement to modern iridescent feathers. The silver sheen visible to the naked eye is caused by the three-dimensional arrangement of melanosome layers and a thin layer of sediment (inset diagram). Scale bars represent 5 mm in (**a**), 1 mm in (**b**) and 10 μm in (**c**)

melanosomes were present within a presumed filamentous keratin matrix. Issues surrounding the immunohistochemistry techniques have however been raised, including their propensity to provide false positives (e.g. from consolidants used to conserve fossils) and statistical artefacts (Saitta et al. 2017, 2018). Other studies have maintained the possibility of keratin protein preservation using antibody experiments (Lindgren et al. 2017). However, the observation that these fossils preserve melanosomes liberated from their keratin matrix and that unmelanised keratinous tissues preserve no organic residues (Vinther et al. 2008) in addition to previous knowledge on protein stability (Eglinton and Logan 1991; Demarchi et al. 2016; Saitta et al.

2017) goes against the claimed presence of these highly unstable molecules in such ancient samples. Additionally, the mass spectroscopic methods employed fail to recover a protein signal. Antibody experiments are therefore demonstrably unsuitable for analysing fossil samples due to the well-known problems with frequent unspecific binding of antibodies (Saitta et al. 2017, 2018).

While organically preserved keratin is controversial, some originally keratinous structures can survive in the fossil record under certain circumstances through authigenic mineralisation. Claw sheaths, for example, are relatively common in deposits such as Jehol (e.g. Gong et al. 2012; Smithwick et al. 2017a, b) and Solnhofen (e.g. Frey et al. 2003; Fig. 11.2b), and fossil baleen has been shown to be prevalent in whales from the Neogene Pisco Formation of Peru (Esperante et al. 2008; Gioncada et al. 2016). The preservation of these features is likely due to the presence of hardening calcium phosphate salts (apatite) (Saitta et al. 2017). Hard keratinous tissues in living animals can contain up to 15% calcium phosphate by dry weight (O'Connor et al. 2015; Gioncada et al. 2016). Apatite is one of the most important minerals in both hard and soft part preservation in vertebrate fossils (Briggs and Kear 1993; Briggs et al. 1993; Briggs and Wilby 1996; Parry et al. 2017 and references therein), and its presence in keratinous tissues likely aids its mineralisation and preservation in fossils. These mineralised keratinous tissues can retain the structure's original morphology but are highly unlikely to preserve any organic traces of the original decay-prone and labile proteins (Saitta et al. 2017). The only part of feathers that has been shown to be hardened in this manner is the rachis (Blakey et al. 1963; Saitta et al. 2017), but none have been found preserved in fossils with intact keratinous ultrastructure. While it is plausible that apatite is present in other parts of feathers (Blakey et al. 1963), it seems that levels are not sufficient to allow for authigenic mineralisation of feather keratin in most circumstances. An alternative proposed pathway for mineralisation of organic features (mineral preserved organics—MPO) such as keratin has been proposed based on close association with metals that could coat and/or invade the tissues, promoting preservation (O'Connor et al. 2015). As yet however, this has not been identified in fossil feathers and is more relevant to much younger archaeological remains.

11.4 Bringing the Past to Life: Palaeocolour Reconstructions of Extinct Dinosaurs

Alongside research into how and why melanin may preserve, much work has been carried out since the initial studies of fossil melanosomes (Vinther et al. 2008, 2009) to better understand what the preservation of melanin and other pigments can tell us about the colouration of extinct animals, particularly birds and other dinosaurs, and how we can use that information to better inform understanding of past ecologies.

Once it had been established that melanosomes could be found preserved with high fidelity in fossil feathers, attention turned to whether the original arrangement of the melanosomes within a fossil feather could be found and thus provide information about potential structural colouration in extinct taxa. In 2009, Vinther et al. looked at isolated fossil feathers from Messel, some of which showed a silvery sheen in the barbules visible to the naked eye (Fig. 11.8a, b). The aim was to determine whether this was a preservational artefact or remnants of original structural arrangements of melanosomes. In one particular contour feather, which became the focus of the study, the arrangement of the melanosomes distinctly varied with a contrast being apparent between the proximal and distal portions. This variation matched the visible differences in the feather. Proximally, melanosomes were arranged in an aligned but loosely packed configuration in the barbs and barbules as well as the rachis. In the distal portion however, the barbs formed an open and pennaceous arrangement with the barbule melanosomes forming solid, smooth and continuous dense external layers. Underlying these layers, further melanosomes were more loosely packed and less densely arranged. This arrangement is similar to the single thin-film nanostructural array seen in many modern bird feathers exhibiting iridescence (Prum 2006; Vinther et al. 2009; Vitek et al. 2013).

Subsequent work on the same originally iridescent feathers from Messel revealed that the silvery sheen was related to the original structural arrangement of melanosomes in the feather (Fig. 11.8c). Focused ion beam scanning electron microscopy (FIB-SEM), a technique that cuts a micrometre-scale trench in a sample allowing three-dimensional structures to be observed, was used to examine this. Thin wedges of sediment were found to be acting in place of the original keratin as a material with a different refractive index than the underlying melanosomes (Vitek et al. 2013). Due to the variable thickness of the wedge of sediment, all light waves were scattered, like in a chirped mirror, hence creating the observed silvery sheen. This showed that the feather was most likely strongly iridescent in life. The presence of structural colour in an Eocene bird feather provided novel information about the Messel ecosystem.

Messel has the richest avifauna of any known fossil location, with a level of diversity and ecological disparity rivalling modern forested ecosystems (Mayr 2017). While skeletal remains of these birds can inform us about their potential habits, such as feeding strategies and perching ability (Mayr 2017), being able to infer plumage colouration has the potential to deepen our understanding of social and behavioural interactions of this extinct ecosystem. In modern birds, iridescence is often associated with social signalling (Cuthill et al. 1999; Prum 2006; Doucet and Meadows 2009; Stavenga et al. 2010; Vukusic 2011). Due to the exceptional preservation of original structural colours in Messel feathers, we can infer that such interactions were likely occurring in the Early Eocene bird assemblages (and indeed earlier).

Most fossil feathers do not show evidence of original structural colours. Instead, we must rely on the morphology, density and chemistry of the melanosomes to infer original colouration. Since melanosome morphology provides the ability to distinguish pigmentary colours in extant taxa (Fig. 11.1), methods for predicting colour from melanosome shape in fossils have been investigated. In 2010, two research groups independently analysed the integument of Mesozoic feathered dinosaurs. Zhang et al. (2010) were able to demonstrate the preservation of both elongate and smaller ovoid melanosomes in theropods and pygostylians from the Jehol Biota. Li et al. (2010) described melanosomes from the Late Jurassic paravian *Anchiornis huxleyi* from the Tiaojishan Formation of China. SEM imaging of 29 samples from feathers across the body of *Anchiornis* revealed abundant impressions of oblong structures again identical to modern melanosomes. These melanosomes preserved the original alignment along feather structures in most places. Some light-coloured feather impression samples were barren of melanosomes and organic remains and hence were considered unpigmented. From the preserved melanosomes and impressions in the rock matrix, measurements were taken of the long and short axis of each structure and added to a database of measurements from modern bird feathers with known associated colours (10 black, 10 brown and 10 grey samples). A canonical discriminant analysis was performed on this database which predicted the likely colour of *Anchiornis*. An example of how this method can be done is presented for a Messel feather in Fig. 11.6. Samples from different areas of *Anchiornis* were predicted as black, brown and grey with different but generally high P-values. This was used to create a complete reconstruction of the plumage of this dinosaur (Fig. 11.9a, b). The body of the animal was predicted as a mixture of grey and black with unpigmented white bands on the fore and hind wings with spangled, black tips. Melanosomes taken from the feathers on the distal crown feathers were particularly small, clustering with rufous-red feathers in the brown category in the canonical analysis. This contrast to the body melanosomes suggests a display function. This method of statistically comparing fossil melanosome morphologies to extant feather melanosomes with known colours has become the standard for many palaeocolour predictions (Fig. 11.6).

Further studies of dinosaur integumentary structures expanded the extant colour database to include iridescence as a category. Generally,

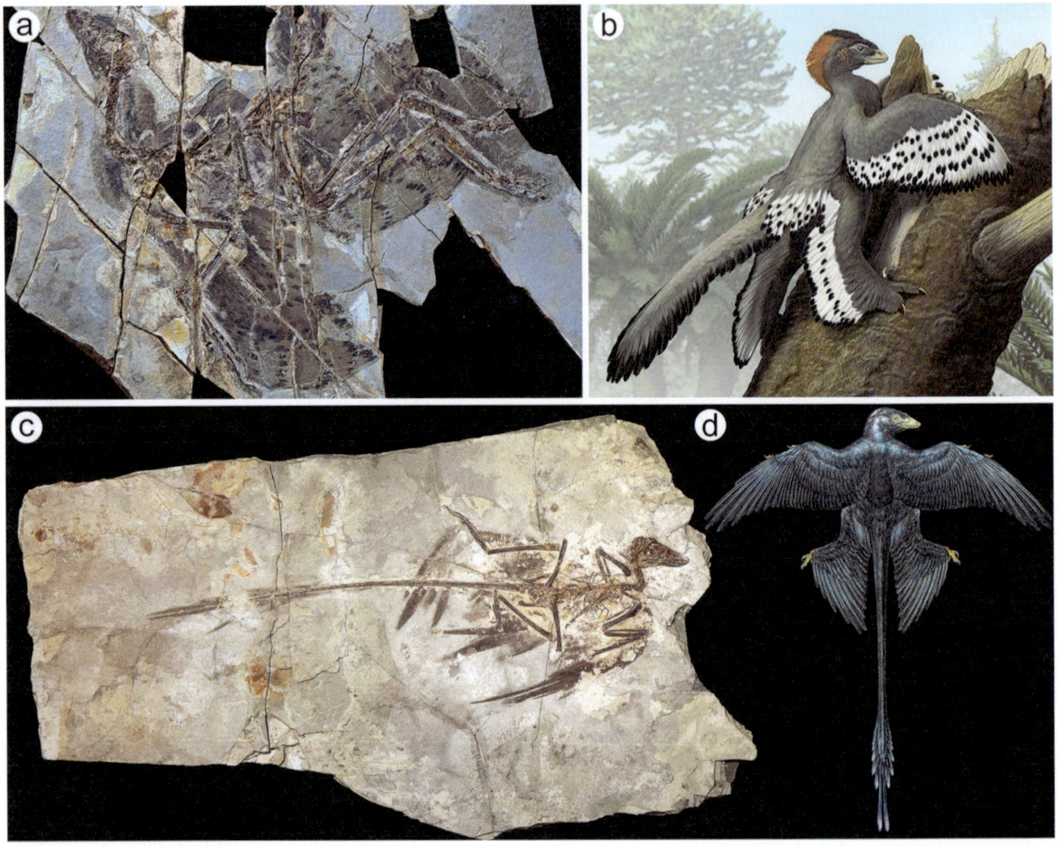

Fig. 11.9 Palaeocolour reconstructions of paravian dinosaurs. (**a**) *Anchiornis* from the Middle Jurassic of Liaoning, China. (**b**) Palaeocolour reconstruction of *Anchiornis* based on melanosome sampling showing a black and grey body with mottled wings and a rufous head crest. (**c**) *Microraptor*, a four-winged dinosaur from the Early Cretaceous of Liaoning, China. (**d**) Palaeocolour reconstruction of *Microraptor* showing subtle corvid-like iridescence based on melanosome sampling. Reconstructions courtesy of Robert Clark (**a**), Carl Zimmer (Commissioner) and Carl Buell (Illustrator), from 'Evolution—making sense of life' (Zimmer and Emlen, Roberts and Co. Publishers) (**b**) and Mick Ellison (**c**, **d**)

iridescent melanosomes are consistently longer and more skinny than black melanosomes apart from in certain clades such as hummingbirds (Fig. 11.1h, j). These works have shown that some dinosaurs likely exhibited iridescence in a similar fashion to modern birds. In 2012, Li et al. looked at the paravian *Microraptor gui*, an unusual 'four-winged' dinosaur from the Jehol Biota. The morphology of the melanosomes in *Microraptor*, being relatively longer and thinner and preserved as aligned impressions, predicted them as iridescent. The limited evidence of the original keratin nanostructure, and thus structural colouration, but with the dense end-to-end arrangement of the melanosomes in *Microraptor* allowed for a conservative interpretation that it would have most likely exhibited thin-film iridescence. This type of iridescence is common to some extant birds such as members of the family Corvidae (Li et al. 2012; Lee et al. 2016), which only show weak iridescence (Fig. 11.9c, d). As aforementioned, since the overlying keratin layer which is integral in determining exact iridescent hues does not fossilise, this cannot be inferred from the fossils. From this reconstruction, inferences were also made about the potential ecology of *Microraptor*. Previous analysis of the scleral ring morphology of the dinosaur indicated a nocturnal lifestyle; however, dark glossy plumage is more common in diurnal modern birds (Li et al. 2012). More recently, another paravian, *Caihong juji* (Hu et al. 2018), was described with solid platelet-shaped melanosomes, which is a feature only known from brightly iridescent extant birds such as hummingbirds and tree swifts (Fig. 11.1j), thus extending this colour-producing feature back into the Jurassic.

More recent studies utilising palaeocolour have started to further explore the intricacies of behaviour and ecology in dinosaurs outside of Maniraptora. Integumentary structures suggested as potentially homologous to feathers are present in certain ornithischian taxa (Zheng et al. 2009; Godefroit et al. 2014) as well as the so-called protofeathers present in basal theropods (Chen et al. 1998; Rauhut et al. 2012). These structures have also been contended as possible decayed scales or collagen fibres rather than feathers of feather antecedents (Lingham-Soliar et al. 2007). The presence of melanosomes preserved in these features (Zhang et al. 2010; Godefroit et al. 2014) however makes it possible to characterise them as genuine integumentary appendages rather than dermal collagen (which does not contain melanosomes). Further, significant issues surrounding the identification of the structures as dermal in origin have been highlighted (Smithwick et al. 2017b). Along with likely feathers (or at least feather homologues) in some early theropod dinosaurs, certain cases of exceptional preservation of genuine scales with original melanised colour patterns have been described.

Fossil colour patterns in these ancient and ground-dwelling dinosaurs are able to provide important clues to the nature of the terrestrial predator-prey landscape and how these adapted to it. A well-preserved specimen of the small ceratopsian *Psittacosaurus* sp. from the Jehol Biota preserves visible colour patterns (Lingham-Soliar and Plodowski 2010; Mayr et al. 2016; Vinther et al. 2016). SEM imaging of samples of dark patches on the externally scaled integument of *Psittacosaurus* revealed abundant melanosome impressions which resembled phaeomelanosomes and were predicted as being brown in quadratic discriminant analysis (Vinther et al. 2016). The distribution of the melanosome-containing organic material exhibits distinct patterns that resemble those seen in extant animals, such as stripes, spots and countershading (Fig. 11.10a). Countershading is a common form of camouflage in modern animals which acts to reduce the three-dimensionality of an object by optically flattening the appearance of the body by reducing self-shadowing (Thayer 1896; Rowland 2009; Allen et al. 2012; Vinther et al. 2016). Self-shadowing is an important visual cue (shape from shading) in practically all vertebrate visual systems (Allen et al. 2012). Unpigmented scales are also discernible across the whole body of *Psittacosaurus* by their modest relief, but they preserve as films of calcium phosphate (see above) that fluoresce in UV- and laser-stimulated fluorescence (LSF) imaging (Vinther et al. 2016). To understand the distribution of the observed colour patterns, an anatomically accurate 3D model was made of dinosaur which had the pigment distribution carefully projected onto it (Fig. 11.10b). This allowed further investigation of the countershading pattern and in turn the likely habitat that the animal live in.

Studies have shown that there is distinct relationship between extant animals living in closed versus open habitats and the transition in countershading, being sharper and higher on the body when directly illuminated versus more gradual and further down the body in closed habitats (Allen et al. 2012). To understand the light environment that the countershading gradient of *Psittacosaurus* would have been best adapted for, a further 3D model (painted grey) was imaged under different lighting conditions found in open and closed habitats. By comparing the shadows produced under each lighting condition with the actual preserved colour patterns, it was shown that *Psittacosaurus* would have been best adapted to living in a closed forested habitat.

The principals behind this study were extended to a ground-dwelling feathered non-avian theropod dinosaur from the same deposits by Smithwick et al. (2017a). *Sinosauropteryx prima*, one of the first dinosaurs to have its colour deduced from melanosome imaging (Zhang et al. 2010), also shows a darker dorsum and lighter (likely unpigmented) ventrum indicating a countershaded pattern (Fig. 11.10c). Reconstructions of the colour pattern across the

Fig. 11.10 Palaeocolour reconstructions of non-maniraptoran dinosaurs. (**a**) *Psittacosaurus*, a ceratopsian from the Early Cretaceous of Liaoning, China. (**b**) 3D palaeocolour model of *Psittacosaurus* based on melanosome sampling and pigment distributions showing a low countershaded pattern and dark brown hue. (**c**) *Sinosauropteryx*, a non-maniraptoran theropod from the Early Cretaceous of Liaoning, China. (**d**) Palaeocolour reconstruction of *Sinosauropteryx* showing countershading, a striped tail and a 'bandit' eye mask, based on the distribution of pigmented feathers across the body. (**e**) *Borealopelta*, a large armoured ankylosaur from the Early Cretaceous of Alberta, Canada. (**f**) Palaeocolour reconstruction of *Borealopelta* showing a phaeomelanised countershaded pattern, based on the distribution of pigment across the body and the chemistry of the preserved melanin. Reconstructions courtesy of Robert Nicholls

dinosaur (Fig. 11.10d) were again compared to predicted optimal countershading patterns based on 3D models imaged under different conditions. This time however, a countershading pattern more closely matching that predicted for an animal living in an open habitat was seen, despite the fact that both *Sinosauropteryx* and *Psittacosaurus* came from the same deposits. This suggests that the Jehol Biota was made up of a range of different habitat types with different animals adapted to different conditions. Other important features of the plumage of *Sinosauropteryx* were also described in detail in this work. These included a previously noted banded tail (Zhang et al. 2010), which is a very common plumage pattern seen in modern birds and is associated with both social signalling and camouflage (Marques et al. 2016), and an eye stipe across the face (Fig. 11.10c, d). This feature is informative with regard to bird evolution, as it is a common feature seen in many modern bird taxa and is known to be associated with both camouflage (hiding the presence of the eye) and as an anti-glare device to protect the eye from the sun (Ficken and Wilmot 1968; Bortolotti et al. 2006).

Countershading has also recently been observed in a large Cretaceous nodosaurid ankylosaur *Borealopelta markmitchelli* (Brown et al. 2017). This provided an opportunity for understanding the non-actualistic nature of Mesozoic predator-prey landscapes. The frequency of countershading occurring in living terrestrial mammals drops with increasing body size and is lost above 1000 kg. This is due to the safe haven provided by gigantism at this threshold. Showing that a heavily armoured ornithischian dinosaur, estimated to have weighed more than 1300 kg, was countershaded (Fig. 11.10e, f) demonstrates the difference in the nature of the predator-prey balance in the Mesozoic to today. This was likely due to the presence of large theropodan predators that meant the safety of large body sizes was only effective at even greater magnitudes than would be necessary today, as exemplified by the giant sauropods (Brown et al. 2017).

Fossil colour patterns therefore provide important insights to extinct ecologies, which would be limited from conventional lines of evidence, such as osteology, trace fossils and stable isotopes. As our ability to reconstruct colour in extinct taxa improves, a more comprehensive picture of the past landscape is becoming clearer.

11.5 Limitations

While a wealth of information on past animal colour has been revealed since the discovery of melanin preservation in birds and extinct dinosaurs, there are currently limitations. These include inferences of specific hues, detection of other pigments and some taphonomic considerations.

Although melanin is the most common pigment utilised by vertebrates for colouration, the myriad of less common pigments contributes a significant extra gamut of possible colours (McGraw 2006a, c). By contrast, melanised colours are limited to black, browns, rufous reds and greys (McGraw 2006b).

Carotenoids are the most widespread pigment in extant avian clades after melanin. This pigment is taken up through the diet and appears with little phylogenetic constraint in different groups. Passerines most commonly exhibit carotenoid-based patches of plumage in about 40% of taxa (Thomas et al. 2014b), while in non-passerines it is much more restricted to only 13% (Thomas et al. 2014b). It is likely that dinosaurs could have exhibited carotenoid-based plumage and integument. It is a common feature in other diapsids. The pigment does have a preservation potential (Damsté and Koopmans 1997; Summons 2014), but is not hosted inside organelles with a preservation potential as the pigment does not form polymerised macromolecules like melanin does. Carotenoid preservation, or the vesicles containing them, has been proposed in a Late Miocene snake (McNamara et al. 2016b), but no evidence of the pigment has been found in any fossil feather (Thomas et al. 2014a; Vinther 2015a), which may be due to preservation and its utmost rarity. A number of other non-melanin pigments found in modern birds are clade specific (e.g. psittacofulvins in the Psittaciformes and green turacoverdin in turacos) (McGraw 2006c) and are therefore unlikely to have been present in any birds outside of these groups. Whether any extinct birds or other dinosaurs had their own unique class of pigments is an open question, but given the rarity in modern birds, it is unlikely.

In order to detect non-melanin pigments, mass spectroscopic or other chemical methods would have to be employed, and most of these are highly destructive. Porphyrins have been characterised compellingly with the otherwise less diagnostic ToF-SIMS (Greenwalt et al. 2013). How much of a concern is this when proposing broad colour patterns from a non-avian dinosaur?

First, these pigments are very rare as previously mentioned. Hence, the likelihood of having to entertain non-melanin pigments outside of passerines, turacos, owls and parrots is small. Second, co-occurring melanin pigments mask the colour of these pigments (McGraw 2006a, b, c; Vinther 2015a). Hence, only feathers lacking melanosomes are likely to have been either white or patterned with alternative pigments. It is possible that labile non-melanin pigments could have been present in these presumed unpigmented areas and have since been lost through diagenesis. However, the most parsimonious interpretation would be that these regions could have been white, given its higher abundance than these pigments. Alternatively, one can entertain exploring for these pigments in these particular regions of the body. However, if the white region forms a dorsoventral gradient, it is most likely that it represents countershading transition as this is one of the commonest colour patterns in modern animals (Rowland 2009).

While pigments other than melanin have been found in certain fossils and sediments [e.g. flavonoids in leaves (Rieseberg and Soltis 1987) and geoporphyrins, derived from haem, in a mosquito (Greenwalt et al. 2013)], assigning them as endogenous to a specific fossil is often problematic due to the propensity of the pigments to remobilise during decay and diagenesis (Vinther 2015a). While it may be possible to find other pigments in fossil birds and dinosaurs, ruling out contamination from remobilisation (e.g. from decaying algae) would require careful comparison of integumentary features to surrounding sediments (Vinther 2015a).

Another limitation of palaeocolour reconstructions is the preservation potential of keratin. As keratin is lost early on in diagenesis [e.g. within decades to millennia in archaeological sites of otherwise promising preservation potential (O'Connor et al. 2015)], original non-iridescent structural colouration that is formed via light-scattering air bubbles inside the keratin is also lost (Saitta et al. 2017). As outlined previously, iridescence, which is generated by organised melanosomes, can be identified, however, through either the preserved arrangement of the melanosomes (Vitek et al. 2013) or through their characteristic shape (Li et al. 2012; Hu et al. 2018).

As it stands, palaeocolour can only provide information on broad hues and iridescence. Non-iridescent structural colour cannot be identified through melanosome preservation (Babarović et al. 2019), and detecting albeit rare non-melanin pigments is complicated. However, inferences about distinct colouration strategies have been performed from fossils such as display (Li et al. 2010, 2012) and camouflage (Vinther et al. 2016; Brown et al. 2017; Smithwick et al. 2017a). In addition, broad-scale colour patterns such as countershading, stripes and spots can be highly informative as to an animal's ecology and behaviour irrespective of the precise hues.

11.6 Conclusions

Over the last decade, palaeocolour has evolved significantly as a discipline. The preservation potential of melanin and other pigments under exceptional circumstances has allowed for inferring aspects of dinosaurian appearance and ecology that was thought to be impossible. While palaeocolour is limited to a few fossils from few localities, it has shown its potency for contributing crucial input to evolutionary and ecological studies of extinct ecosystems.

Palaeocolour has contributed significantly to our understanding of the evolution and origin of avian plumage, and its colour gamut and many discoveries are still to be made. Colour reconstructions have helped to advance our knowledge of the predator-prey landscape in the Mesozoic, further highlighting major differences to today, but also some important similarities. Understanding which types of camouflage were present and in which groups helps to elucidate how Jurassic Park would have played out, if the book were written today.

Acknowledgements We thank the editors for their encouragement and patience with us in producing this book chapter as well as Gerald Mayr and others at the Senckenberg Museum, Frankfurt, for access of specimens.

References

Allen WL, Baddeley R, Cuthill IC, Scott-Samuel NE (2012) A quantitative test of the predicted relationship between countershading and lighting environment. Am Nat 180:762–776

Allison PA, Briggs DE (1993) Exceptional fossil record: distribution of soft-tissue preservation through the Phanerozoic. Geology 21:527–530

Babarović F, Puttick MN, Zaher M, Learmonth E, Gallimore E-J, Smithwick FM, Mayr G, Vinther J (2019) Characterization of melanosomes involved in the production of non-iridescent structural feather colours and their detection in the fossil record. J R Soc Interface 16(155):20180921

Baker RR, Parker G (1979) The evolution of bird coloration. Philos Trans R Soc B 287:63–130

Barden HE, Wogelius RA, LI D, Manning PL, Edwards NP, Van Dongen BE (2011) Morphological and geochemical evidence of eumelanin preservation in the feathers of the early cretaceous bird, Gansus yumenensis. PLoS One 6:e25494

Beyermann K, Hasenmaier D (1973) Identifizierung 180 Millionen Jahre alten, wahrscheinlich unverändert erhaltenen Melanins (identification of 180 million years old, probably unchanged melanine). Fresenius' Zeitschrift für analytische Chemie 266(3):202–205

Blakey PR, Earland C, Stell JGP (1963) Calcification of keratin. Nature 198:481

Bortolotti GR, Hill G, Mcgraw K (2006) Natural selection and coloration: protection, concealment, advertisement, or deception. Bird Coloration 2:3–35

Briggs DE, Kear AJ (1993) Fossilization of soft tissue in the laboratory. Science 259:1439–1442

Briggs DE, Wilby PR (1996) The role of the calcium carbonate-calcium phosphate switch in the mineralization of soft-bodied fossils. J Geol Soc 153:665–668

Briggs DEG, Kear AJ, Martill DM, Wilby PR (1993) Phosphatization of soft-tissue in experiments and fossils. J Geol Soc 150:1035–1038

Brown CM, Henderson DM, Vinther J, Fletcher I, Sistiaga A, Herrera J, Summons RE (2017) An exceptionally preserved three-dimensional Armored dinosaur reveals insights into coloration and cretaceous predator-prey dynamics. Curr Biol 27: 2514–2521

Brusatte SL, O'connor JK, Jarvis ED (2015) The origin and diversification of birds. Curr Biol 25: R888–R898

Carney RM, Vinther J, Shawkey MD, D'alba L, Ackermann J (2012) New evidence on the colour and nature of the isolated archaeopteryx feather. Nat Commun 3:637

Chang M-M (2003) The Jehol biota: the emergence of feathered dinosaurs, beaked birds and flowering plants. Scientific & Technical Publ, Shanghai

Chen PJ, Dong ZM, Zhen SN (1998) An exceptionally well-preserved theropod dinosaur from the Yixian formation of China. Nature 391:147

Chittenden RH, Albro AH (1899) The formation of melanins or melanin-like pigments from proteid substances. Am J Physiol Legacy Cont 2:291–305

Clarke JA, Ksepka DT, Salas-Gismondi R, Altamirano AJ, Shawkey MD, D'alba L, Vinther J, Devries TJ, Baby P (2010) Fossil evidence for evolution of the shape and color of penguin feathers. Science 330:954–957

Clements T, Dolocan A, Martin P, Purnell MA, Vinther J, Gabbott SE (2016) The eyes of Tullimonstrum reveal a vertebrate affinity. Nature 532:500–503

Colleary C, Dolocan A, Gardner J, Singh S, Wuttke M, Rabenstein R, Habersetzer J, Schaal S, Feseha M, Clemens M (2015) Chemical, experimental, and morphological evidence for diagenetically altered melanin in exceptionally preserved fossils. Proc Natl Acad Sci 112:12592–12597

Cuthill IC (2006) Color perception. Bird Coloration 1:3–40

Cuthill I, Bennett A, Partridge J, Maier E (1999) Plumage reflectance and the objective assessment of avian sexual dichromatism. Am Nat 153:183–200

D'alba L, Shawkey MD (2018) Melanosomes: biogenesis, properties, and evolution of an ancient organelle. Physiol Rev 99:1–19

Damsté JS, Koopmans MP (1997) The fate of carotenoids in sediments: an overview. Pure Appl Chem 69:2067–2074

Darwin C (1859) The origin of species by means of natural selection: or, the preservation of favored races in the struggle for life

Darwin C (1871) The descent of man and selection in relation to sex. Рипол Классик, Moscow

Davis PG, Briggs DE (1995) Fossilization of feathers. Geology 23:783–786

Demarchi B, Hall S, Roncal-Herrero T, Freeman CL, Woolley J, Crisp MK, Wilson J, Fotakis A, Fischer R, Kessler BM, Jersie-Christensen RR (2016) Protein sequences bound to mineral surfaces persist into deep time. elife 5:e17092

Doguzhaeva L, Mapes R, Mutvei H (2004) Occurrence of ink in Paleozoic and Mesozoic coleoids (Cephalopoda). Mitteilungen aus dem Geologisch-Paläontologischen Institut der Universität Hamburg 88:145–155

Doucet SM, Meadows MG (2009) Iridescence: a functional perspective. J R Soc Interface 6:S115–S132
Eglinton G, Logan GA (1991) Molecular preservation. Phil Trans R Soc Lond B 333:315–328
Esperante R, Brand L, Nick KE, Poma O, Urbina M (2008) Exceptional occurrence of fossil baleen in shallow marine sediments of the Neogene Pisco formation, southern Peru. Palaeogeogr Palaeoclimatol Palaeoecol 257:344–360
Ficken RW, Wilmot LB (1968) Do facial eye-stripes function in avian vision? Am Midl Nat:522–523
Field DJ, D'alba L, Vinther J, Webb SM, Gearty W, Shawkey MD (2013) Melanin concentration gradients in modern and fossil feathers. PLoS One 8:e59451
Foth C, Tischlinger H, Rauhut OW (2014) New specimen of archaeopteryx provides insights into the evolution of pennaceous feathers. Nature 511:79–82
Frey E, Tischlinger H, Buchy MC, Martill DM (2003) New specimens of Pterosauria (Reptilia) with soft parts with implications for pterosaurian anatomy and locomotion. Geol Soc Lond, Spec Publ 217:233–266
Gabbott SE, Donoghue PC, Sansom RS, Vinther J, Dolocan A, Purnell MA (2016) Pigmented anatomy in carboniferous cyclostomes and the evolution of the vertebrate eye. Proc Biol Sci 283:1836
Galván I, Solano F (2016) Bird integumentary melanins: biosynthesis, forms, function and evolution. Int J Mol Sci 17:520
Galván I, Wakamatsu K (2016) Color measurement of the animal integument predicts the content of specific melanin forms. RSC Adv 6:79135–79142
Gioncada A, Collareta A, Gariboldi K, Lambert O, Di Celma C, Bonaccorsi E, Urbina M, Bianucci G (2016) Inside baleen: exceptional microstructure preservation in a late Miocene whale skeleton from Peru. Geology 44:839–842
Glass K, Ito S, Wilby PR, Sota T, Nakamura A, Bowers CR, Vinther J, Dutta S, Summons R, Briggs DE (2012) Direct chemical evidence for eumelanin pigment from the Jurassic period. Proc Natl Acad Sci 109:10218–10223
Glass K, Ito S, Wilby PR, Sota T, Nakamura A, Bowers CR, Miller KE, Dutta S, Summons RE, Briggs DE (2013) Impact of diagenesis and maturation on the survival of eumelanin in the fossil record. Org Geochem 64:29–37
Godefroit P, Sinitsa SM, Dhouailly D, Bolotsky YL, Sizov AV, Mcnamara ME, Benton MJ, Spagna P (2014) A Jurassic ornithischian dinosaur from Siberia with both feathers and scales. Science 345:451–455
Gong EP, Martin LD, Burnham DA, Falk AR, Hou LH (2012) A new species of microraptor from the Jehol Biota of northeastern China. Palaeoworld 21:81–91
Gottfried MD (1989) Earliest fossil evidence for protective pigmentation in an actinopterygian fish. Hist Biol 3:79–83
Greenwalt DE, Goreva YS, Siljeström SM, Rose T, Harbach RE (2013) Hemoglobin-derived porphyrins preserved in a middle Eocene blood-engorged mosquito. Proc Natl Acad Sci 110(46): 18496–18500
Gren JA, Sjövall P, Eriksson ME, Sylvestersen RL, Marone F, Sigfridsson Clauss KG, Taylor GJ, Carlson S, Uvdal P, Lindgren J (2017) Molecular and microstructural inventory of an isolated fossil bird feather from the Eocene Fur formation of Denmark. Palaeontology 60:73–90
Griffiths P (1996) The isolated archaeopteryx feather. Archaeopteryx 14:1–26
Hill G, McGraw K (2006a) Function and evolution, vol 2. Harvard University Press, Cambridge, MA
Hill G, McGraw K (2006b) Mechanisms and measurements, vol 1. Harvard University Press, Cambridge, MA
Hu D, Clarke JA, Eliason CM, Qiu R, LI Q, Shawkey MD, Zhao C, D'alba L, Jiang J, Xu X (2018) A bony-crested Jurassic dinosaur with evidence of iridescent plumage highlights complexity in early paravian evolution. Nat Commun 9:217
Igic B, D'alba L, Shawkey MD (2016) Manakins can produce iridescent and bright feather colours without melanosomes. J Exp Biol 219:1851–1859
Kellner AW (2002) A review of avian Mesozoic fossil feathers. In: Mesozoic Birds: above the heads of Dinosaurs. University of California Press, Berkeley CA, pp 389–404
Koschowitz MC, Fischer C, Sander M (2014) Beyond the rainbow. Science 346:416–418
Lafountain AM, Prum RO, Frank HA (2015) Diversity, physiology, and evolution of avian plumage carotenoids and the role of carotenoid–protein interactions in plumage color appearance. Arch Biochem Biophys 572:201–212
Lee S-I, Kim M, Choe JC, Jablonski PG (2016) Evolution of plumage coloration in the crow family (Corvidae) with a focus on the color-producing microstructures in the feathers: a comparison of eight species. Anim Cells Syst 20:95–102
Li Q, Gao K-Q, Vinther J, Shawkey MD, Clarke JA, D'alba L, Meng Q, Briggs DE, Prum RO (2010) Plumage color patterns of an extinct dinosaur. Science 327:1369–1372
Li Q, Gao K-Q, Meng Q, Clarke JA, Shawkey MD, D'alba L, Pei R, Ellison M, Norell MA, Vinther J (2012) Reconstruction of microraptor and the evolution of iridescent plumage. Science 335:1215–1219
Lindgren J, Uvdal P, Sjövall P, Nilsson DE, Engdahl A, Schultz BP, Thiel V (2012) Molecular preservation of the pigment melanin in fossil melanosomes. Nat Commun 3:824
Lindgren J, Sjövall P, Carney RM, Uvdal P, Gren JA, Dyke G, Schultz BP, Shawkey MD, Barnes KR, Polcyn MJ (2014) Skin pigmentation provides evidence of convergent melanism in extinct marine reptiles. Nature 506(7489):484–488
Lindgren J, Moyer A, Schweitzer MH, Sjövall P, Uvdal P, Nilsson DE, Heimdal J, Engdahl A, Gren JA, Schultz BP (2015) Interpreting melanin-based coloration

through deep time: a critical review. Proc R Soc B 282:1813

Lindgren J, Kuriyama T, Madsen H, Sjövall P, Zheng W, Uvdal P, Engdahl A, Moyer AE, Gren JA, Kamezaki N, Ueno S (2017) Biochemistry and adaptive colouration of an exceptionally preserved juvenile fossil sea turtle. Sci Rep 7:13324

Lingham-Soliar T, Plodowski G (2010) The integument of Psittacosaurus from Liaoning Province, China: taphonomy, epidermal patterns and color of a ceratopsian dinosaur. Naturwissenschaften 97:479–486

Lingham-Soliar T, Feduccia A, Wang X (2007) A new Chinese specimen indicates that 'protofeathers' in the early cretaceous theropod dinosaur Sinosauropteryx are degraded collagen fibres. Proc R Soc Lond B Biol Sci 274:1823–1829

Liu Y, Simon JD (2005) Metal–ion interactions and the structural organization of Sepia eumelanin. Pigment Cell Melanoma Res 18:42–48

Mapes RH, Davis RA (1996) Color patterns in ammonoids. *Ammonoid paleobiology*. Springer, Boston, MA

Margalida A, Negro JJ, Galván I (2008) Melanin-based color variation in the bearded vulture suggests a thermoregulatory function. Comp Biochem Physiol A Mol Integr Physiol 149:87–91

Marques CI, Batalha HR, Cardoso GC (2016) Signalling with a cryptic trait: the regularity of barred plumage in common waxbills. R Soc Open Sci 3:160195

Mayr G (2017) The early Eocene birds of the Messel fossil site: a 48 million-year-old bird community adds a temporal perspective to the evolution of tropical avifaunas. Biol Rev Camb Philos Soc 92 (2):1174–1188

Mayr G, Pittman M, Saitta E, Kaye TG, Vinther J (2016) Structure and homology of Psittacosaurus tail bristles. Palaeontology 59:793–802

McGraw KJ (2006a) Mechanics of carotenoid-based coloration. Bird Coloration 1:177–242

McGraw KJ (2006b) Mechanics of melanin-based coloration. In: Bird coloration, Mechanisms and measurements, vol 1. Harvard University Press, Cambridge MA

McGraw KJ (2006c) Mechanics of uncommon colors: pterins, porphyrins, and psittacofulvins. Bird Coloration 1:354–398

McGraw KJ (2008) An update on the honesty of melanin-based color signals in birds. Pigment Cell Melanoma Res 21:133–138

McGraw KJ, Nogare MC (2004) Carotenoid pigments and the selectivity of psittacofulvin-based coloration systems in parrots. Comp Biochem Physiol B: Biochem Mol Biol 138:229–233

McGraw K, Safran R, Wakamatsu K (2005) How feather colour reflects its melanin content. Funct Ecol 19:816–821

McNamara ME, Orr PJ, Kearns SL, Alcalá L, Anadón P, Peñalver-Mollá E (2006) High-fidelity organic preservation of bone marrow in ca. 10 ma amphibians. Geology 34:641–644

McNamara M, Orr PJ, Kearns SL, Alcalá L, Anadón P, Peñalver-Mollá E (2010) Organic preservation of fossil musculature with ultracellular detail. Proc R Soc Lond B Biol Sci 277:423–427

McNamara ME, Briggs DE, Orr PJ, Field DJ, Wang Z (2013) Experimental maturation of feathers: implications for reconstructions of fossil feather colour. Biol Lett 9:20130184

McNamara ME, Dongen BE, Lockyer NP, Bull ID, Orr PJ (2016a) Fossilization of melanosomes via sulfurization. Palaeontology 59:337–350

McNamara ME, Orr PJ, Kearns SL, Alcalá L, Anadón P, Peñalver E (2016b) Reconstructing carotenoid-based and structural coloration in fossil skin. Curr Biol 26:1075–1082

McNamara ME, Briggs DE, Orr PJ, Field DJ, Wang Z (2017) Correction to 'experimental maturation of feathers: implications for reconstructions of fossil feather colour'. Biol Lett 13:3

McNamara ME, Kaye JS, Benton MJ, Orr PJ, Rossi V, Ito S, Wakamatsu K (2018) Non-integumentary melanosomes can bias reconstructions of the colours of fossil vertebrates. Nat Commun 9:2878

Morris SC, Caron JB (2014) A primitive fish from the Cambrian of North America. Nature 512:419

Moyer AE, Zheng W, Johnson EA, Lamanna MC, LI D-Q, Lacovara KJ, Schweitzer MH (2014) Melanosomes or microbes: testing an alternative hypothesis for the origin of microbodies in fossil feathers. Sci Rep 4:4233

Moyer AE, Zheng W, Schweitzer MH (2016) Microscopic and immunohistochemical analyses of the claw of the nesting dinosaur, *Citipati osmolskae*. Proc R Soc B 283:20161997

O'Connor S, Solazzo C, Collins M (2015) Advances in identifying archaeological traces of horn and other keratinous hard tissues. Stud Conserv 60:393–417

Olson VA, Owens IP (1998) Costly sexual signals: are carotenoids rare, risky or required? Trends Ecol Evol 13:510–514

Ostrom JH (1974) *Archaeopteryx* and the origin of flight. Q Rev Biol 49:27–47

Owen P (1863) On the *Archeopteryx* of von Meyer, with a description of the fossil remains of a long-tailed species, from the lithographic stone of Solnhofen. Philos Trans R Soc Lond 153:33–47

Pan Y, Zheng W, Moyer AE, O'connor JK, Wang M, Zheng X, Wang X, Schroeter ER, ZHOU Z, Schweitzer MH (2016) Molecular evidence of keratin and melanosomes in feathers of the early cretaceous bird Eoconfuciusornis. Proc Natl Acad Sci 113: E7900–E7907

Parry L, Smithwick FM, Briggs DE, Vinther J (2017) Soft-bodied fossils are not simply rotten carcasses—towards a holistic understanding of exceptional fossil preservation. BioEssays 40:1

Prado GM, Anelli LE, Petri S, Romero GR (2016) New occurrences of fossilized feathers: systematics and taphonomy of the Santana formation of the Araripe Basin (cretaceous), NE, Brazil. PeerJ 4:e1916

Prum RO (2006) Anatomy, physics, and evolution of structural colors. Bird Coloration 1:295–353

Rauhut OW, Foth C, Tischlinger H, Norell MA (2012) Exceptionally preserved juvenile megalosauroid theropod dinosaur with filamentous integument from the late Jurassic of Germany. Proc Natl Acad Sci 109:11746–11751

Rieseberg LH, Soltis DE (1987) Flavonoids of fossil Miocene Platanus and its extant relatives. Biochem Syst Ecol 15:109–112

Rowland HM (2009) From Abbott Thayer to the present day: what have we learned about the function of countershading? Philos Trans R Soc Lond B Biol Sci 364:519–527

Saitta ET, Rogers C, Brooker RA, Abbott GD, Kumar S, O'reilly SS, Donohoe P, Dutta S, Summons RE, Vinther J (2017) Low fossilization potential of keratin protein revealed by experimental taphonomy. Palaeontology 60:547–556

Saitta ET, Fletcher I, Martin P, Pittman M, Kaye TG, True LD, Norell MA, Abbott GD, Summons RE, Penkman K, Vinther J (2018) Preservation of feather fibers from the late cretaceous dinosaur Shuvuuia deserti raises concern about immunohistochemical analyses on fossils. Org Geochem 125:142–151

Schweitzer MH, Watt JA, Avci R, Forster CA, Krause DW, Knapp L, Rogers RR, Beech I, Marshall M (1999) Keratin immunoreactivity in the late cretaceous bird *Rahonavis ostromi*. J Vertebr Paleontol 19:712–722

Schweitzer MH, Lindgren J, Moyer AE (2015) Melanosomes and ancient coloration re-examined: a response to Vinther 2015 (doi:10.1002/bies.201500018). BioEssays 37:1174–1183

Shu DG, Morris SC, Han J, Zhang ZF, Yasui K, Janvier P, Chen L, Zhang XL, Liu JN, Li Y, Liu HQ (2003) Head and backbone of the early Cambrian vertebrate Haikouichthys. Nature 421:526

Sinninghe Damsté JS, Rijpstra WIC, Kock-van Dalen AC, De Leeuw JW, Schenck PA (1989) Quenching of labile functionalised lipids by inorganic Sulphur species: evidence for the formation of sedimentary organic Sulphur compounds at the early stages of diagenesis. Geochim Cosmochim Acta 53:1343–1355

Smithwick FM, Nicholls R, Cuthill I, Vinther J (2017a) Countershading and stripes in the theropod dinosaur *Sinosauropteryx* reveal heterogeneous habitats in the early cretaceous Jehol biota. Curr Biol 27 (21):3337–3343.e2

Smithwick FM, Mayr G, Saitta ET, Benton MJ, Vinther J (2017b) On the purported presence of fossilized collagen fibres in an ichthyosaur and a theropod dinosaur. Palaeontology 60:409–422

Stavenga DG, Leertouwer HL, Marshall NJ, Osorio D (2010) Dramatic colour changes in a bird of paradise caused by uniquely structured breast feather barbules. Proc R Soc Lond B Biol Sci 278(1715). https://doi.org/10.1098/rspb.2010.2293

Summons RE (2014) The exceptional preservation of interesting and informative biomolecules. Paleontol Soc Pap 20:217–236

Thayer AH (1896) The law which underlies protective coloration. Auk 13:124–129

Thomas DB, Nascimbene PC, Dove CJ, Grimaldi DA, James HF (2014a) Seeking carotenoid pigments in amber-preserved fossil feathers. Sci Rep 4:5226

Thomas DB, McGraw KJ, Butler MW, Carrano MT, Madden O, James HF (2014b) Ancient origins and multiple appearances of carotenoid-pigmented feathers in birds. Proc R Soc Lond B Biol Sci 281:20140806

Vickerman JC, Briggs D (2001) ToF-SIMS: surface analysis by mass spectrometry. IM, Chichester

Vinther J (2015a) A guide to the field of palaeo colour. BioEssays 37:643–656

Vinther J (2015b) Fossil melanosomes or bacteria? A wealth of findings favours melanosomes. BioEssays 38:220–225

Vinther J, Briggs DE, Prum RO, Saranathan V (2008) The colour of fossil. Feathers. Biol Lett 4:522–525

Vinther J, Briggs DE, Clarke J, Mayr G, Prum RO (2009) Structural coloration in a fossil feather. Biol Lett 6:1. https://doi.org/10.1098/rsbl.2009.0524

Vinther J, Nicholls R, Lautenschlager S, Pittman M, Kaye TG, Rayfield E, Mayr G, Cuthill IC (2016) 3D camouflage in an ornithischian dinosaur. Curr Biol 26:2456–2462

Vitek NS, Vinther J, Schiffbauer JD, Briggs DE, Prum RO (2013) Exceptional three-dimensional preservation and coloration of an originally iridescent fossil feather from the middle Eocene messel oil shale. Paläontol Z 87:493–503

Voigt E (1936) Über das Haarkleid einiger Säugetiere aus der mitteleozänen Braunkohle des Geiseltales. Nova Acta Leopoldina 4:317–334

Voigt E (1988) Preservation of soft tissues in the Eocene lignite of the Geiseltal near Halle/S. Cour Forschungsinst Senck 107:325–343

von Meyer H (1861a) Archaeopteryx lithographica. Neues Jahrbuch für Mineralogie, Geologie und Paläontologie 1861:678–679

von Meyer H (1861b) Archaeopteryx lithographica (Vogel-Feder) und Pterodactylus von Solnhofen. Neues Jahrbuch für Mineralogie, Geognosie, Geologie und Petrefakten-Kunde 1861:678–679

von Meyer H (1862) Archaeopteryx lithographica aus dem lithographischen Schiefer von Solenhofen. Palaeontographica (1846-1933) 10:53–56

Vorobyev M, Osorio D, Bennett AT, Marshall NJ, Cuthill IC (1998) Tetrachromacy, oil droplets and bird plumage colours. J Comp Physiol A 183:621–633

Vukusic P (2011) Structural colour: elusive iridescence strategies brought to light. Curr Biol 21:R187–R189

Whitear M (1956) XCVII.—on the colour of an ichthyosaur. J Nat Hist 9:742–744

Wilby PR, Briggs DE (1997) Taxonomic trends in the resolution of detail preserved in fossil phosphatized soft tissues. Geobios 30:493–502

Wogelius R, Manning P, Barden H, Edwards N, Webb S, Sellers W, Taylor K, Larson P, Dodson P, You H (2011) Trace metals as biomarkers for eumelanin pigment in the fossil record. Science 333:1622–1626

Wolfenden RN (1884) On certain constituents of the eggs of the common frog. J Physiol 5:91–97

Wuttke M (1983) Weichteil-Erhaltung'durch lithifizierte Mikroorganismen bei mittel-eozänen Vertebraten aus den Ölschiefern der 'Grube Messel'bei Darmstadt. Senckenb Lethaea 64:509–527

Xing L, Mckellar RC, Wang M, Bai M, O'connor JK, Benton MJ, Zhang J, Wang Y, Tseng K, Lockley MG, Li G (2016) Mummified precocial bird wings in mid-cretaceous Burmese amber. Nat Commun 7:12089

Zhang F, Kearns SL, Orr PJ, Benton MJ, Zhou Z, Johnson D, Xu X, Wang X (2010) Fossilized melanosomes and the colour of cretaceous dinosaurs and birds. Nature 463:1075–1078

Zheng X-T, You H-L, Xu X, Dong Z-M (2009) An early cretaceous heterodontosaurid dinosaur with filamentous integumentary structures. Nature 458:333–336

On the Ancestry of Feathers in Mesozoic Dinosaurs 12

Nicolás E. Campione, Paul M. Barrett, and David C. Evans

12.1 Introduction

Prior to the late twentieth century, most, if not all, non-avian dinosaurs were considered to have been covered in scales (e.g. Charig 1979; Norman 1985). This model was based on interpretation of dinosaur biology within a classical 'reptilian' paradigm and was supported by the recovery of rare skin impressions from a diverse range of taxa, including ankylosaurs, ceratopsians, hadrosaurids, and sauropods, whose preserved integument provided clear evidence for a covering of non-imbricating, polygonal scales (reviewed in Czerkas 1994; Xu and Guo 2009; Arbour et al. 2014; Bell 2014; Davis 2014; Barrett et al. 2015). This consensus was challenged by the recognition that birds are dinosaurs, lying in a deeply nested position within Theropoda (e.g. Ostrom 1976; Gauthier 1986), leading to speculation that the most talismanic of avian features, feathers, would eventually be discovered in their non-avian relatives (e.g. Bakker and Galton 1974; Paul 1988). Spectacular discoveries of exquisitely preserved non-avian theropods from the Early Cretaceous Jehol Biota of China confirmed this prediction (Chen et al. 1998), and numerous subsequent finds showed that feathers and other feather-like structures were widespread amongst coelurosaurs and potentially primitive for theropods as a whole (e.g. Norell and Xu 2005; Xu 2006; Xu and Guo 2009; Rauhut et al. 2012; see below). Moreover, the recognition of feather-like features in the ornithischian dinosaurs *Kulindadromeus* (Godefroit et al. 2014a; Saveliev and Alifanov 2014), *Psittacosaurus* (Mayr et al. 2002, 2016), and *Tianyulong* (Zheng et al. 2009; Sereno 2012) led to the proposal that all dinosaurs might have been primitively feathered (Witmer 2009; Zheng et al. 2009; Brusatte 2012; Godefroit et al. 2014a). However, this proposal remains controversial as (1) direct evidence for feathers is lacking in many non-avian dinosaurs, (2) inferring the presence/absence of feathers in the common ancestor of dinosaurs is based on numerous unproven assumptions regarding the antiquity and ubiquity of feathers amongst theropods and

Electronic supplementary material: The online version of this chapter (https://doi.org/10.1007/978-3-030-27223-4_12) contains supplementary material, which is available to authorized users.

N. E. Campione (✉)
Palaeoscience Research Centre, School of Environmental and Rural Science, University of New England, Armidale, NSW, Australia

Palaeobiology Programme, Department of Earth Sciences, Uppsala University, Uppsala, Sweden

Subdepartment of Evolution and Development, Department of Organismal Biology, Uppsala University, Uppsala, Sweden
e-mail: ncampion@une.edu.au

P. M. Barrett
Department of Earth Sciences, Natural History Museum, London, UK

D. C. Evans
Department of Natural History, Royal Ontario Museum, Toronto, ON, Canada

C. Foth, O. W. M. Rauhut (eds.), *The Evolution of Feathers*, Fascinating Life Sciences,
https://doi.org/10.1007/978-3-030-27223-4_12

ornithischians and (3) phylogenetic optimisations for the presence/absence of feathers and feather-like structures are sensitive to the proposed outgroup condition in the earliest pterosaurs, which is also unresolved (Barrett et al. 2015).

New discoveries from the fossil record of feathered dinosaurs were paralleled by numerous advances in developmental biology and led to the proposal of several different, but overlapping, models for feather evolution (Prum 1999; Brush 2000; Maderson and Alibardi 2000; Prum and Brush 2002; Sawyer and Knapp 2003; Sawyer et al. 2003a, b; Dhouailly 2009; Xu and Guo 2009). One such model proposes that feathers originated following the development of the follicle and that many of the structural β-keratins found in feathers appeared after the origin of feather precursors, so that feathers should not be considered direct homologues of scales (Prum 1999; Prum and Brush 2002). Under this model, filaments arising from within follicles represent the first stage of feather evolution, which become more elaborate during theropod evolution via the addition of barbs and barbules (Prum 1999; Prum and Brush 2002). This model proved attractive to palaeontologists as it predicts that filamentous 'protofeathers' should appear in theropod phylogeny prior to the appearance of more complex branched structures, which was supported by fossil evidence—early diverging coelurosaurians, such as *Sinosauropteryx*, possess simple, unbranched, filamentous integumentary structures (Chen et al. 1998), whereas more complex, branched structures are present in more deeply nested maniraptorans (e.g. dromaeosaurs and birds: Xu et al. 1999a). By contrast, an alternative model posits that the β-keratins found in the avian epidermis have a wider distribution and deeper origin within Amniota (Sawyer et al. 2000), suggesting the deep homology of feathers and scales, and that feathers first appeared as filamentous epidermal outgrowths, similar to the bristles found in turkey 'beards' or the embryonic filaments of 'true' feathers, prior to the appearance of follicles (Sawyer et al. 2003a, b). A third class of model suggests that feathers originated directly from flattened scales similar to those present in the Triassic archosauromorph *Longisquama* (Maderson and Alibardi 2000).

Regardless of which model is preferred, all agree that, at some stage in feather evolution, a follicle forms by the invagination of the epidermis around a dermal papilla, the latter being formed from an epidermal placode that produces the population of cells that goes on to develop into the feather bud. Also, changes in cellular signalling, developmental timing, and interactions between the follicle and the developing feather bud are responsible for generating the morphological variety present in more complex feathers (Chen et al. 2015). Other recent work indicated that the ectodermal placodes of birds are homologous to those present in non-avian reptiles (Musser et al. 2015; Di-Poï and Milinkovitch 2016) and further suggest that scales and feathers should be considered homologous at a developmental level. Following the appearance of the follicle, feathers are hypothesised to have passed through a variety of evolutionary and developmental stages, with the incremental, stepwise appearance of a collar, barb ridges, the presence of unbranched barbs, and a rachis, through to the development of branched barbs and pennaceous feathers with vanes (Prum 1999; Prum and Brush 2002; Dhouailly 2009; Xu and Guo 2009). Finally, it was also suggested that bird scales, as found on their legs, may not be directly homologous with those in other reptiles (including other non-avian dinosaurs), but might represent secondarily modified feathers (Sawyer and Knapp 2003; Dhouailly 2009).

A number of terms have been used to describe the integumentary features present in Mesozoic archosaurs, including (but not limited to) scales, filaments, protofeathers, quills, and feathers (unbranched, branched, or pennaceous). Here, we use the general term 'filament' for any non-squamous integumentary structure in recognition of the fact that these may have different developmental origins (as either an epidermal outgrowth, modification of an epidermal scale, or from a true follicle) and may not be homologous, thus avoiding the epistemological pitfall of using the term 'protofeather', which implies direct homology with the varied feather types found in extant dinosaurs. It is worth noting that without direct evidence from developmental observations, it is currently impossible to distinguish between filaments that might have arisen from either epidermal outgrowths or from within follicles in fossils. We reserve the use of the term 'feather' for structures comparable to those seen in living birds, which show definitive evidence of barbs and/or barbules, and which are inferred to have emerged from a follicle. Hence, we would not regard turkey 'beard' bristles as feathers, given their differing developmental origin, but rather as a distinct subcategory of 'filament' (e.g. Mayr et al. 2016).

Determining the phylogenetic distribution of the varied integumentary structures present amongst Mesozoic archosaurs represents the first step in understanding the origins and potential selective pressures driving the evolution of these features and also offers insights into the palaeobiology of these taxa. Here, we review the distribution of preserved integumentary structures amongst Mesozoic archosauromorphs to provide a context for understanding their distribution within a phylogenetic framework. We then use maximum likelihood methods to recover the probabilities for the presence of varied integumentary structures at different nodes within archosaur phylogeny, building upon and testing the results presented in Barrett et al. (2015). Our analyses examine the effects of different branch-scaling methods, character construction, and tree topologies upon our understanding of archosaur skin evolution and also test the impact of adding hypothetical 'feathered' taxa from poorly sampled timed intervals to dinosaur phylogenies. We also incorporate additional information and discussion on pterosaur integument. Finally, we compare the phylogenetic distribution of dinosaur integumentary structures against several evolutionary models to determine which, if any, have the most explanatory power to account for the observed pattern. We then discuss the implications of these results for understanding dinosaur skin evolution and highlight ways in which current controversies can be resolved.

12.2 Review of Fossil Archosaur Skin

Currently, there are no overviews of the epidermal structures present in extinct archosauromorphs [although there are overviews of dinosaur skin structures: e.g. Xu and Guo (2009), Arbour et al. (2014), Bell (2014), Barrett et al. (2015)] nor has there been any attempt to discuss the evolution of these features within the clade as a whole. This is surprising given the recent explosion of interest in feather evolution, and controversies over purported feather-like structures, which have both been studied intensively on the basis of palaeontological, taphonomic, and developmental evidence (see above). This deficit might stem, in part, from a dearth of information on the skin structure for large swathes of the archosauromorph tree. For example, although there is extensive literature on the skin of extant birds and non-avian dinosaurs, there is limited or no published information on the skin of most pseudosuchians and non-dinosaurian avemetatarsalians. This lack of basic anatomical data is problematic as it obscures the phylogenetic distributions of key features and frustrates attempts to reconstruct the evolutionary history of epidermal structure.

Only one example of skin preservation has been documented in a non-archosaurian archosauromorph, from a single individual of the Early Triassic South African taxon *Proterosuchus* (Thornley 1970; Cruickshank 1972). This specimen included a number of poorly preserved epidermal structures termed 'scutes', which were embedded in the matrix surrounding the skeleton. Histological analysis of these features suggested that they were structurally similar to the scales of extant reptiles, but no other details were provided (Thornley 1970).

In stark contrast to the rich record of Mesozoic pseudosuchian (especially crocodilian) body fossils, non-ornithodiran skin impressions are rare and poorly documented. Preserved body outlines are available for some marine Jurassic thalattosuchians, but these lack clear evidence of squamation and, instead, have a wrinkled appearance (Andrews 1913). Other available evidence includes rare scale impressions within tracks referred to the widespread Triassic ichnotaxa *Chirotherium* and *Synaptichium* (see review in Avanzini 2000). The precise identities of these trackmakers are currently unknown, but they are generally considered to be crown-group archosaurs (e.g. Klein et al. 2011). Preserved scale impressions are polygonal to rectangular or subelliptical in outline, non-imbricating and arranged in a 'chequer board' pattern of alternating rows (Avanzini 2000). They vary in size depending on their position within the print, with smaller scales at the junctions between phalangeal pads and larger scales within the pads, on the more proximal portions of the digits and on the sole of the foot (Avanzini 2000). Scale impressions are also present within tracks referred to *Apatopus* (Foster et al. 2003), which are attributed to a phytosaur (Padian et al. 2010). These scales are small, polygonal, and arranged in a 'honeycomb-like' pattern (Foster et al. 2003). Most Mesozoic crocodylomorph footprints lack scale impressions but, where present, their squamation patterns are similar to those of extant crocodilians (Ellenberger 1974; Bernardi et al. 2010).

Several Lagerstätten, notably those of the Solnhofen beds (Late Jurassic), Yanliao Biota (Late Jurassic), the Crato and Santana formations (Early Cretaceous), and the Jehol Biota (Early Cretaceous), together with a handful of other sites, have yielded a number of exceptional pterosaur specimens that include a variety of preserved soft tissues (e.g. Kellner et al. 2010; Sullivan et al. 2014). These indicate that pterosaur skin was, at least in large part, naked and leathery, and that it lacked scales over most of the body surface (Martill and Unwin 1989; Kellner 1996; Lü 2002; Wang et al. 2007; Kellner et al. 2010). The absence of scales appears genuine, as many of the specimens with soft tissue preservation possess other features with similar preservation potential as scales: details of actinofibres in the wing membranes, keratinous crests, rhamphothecae, claw sheaths, shell membranes, and pycnofibres (e.g. Kellner et al. 2010). The abdominal skin of *Gegepterus* was described as 'rugose' (Wang et al. 2007), but it is not clear if this represents the degraded remains of scales or a different, wrinkled epidermal structure. Several specimens demonstrate that scales were not completely lost in pterosaurs, however, as they were recorded in footpads around the heel (Frey et al. 2003). Some pterosaur tracks also include scale impressions (e.g. Mazin et al. 1995). Padian and Rayner (1993) proposed that the actinofibrils within the wing membranes might be homologous with epidermal scales, but this interpretation has been questioned (Kellner et al. 2010).

Pterosaurs from many different clades possess small, branched or unbranched, bristle-like filaments that were presumably for display and/or insulation (e.g. Unwin 2006). These were termed 'hair' or 'fur' (e.g. Sharov 1971), but this implies homology with mammalian fur, which was rejected on both phylogenetic and structural grounds: the term pycnofibre was proposed as an alternative (Kellner et al. 2010). These structures have a wide distribution within the clade and have been documented in a variety of taxa, including the Late Triassic *Eudimorphodon* (Wild 1993), anurognathids (Sharov 1971; Wang et al. 2002; Kellner et al. 2010), ctenochasmatids (Lü 2002; Wang et al. 2007; Jiang and Wang 2011; Lü et al. 2012), rhamphorhynchids (e.g. Padian 2008; Bennett 2015), pterodactylids (e.g. Frey and Martill 1998), and the pterodactyloid *Ningchengopterus* (Lü 2009), as well as several taxa of unresolved phylogenetic position (e.g. *Daohugoupterus*: Cheng et al. 2015; *Pterorhynchus*: Czerkas and Ji 2002). It is noteworthy that the full extent of pycnofibre distribution over the body is currently unknown for any pterosaur taxon (e.g. Kellner 1996). In most specimens, preservation is limited to small areas of the skeleton, and it is unclear if pterosaurs were covered in a continuous pelage or if pycnofibres were restricted to particular body regions. Additional information on this point will be essential in determining the function of these structures: for example, small tufts of pycnofibres are unlikely to be effective insulators, whereas a complete coat might imply an important thermoregulatory function.

Although it is tempting to speculate that the presence of pycnofibres might be primitive for pterosaurs, due to their presence in so many pterosaur clades (including at least one early diverging Late Triassic taxon), this issue has yet to receive rigorous scrutiny. Some well-preserved Late Triassic and Early Jurassic pterosaur specimens, such as *Carniadactylus*, lack evidence for pycnofibres, even though they include preservation of other soft tissues (e.g. wing membranes: Dalla Vecchia 2009). Moreover, at present, there is no direct evidence for the presence of pycnofibres in the majority of pterosaur species. However, it is possible that pycnofibres were subject to early taphonomic loss and/or that they are preserved in only a small number of depositional settings: all available pterosaur specimens with soft tissue preservation are derived from a small subset of the total number of pterosaur localities worldwide (see Barrett et al. (2008) for an overview of pterosaur distribution). For the time being, it seems most parsimonious to assume that all pterosaurs lacked scales on the body surface (with the exception of the foot), as these are absent from all known specimens for which appropriate material is available, and it seems plausible that pycnofibres were also primitively present, at least on some regions of the body, although more information on early pterosaurs is required to validate this proposal.

Pycnofibres might be homologous with the filaments/feathers of dinosaurs (e.g. Ji and Yuan 2002; Wang et al. 2002), but this proposal has yet to be confirmed on the basis of either detailed structural/biogeochemical or phylogenetic evidence (Wang et al. 2002; Kellner et al. 2010). For example, it was suggested that pycnofibres were superficial structures, borne on the outer surface of the skin, whereas feathers arise from follicles that are much more deeply embedded within it (Unwin 2006), though interpreting the detailed structure of these small, filamentous structures in fossils can be problematic (Foth 2012a).

Skin impressions are currently unknown for all non-dinosaurian dinosauromorphs, so these taxa offer no information on the primitive condition of dinosaur skin. Similarly, with the exception of scale impressions within footprints (Kundrát 2004), there is no direct evidence for dinosaur skin morphology prior to the Jurassic (Barrett et al. 2015). Although dinosaur skin impressions are now known for almost every major clade, the majority of examples are of Cretaceous age (see compilation in Barrett et al. (2015) and discussion, below; Table 12.1).

Table 12.1 Temporal sampling of fossilised skin in Mesozoic dinosaurs

Epoch	Period	Sample size
Upper	Cretaceous	34
Lower	Cretaceous	20
Upper	Jurassic	21
Middle	Jurassic	1
Lower	Jurassic	1
Upper	Triassic	0

Sauropodomorph skin impressions are surprisingly rare and are currently unknown for 'prosauropods' (Barrett et al. 2015). They are known primarily from neosauropods, with the exception of a single occurrence in the non-neosauropod *Mamenchisaurus* (Ouyang and Ye 2002). All known examples of sauropod skin consist of non-overlapping, flattened, polygonal scales that vary in size depending on position (e.g. Mantell 1852; Czerkas 1994; Coria and Chiappe 2007). There is some evidence that scales might be arranged in regular rosette-like patterns over some body regions (Czerkas 1992, 1994; Coria and Chiappe 2007). There is no direct evidence for any other epidermal structures in the vast majority of sauropodomorphs (e.g. elongated fringing scales, filaments), although these have been incorporated into life reconstructions by some palaeoartists (e.g. Witton 2016). 'Dermal spines', which appear to be hypertrophied scales, were reported along the dorsal midline of one indeterminate diplodocoid specimen (Czerkas 1992).

A wider variety of skin structures are present in ornithischians and samples are available for most of the major lineages (with the exception of pachycephalosaurs), although they are rare prior to the Late Cretaceous (Barrett et al. 2015). Most of our information comes from hadrosaurids, ceratopsians, and ankylosaurs, with rare examples from non-hadrosaurid ornithopods, heterodontosaurids, and stegosaurs. The majority of taxa possess flattened, non-overlapping, polygonal scales over most of the body surface, exemplified by numerous hadrosaurid specimens with extensive skin preservation, including almost complete skin envelopes (see reviews in Bell 2012, 2014; Davis 2014), but as also seen in a range of thyreophoran and ceratopsian taxa (e.g. Brown 1917; Christiansen and Tschopp 2010; Arbour et al. 2014). Skin preservation is of high enough quality and extent in some hadrosaurid taxa that differences in squamation patterns can be used for taxonomic purposes (Bell 2012). Some hadrosaurids, such as *Gryposaurus*, *Saurolophus*, and *Edmontosaurus*, possess flattened, tabular scales that are situated along the dorsal midline, forming a fringe-like structure similar to those of some extant lizards (e.g. Horner 1984; Bell 2012).

Elongate, unbranched, bristle-like structures are present on the tail of one specimen of the early ceratopsian *Psittacosaurus* (Mayr et al. 2002, 2016). These features probably did not arise from follicles, but it was proposed that they might be homologous to the filaments present in some theropod dinosaurs on the basis of gross morphological and structural evidence (it should be noted that polygonal scales are also present in virtually all other body regions of the same individual: Mayr et al. 2002, 2016). Similar filamentous structures have also been reported on the neck, back, and tail of the heterodontosaurid *Tianyulong*, but in this taxon, at least some of these features appear to be branched (Zheng et al. 2009). *Kulindadromeus,* from the ?Middle–Late Jurassic of Russia, possesses the most diverse epidermal structures found in any ornithischian, including three-scale morphotypes (found on different regions of the limbs and tail), monofilaments along the neck and back, ribbon-like structures on the distal hind limb, and unusual compound structures that consist of a basal plate supporting a number of unbranched filaments (Godefroit et al. 2014a). The monofilaments, ribbon-like structures, and compound structures were interpreted as feather homologues (Godefroit et al. 2014a). It was also proposed that at least some of the structures present in *Kulindadromeus*, *Psittacosaurus*, and *Tianyulong* might represent degraded collagen fibres from within the skin or the remains of extensive membranous frills (Lingham-Soliar 2014), though these interpretations were heavily disputed (Godefroit et al. 2014b; Mayr et al. 2016). It was also suggested that they might represent highly modified scales, rather than true feather homologues (Saveliev and Alifanov 2014; Barrett et al. 2015). Other epidermally derived soft tissue structures, such as rhampthothecae and fleshy cranial crests, are also known to have been present in ornithischians (e.g. Morris 1976; Bell et al. 2014).

Considerable variation exists in the epidermal structures of theropod dinosaurs, which include taxa that appear to be extensively covered in scales, a diversity of maniraptorans that are completely feathered and include varied feather types, as well as those seen in Neornithes, the avian crown group. In the last two decades, two Chinese Lagerstätten, the Yanliao Biota (Late Jurassic) and the Jehol Biota (Early Cretaceous), have produced a large number of exceptional small theropod specimens with soft tissue preservation that dramatically increased knowledge of the integumentary diversity of this group and considerably illuminated the origins of feathers and the avian *Bauplan* (Norell and Xu 2005; Xu et al. 2014; Sullivan et al. 2017). More recently, several spectacular coelurosaur specimens preserved in amber from the 'middle' Cretaceous of Myanmar (Xing et al. 2016) and Canada (McKellar et al. 2011) provide incredibly detailed, three-dimensional views of Cretaceous non-avian and avian theropod integument.

Few examples of fossilised integumentary structures are known in non-coelurosaurian, or 'basal' theropod lineages. Filamentous impressions in the abdominal region (prepubic and ischiadic imprints) were associated with sitting traces of an unknown Early Jurassic theropod from the Turners Falls Formation of Massachusetts, USA (Kundrát 2004); small trapezoidal scales are reported in the pedal impressions of the same individual. Skin impressions for Ceratosauria are limited to abelisaurids, with a detailed description available for those of *Carnotaurus sastrei* from the Upper Cretaceous (Maastrichtian) La Colonia Formation of Argentina (Bonaparte et al. 1990; Czerkas and Czerkas 1997). The largely complete articulated skeleton of the *Carnotaurus* holotype was found with skin impressions from the back of the lower jaw, neck, shoulder region, rib cage, and anterior part of the tail, which consists of non-imbricating scales from 5 to 12 mm in diameter arranged as individual background tubercles and in rosette patterns (Czerkas and Czerkas 1997). Larger feature scales of low conical morphology 40–50 mm in diameter occur intermittently along the flank, with smaller conical feature scales occurring ventrally; the height of the larger conical scales was suggested to be up to 50 mm (Czerkas and Czerkas 1997). Polygonal scales have been reported in an undescribed juvenile specimen of *Allosaurus* from the Late Jurassic Morrison Formation (Pinegar et al. 2003) and in the carcharodontosaurid *Concavenator corcovatus* from the Lower Cretaceous (Barremian) Calizas de La Huérguinahas Formation (Ortega et al. 2010). Interestingly, pennaceous feathers have been inferred in *Concavenator*, *Dakotaraptor*, and *Apatoraptor* based on structures resembling quill knobs on the ulna (Ortega et al. 2010; DePalma et al. 2015; Funston and Currie 2016), though these structures were also interpreted as representing an intramuscular line (Foth et al. 2014).

The holotype of the possible megalosauroid *Sciurumimus albersdoerferi* (but see Godefroit et al. (2013), where it has been inferred to be a basal coelurosaurian) preserves large patches of skin on the tail and smaller patches or soft tissue in other regions of the body (Rauhut et al. 2012; Foth et al. 2020). The skin is smooth, with no clear scalation, and long, thin monofilaments filaments are present on the tail. Shorter filaments are also found above the mid-dorsal vertebrae and on the ventral part of the body. Follicle-like structures were also noted in this taxon and used to suggest that follicles were associated with the origin of filamentous feathers, rather than the origin of a rachis or barb ridges (Rauhut et al. 2012).

The epidermal structures in coelurosaurian theropods have been extensively studied and described. Although scales are present in some taxa (see below), the integument in this clade also includes filamentous (mono- or multifilament) and shafted structures that are doubtless homologous to the remiges of living birds. Xu and Guo (2009) reviewed the integumentary structures in coelurosaurs and identified eight feather morphotypes, ranging from simple monofilaments to shafted flight feathers with barbs and barbules (see below; Levèfre et al. 2020; Xu 2020). Body scales are known only in several non-maniraptoran coelurosaurs; to date, known tyrannosaurid integument consists only of scales (Bell et al. 2017), and both body scales and monofilaments were reported in the compsognathid *Juravenator starki* (Chiappe and Göhlich 2010; Foth et al. 2020). Pennebrachia, or wing-like feathered forelimbs with shafted feathers, are widespread amongst coelurosaurs, occurring in the least inclusive clade composed of Ornithomimidae and Maniraptora (Zelenitsky et al. 2012).

Filamentous structures consisting of simple monofilaments (Morphotype 1) are present in early tyrannosauroids (*Dilong* and *Yutyrannus*: Xu et al. 2004, 2012), compsognathids (*Sinosauropteryx*: Chen et al. 1998), therizinosauroids (*Beipiaosaurus*: Xu et al. 1999b, 2009) and alvarezsauroids (*Shuvuuia*: Schweitzer et al. 1999), but have not been recognised in paravians (Xu and Guo 2009). Basally arrayed multifilaments (Morphotype 2) are present in numerous coelurosaurs, including therizinosauroids (*Beipiaosaurus*: Xu et al. 2009), scansoriopterygids (*Epidexipteryx*: Zhang et al. 2008), troodontids (*Anchiornis*: Hu et al. 2009), and dromaeosaurids (e.g. *Microraptor* spp., *Sinornithosaurus*: Xu et al. 1999a, 2000, 2003). Elongated broad filamentous, unbranched structures that differ from 'typical' monofilaments were reported in *Beipiaosaurus* (Xu et al. 2009).

Branching feather types that have a central shaft (rachis) supporting multiple filaments (barbs) (morphotypes 3, 4, and 6–8) occur in Ornithomimosauria (Zelenitsky et al. 2012; Van der Reest et al. 2016), Oviraptorosauria (e.g. *Protarchaeopteryx*, *Caudipteryx*: Ji et al. 1998), scansoriopterygids (e.g. *Epidexipteryx*, *Yi*: Zhang et al. 2008; Xu et al. 2015), troodontids (e.g. *Anchiornis*, *Jinfengopteryx*: Ji et al. 2005;

Hu et al. 2009), dromaeosaurids (e.g. *Microraptor* spp., *Sinornithosaurus*: Xu et al. 1999a, 2000, 2003), and avialians, including Neornithes. The scansoriopterygid *Yi qi* has filamentous integument over large portions of the body, as well as a membranous integument around the styliform element and digits in both hands (Xu et al. 2015). Planar, pennaceous feathers with barbules (morphotypes 6 and 7) occur in oviraptorosaurs and paravians, including birds. Most Mesozoic birds (Avialae) have the same complement of feather morphotypes (morphotypes 4, 6 and 7) as modern birds, and lack more primitive feather types including monofilaments (morphotype 1) and non-planar multifilaments (morphotypes 2 and 3) (also see O'Connor 2020).

12.3 Quantitative Framework

Based on the review of archosaurian integumentary structures provided here, we present an elaborated analysis of the results presented in Barrett et al. (2015), with the goal of generating a variety of ancestral state estimates to infer the probable ancestry of dinosaur integumentary structures. Techniques for reconstructing the distributions of discrete ancestral states have advanced beyond the ACCTRAN/DELTRAN optimisation criteria introduced by Swofford and Maddison (1987) and, although these methods are still utilised to visualise the evolution of discrete characters following maximum parsimony phylogenetic analyses and to help define clade synapomorphies, the probabilistic framework provided by maximum likelihood offers a more robust approach when attempting to understand the uncertainty associated with a particular ancestral state (Felsenstein 1985; Maddison 1991; Schluter et al. 1997; Cunningham et al. 1998).

The current study follows the methodology outlined in Barrett et al. (2015) in which maximum likelihood is used to reconstruct ancestral integumentary states within a probabilistic framework that we then use to interpret the evolution of dinosaur integument. Here, we expand on the methodological assumptions and analyses presented in our previous study in order to better understand the general nature of the previously described patterns, and the sensitivity of those results to variations in parameters such as tree topology, outgroup conditions, missing data, and the application of alternate evolutionary models. The effects of these parameters are tested using a non-parametric two-way multivariate analysis of variance (npMANOVA; $N_{\text{permutations}} = 999$) implemented through the R package vegan v. 2.4-2 (Oksanen et al. 2017).

Dataset All analyses presented here are based on an updated version of the dataset used in Barrett et al. (2015) that compiled all known species-level occurrences of fossilised integument in non-avian dinosaurs. Since that publication, Bell et al. (2017) presented additional data on the integument of tyrannosaurids and provide evidence that *Tyrannosaurus*, *Daspletosaurus*, *Albertosaurus*, and *Gorgosaurus* were not extensively feathered, contrary to the condition suggested by the presence of extensive plumage in early tyrannosauroids (Xu et al. 2012). Five new feathered dinosaurs from China were also described: the scansoriopterygid *Yi qi* (Xu et al. 2015), the dromaeosaurid *Zhenyuanlong suni* (Lü and Brusatte 2015), the troodontid *Jianianhualong tengi* (Xu et al. 2017), and the paravians *Serikornis sungei* (Lefèvre et al. 2017) and *Caihong juji* (Hu et al. 2018). However, as these five species are deeply nested within Coelurosauria, their inclusion will have a minimal effect on the predicted states at and near the root node; they were not included in the analyses. The current dataset includes a total of 77 species-level occurrences (excluding hypothetical outgroups; see Appendix).

Two integumentary characters were presented in Barrett et al. (2015). These vary in the number of states: character 1 is a multistate character that defines the presence of epidermal scales (state 0), unbranched filaments (state 1), and complex filaments/feathers (state 2), and character 2 is a binary character defining scales (state 0) and all other structures (state 1). In order to address the uncertainty associated with the unknown ancestral state of the outgroup, Pterosauria, Barrett et al. (2015) tested two variations of these characters in which pterosaurs were coded as either scaled or filamentous/feathered, primitively.

Given the ambiguities associated with interpreting the ancestral archosaur integument (see above), we introduce two additional modifications to characters 1 and 2. First, rather than include pterosaurs as the only dinosaurian outgroup, we include an additional crurotarsan terminal unit, coded as scaled (state 0) following the condition in extant and fossil crocodilians (e.g. Ellenberger 1974; Chang et al. 2009; Bernardi et al. 2010). The first appearance date for Crurotarsi was set at 247.2 Ma, as provided by the Paleobiology Database (paleobiodb.org; accessed April 3, 2017). The second modification accommodates for the observation that pterosaur integument can lack both scales and plumage, and hence may have been primitively 'naked' (Kellner et al. 2010; see Sect. 12.2). This state is added to both characters in this iteration as state 3 in character 1 and state 2 in character 2.

Phylogeny Given the relatively low sample of dinosaur taxa known from fossilised skin, there was little phylogenetic ambiguity in the tree topology employed by Barrett et al. (2015), and so alternate topologies were not explored. However, Baron et al. (2017) recently proposed a major restructuring of the dinosaur tree that rejected the traditional saurischian/ornithischian dichotomy. Instead, this new hypothesis places Theropoda as the sister taxon of Ornithischia (united in the clade Ornithoscelida), and Saurischia is redefined to include Sauropodomorpha and Herrerasauridae. Given that this phylogenetic hypothesis alters the interrelationships amongst various early dinosaur lineages, it is possible that these topological changes will impact the results presented in Barrett et al. (2015). All ancestral state reconstructions carried out here will be analysed using both the traditional (Hypothesis 1; H1) and ornithoscelidan (Hypothesis 2; H2) topologies for early dinosaurs (see Supplementary Appendix 12.1). However, given our sampling, the ingroup relationships within more deeply nested clades (e.g. within Tetanurae or Ornithischia) are consistent between these opposing phylogenetic hypotheses and were derived from a variety of sources (Dodson et al. 2004; Upchurch et al. 2004; Butler et al. 2008; Maidment 2010; Prieto-Márquez 2010; Prieto-Márquez et al. 2012; Rauhut et al. 2012; Thompson et al. 2012; Godefroit et al. 2013, 2014a).

Models of Character Evolution In Barrett et al. (2015), we tested two models of character evolution that were identified as unordered and ordered because of the way they treated the rate(s) of character state transitions. The unordered model considers all single transitions, such as between states 0–1, 1–2, and 0–2 and reversals of these transitions, as equal, and hence they are given an equal rate parameter. Such a scenario could be more generally referred to as an equal-rates model (ER) and the transition matrices ($T_{\text{model(states)}}$) given two, three, or four states would be:

$$T_{\text{ER}(0,1)} = \begin{Bmatrix} 0 & 1 \\ 1 & 0 \end{Bmatrix} \text{ or}$$

$$T_{\text{ER}(0,1,2)} = \begin{Bmatrix} 0 & 1 & 1 \\ 1 & 0 & 1 \\ 1 & 1 & 0 \end{Bmatrix} \text{ or}$$

$$T_{\text{ER}(0,1,2,3)} = \begin{Bmatrix} 0 & 1 & 1 & 1 \\ 1 & 0 & 1 & 1 \\ 1 & 1 & 0 & 1 \\ 1 & 1 & 1 & 0 \end{Bmatrix}$$

The values to the right of the diagonal (zero values) represent increases in state numbers, whereas those to the left represent reversals. The numbers in the matrix refer to the rate parameter to be estimated—in this case only a single rate. In comparison, the ordered model (ORD), which is only applicable for multistate characters, assumes that two-step transitions (i.e. between states 0–2 and 2–0) have a different rate parameter, and hence:

$$T_{\mathrm{ORD}(0,1,2)} = \begin{Bmatrix} 0 & 1 & 2 \\ 1 & 0 & 1 \\ 2 & 1 & 0 \end{Bmatrix}$$

However, in this study, we will modify T_{ORD} slightly from that used in Barrett et al. (2015), and remove the option for two-step transitions completely. This should act as a stricter implementation of an ordered character. Accordingly,

$$T_{\mathrm{ORD}(0,1,2)} = \begin{Bmatrix} 0 & 1 & 0 \\ 1 & 0 & 1 \\ 0 & 1 & 0 \end{Bmatrix}$$

We will also consider two other models of character evolution. The first allows all transition rates to vary independently, the unequal-rates (UR) model as follows:

$$T_{\mathrm{UR}(0,1)} = \begin{Bmatrix} 0 & 1 \\ 2 & 0 \end{Bmatrix} \text{ or}$$

$$T_{\mathrm{UR}(0,1,2)} = \begin{Bmatrix} 0 & 1 & 2 \\ 3 & 0 & 4 \\ 5 & 6 & 0 \end{Bmatrix} \text{ or}$$

$$T_{\mathrm{UR}(0,1,2,3)} = \begin{Bmatrix} 0 & 1 & 2 & 3 \\ 4 & 0 & 5 & 6 \\ 7 & 8 & 0 & 9 \\ 10 & 11 & 12 & 0 \end{Bmatrix}$$

The UR model is the most complex, given that as many as 12 rate parameters may be estimated. However, it serves to test an evolutionary model in which all transitions are independent from each other.

The final model attempts to test, in a general sense, the progressive (PROG) evolutionary pattern predicted by various developmental models in which the evolution of feathers is typified by increases in complexity (e.g. Prum 1999; Xu and Guo 2009). In this regard, reversals are not allowed. Ideally, under a strict progressive model, two-step transitions should not be allowed. Unfortunately, the nature of the data set did not permit such a model to be run (i.e. computationally singular). As a result, two variations of the PROG model were adopted when dealing with multistate characters:

$$T_{\mathrm{PROG}(0,1)} = \begin{Bmatrix} 0 & 1 \\ 0 & 0 \end{Bmatrix} \text{ or}$$

$$T_{\mathrm{EPROG}(0,1,2)} = \begin{Bmatrix} 0 & 1 & 1 \\ 0 & 0 & 1 \\ 0 & 0 & 0 \end{Bmatrix} \text{ or}$$

$$T_{\mathrm{UPROG}(0,1,2)} = \begin{Bmatrix} 0 & 1 & 2 \\ 0 & 0 & 3 \\ 0 & 0 & 0 \end{Bmatrix}$$

It should be noted that the PROG models are only used for the purposes of fitting and interpreting models of evolution, but not for interpreting ancestral likelihood values. The PROG models are highly restrictive and do not allow ancestral likelihoods to vary resulting in a forced primitive state at the base of the tree. Ancestral interpretations are only based on those models that allow character states to evolve in various directions. Furthermore, both the ORD and PROG models are only tested within the context of a primitively scaled or filamentous pterosaur, but not the iterations in which pterosaur integument is derived with respect to both (e.g. naked skin). A different condition in pterosaurs will have little bearing on the progressive model of evolution, and, more importantly, it cannot be assumed that such a condition is ordered within the framework of integumentary modifications in dinosaurs. In such cases where pterosaurs are coded differently, we tested an additional model, in which pterosaur integument is given a separate rate from that used for dinosaurs (i.e. a two-rate model [2R]):

$$T_{\mathrm{2R}(0,1,2)} = \begin{Bmatrix} 0 & 1 & 2 \\ 1 & 0 & 2 \\ 2 & 2 & 0 \end{Bmatrix} \text{ or}$$

$$T_{\mathrm{2R}(0,1,2,3)} = \begin{Bmatrix} 0 & 1 & 1 & 2 \\ 1 & 0 & 1 & 2 \\ 1 & 1 & 0 & 2 \\ 2 & 2 & 2 & 0 \end{Bmatrix}$$

Here, character states 2 and 3, respectively, represent the 'naked' pterosaur condition.

All models were compared internally; thus, comparisons are consistent across the various iterations explored here. For example, during each set of model comparisons, the character, the phylogenetic hypothesis, and the branch-scaling method are kept constant. Model fitting was accomplished using the ace function provided in ape (version 4.0: Paradis 2012), and the model log-likelihoods were converted into Akaike information criteria (AIC) in order to accommodate the large variation in the number of parameters between models using the standard equation:

$$\mathrm{AIC} = -2\mathrm{loglik} + 2k,$$

where k is the number of parameters.

Missing Data The fossil record of Mesozoic dinosaur integument is highly incomplete. Of the 1124 currently accepted species (Starrfelt and Liow 2016), only 77 (82 when counting the paravians not included in the analyses) are known that preserve at least some integumentary preservation (Appendix). As a result, studies of integument evolution are limited to just 7% of known non-avian dinosaurian diversity, the provenance of which is almost exclusively from the Cretaceous Period (Table 12.1). However, despite this limitation, the sample of fossilised skin spans almost all major dinosaurian groups, including Theropoda, Sauropoda, Thyreophora, Ceratopsia, and Ornithopoda (see Sect. 12.2). Sampling is particularly poor during the Late Triassic and Early Jurassic, but given that many major dinosaur clades diverge during this interval, it is hypothesised that the recovery of integumentary structures from taxa of this age could have a major bearing on the probabilistic framework adopted in this study. Accordingly, we introduce a series of experiments in which hypothetical 'feathered' operational taxonomic units are inserted into the two main trees discussed above (Table 12.2; Fig. 12.1). The effects of their inclusion are tested, both separately and in tandem, on the resulting ancestral state reconstructions.

Table 12.2 Hypothetical variants explored (Fig. 12.1)

Variants	Details	Age range (Ma)
1	Sauropodomorpha (H1) Saurischia (H2)	237–201.3
2	Theropoda	237–201.3
3	Ornithischia	237–201.3
4	Variations 1, 2, and 3 combined	

All variants are coded as state 1 for all characters

12.4 Results

Several different iterations of these analyses were carried out using both multistate and binary characters. However, the following results will focus on those analyses that deal with a multistate integumentary character only. Overall, the results did not vary substantially between character treatments, and the multistate character has the added benefit of greater interpretive potential, especially in the model-fitting analyses.

12.4.1 Model Fitting

A total of 88 model-fitting exercises were carried out, which encompass 12 comparative sets per branch scaling approach ('mbl' and 'equal'). Each set keeps all criteria constant, varying only the assumed model of character evolution. Overall, the ordered model of evolution (ORD) received overwhelmingly strong support (Table 12.3). In these cases, all other models received Akaike weights <0.1, although the equal-rates model could not be rejected in every set. As a binary character, both the equal- and unequal-rates models receive high Akaike weight values with slightly higher values for the former (0.63–0.73 and 0.24–0.3, respectively). Regardless of whether characters were coded as multistate or bivariate, the strict progressive models (PROG, EPROG and UPROG) received very low support values (Akaike weight = 0–0.14). Finally, support values are more ambiguous if pterosaurs are given a unique character state. The highest support values were recovered for the two-rates model, in which transitions to the

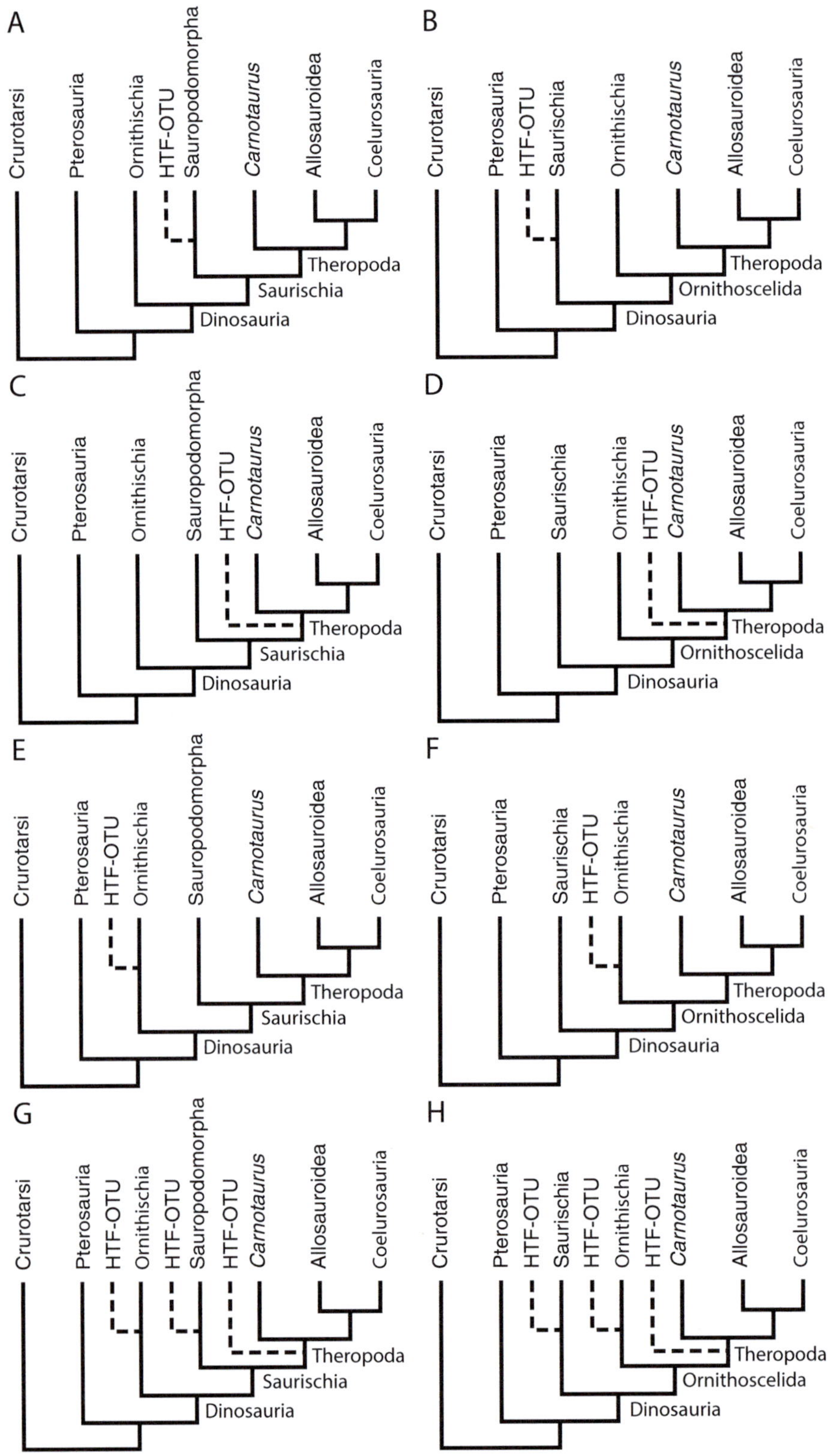

Fig. 12.1 Hypothetical experiments. Variants 1–4 (Table 12.2) explore the effects of a hypothetical Late Triassic feathered operational taxonomic unit (HTF-OUT) placed at the base of Sauropodomorpha/Saurischia (**a**, **b**), Theropoda (**c**, **d**), Ornithischia (**e**, **f**), or at the base of all three (**g**, **h**) on the ancestral dinosaurian state

Table 12.3 Model fitting results

Tree	Character	Model	Pterosaur state	Dinosaur ancestor estimate	log Lik.	Pars.	AIC	A.W.
Set 1								
1	Multistate	ER	Scales	Scales (Scales)	−50.54 (−45.77)	1	103.07 (93.53)	*0.04838* (0.09845)
1	Multistate	ORD	Scales	Scales (Scales)	−47.61 (−43.6)	1	97.22 (89.21)	**0.90152 (0.85369)**
1	Multistate	UR	Scales	Scales (Scales)	−45.5 (−41.51)	6	103 (95.02)	*0.0501 (0.04674)*
1	Multistate	EPROG	Scales	Scales (Scales)	−71.24 (−63.2)	1	144.49 (128.4)	*0 (0)*
1	Multistate	UPROG	Scales	Scales (Scales)	−61.81 (−48.24)	3	129.62 (102.48)	*0 (0.00112)*
Set 2								
1	Binary	ER	Scales	Scales (Scales)	−30.35 (−27.2)	1	62.7 (56.4)	**0.73105 (0.64977)**
1	Binary	UR	Scales	Scales (Scales)	−30.35 (−27.17)	2	64.7 (58.33)	0.26894 (0.24755)
1	Binary	PROG	Scales	Scales (Scales)	−41.36 (−29.04)	1	84.73 (60.09)	*0.00001* (0.10268)
Set 3								
1	Multistate	ER	Filaments	Scales (Scales)	−53.52 (−48.56)	1	109.03 (99.12)	*0.01658 (0.04702)*
1	Multistate	ORD	Filaments	Scales (Scales)	−49.48 (−45.66)	1	100.97 (93.31)	**0.93285 (0.85888)**
1	Multistate	UR	Filaments	Scales (Filaments)	−47.4 (−42.88)	6	106.8 (97.75)	*0.05056* (0.09328)
1	Multistate	EPROG	Filaments	Scales (Scales)	−75.91 (−65.42)	1	153.82 (132.85)	*0 (0)*
1	Multistate	UPROG	Filaments	Scales (Scales)	−66.43 (−50.62)	3	138.85 (107.24)	*0 (0.00081)*
Set 4								
1	Binary	ER	Feathers	Feathers (Scales)	−32.35 (−29.66)	1	66.7 (61.32)	**0.71095 (0.6266)**
1	Binary	UR	Feathers	Feathers (Scales)	−32.25 (−29.63)	2	68.5 (63.27)	0.28905 (0.23635)
1	Binary	PROG	Feathers	Scales (Scales)	−45.85 (−31.18)	1	93.7 (64.36)	*0* (0.13705)
Set 5								
1	Multistate	ER	Unique	Scales (Scales)	−61.01 (−54.14)	1	124.03 (110.28)	0.11844 (0.16623)
1	Multistate	UR	Unique	Filaments (Filaments)	−49.31 (−44.33)	12	122.62 (112.66)	0.2397 *(0.05057)*
1	Multistate	2R	Unique	Scales (Scales)	−58.32 (−51.59)	2	120.65 (107.18)	**0.64187 (0.7832)**
Set 6								
1	Multistate	ER	Unique	Scales (Scales)	−41.4 (−35.93)	1	84.8 (73.87)	0.06277 (0.12252)
1	Multistate	2R	Unique	Scales (Scales)	−38.17 (−33.06)	2	80.33 (70.13)	**0.58668 (0.79494)**
1	Multistate	UR	Unique	Scales (Scales)	−34.68 (−31.33)	6	81.36 (74.66)	0.35054 (0.08254)
Set 7								
2	Multistate	ER	Scales	Scales (Scales)	−50.11 (−45.59)	1	102.22 (93.19)	*0.0352* (0.08275)
2	Multistate	ORD	Scales	Scales (Scales)	−46.86 (−43.25)	1	95.73 (88.5)	**0.90319 (0.86336)**
2	Multistate	UR	Scales	Scales (Scales)	−44.55 (−41.04)	6	101.1 (94.08)	*0.06162 (0.05303)*

(continued)

Table 12.3 (continued)

Tree	Character	Model	Pterosaur state	Dinosaur ancestor estimate	log Lik.	Pars.	AIC	A.W.
2	Multistate	EPROG	Scales	Scales (Scales)	−71.68 (−63.07)	1	145.35 (128.14)	*0 (0)*
2	Multistate	UPROG	Scales	Scales (Scales)	−62.08 (−48.16)	3	130.15 (102.32)	*0 (0.00086)*
Set 8								
2	Binary	ER	Scales	Scales (Scales)	−29.35 (−26.87)	1	60.71 (55.74)	**0.7281 (0.675)**
2	Binary	UR	Scales	Scales (Scales)	−29.34 (−26.87)	2	62.68 (57.74)	0.2719 (0.24832)
2	Binary	PROG	Scales	Scales (Scales)	−41.61 (−29.04)	1	85.22 (60.09)	*0* (0.07669)
Set 9								
2	Multistate	ER	Filaments	Scales (Scales)	−53.18 (−48.41)	1	108.35 (98.81)	*0.01178 (0.04106)*
2	Multistate	ORD	Filaments	Scales (Scales)	−48.81 (−45.35)	1	99.63 (92.7)	**0.92178 (0.87136)**
2	Multistate	UR	Filaments	Filaments (Filaments)	−46.44 (−42.66)	6	104.89 (97.31)	*0.06644* (0.08693)
2	Multistate	EPROG	Filaments	Scales (Scales)	−76.36 (−65.31)	1	154.72 (132.63)	*0 (0)*
2	Multistate	UPROG	Filaments	Scales (Scales)	−66.72 (−50.55)	3	139.44 (107.1)	*0 (0.00065)*
Set 10								
2	Binary	ER	Feathers	Feathers (Scales)	−31.62 (−29.36)	1	65.24 (60.72)	**0.70475 (0.65467)**
2	Binary	UR	Feathers	Feathers (Scales)	−31.49 (−29.36)	2	66.98 (62.72)	0.29525 (0.24084)
2	Binary	PROG	Feathers	Scales (Scales)	−46.13 (−31.2)	1	94.26 (64.39)	*0* (0.10449)
Set 11								
2	Multistate	ER	Unique	Scales (Scales)	−60.68 (−54)	1	123.36 (110)	0.08274 (0.17046)
2	Multistate	UR	Unique	Naked (Scales)	−47.96 (−45.23)	12	119.93 (114.46)	**0.45978** *(0.01833)*
2	Multistate	2R	Unique	Scales (Scales)	−57.97 (−51.44)	2	119.94 (106.88)	**0.45748 (0.81121)**
Set 12								
2	Multistate	ER	Unique	Scales (Scales)	−40.47 (−35.73)	1	82.93 (73.46)	0.06256 (0.11186)
2	Multistate	2R	Unique	Scales (Scales)	−37.27 (−32.76)	2	78.53 (69.51)	**0.56461 (0.8061)**
2	Multistate	UR	Unique	Scales (Scales)	−33.68 (−31.04)	6	79.36 (74.08)	0.37283 (0.08204)

Bracketed values represent results using the 'equal' approach for scaling zero-length branches, else results are based on the 'mbl' approach. Boldface values denote the best-fit model in the set, whereas italics represent very low support models that can be rejected. Tree 1 reflects the traditional dinosaurian topology and Tree 2 follows the topology of Baron et al. (2017). Model acronyms follow those explained in the text

Abbreviations: AIC, Akaike Information Criterion, A.W., Akaike weights, log. Lik., log Likelihoods, Pars., number of possible rate parameters

pterosaur condition are given their own rate parameter compared with other transitions (Akaike weight = 0.46–0.81); however, the equal- and unequal-rates models also receive some support and cannot be rejected (Akaike weights = 0.06–0.17 and 0.05–0.46, respectively; Table 12.3).

12.4.2 Ancestral State Reconstructions

Empirical—Based on the dataset of known fossilised integument occurrences, our analyses provided overwhelming support for the ancestral dinosaur to be scaled (Figs. 12.2 and 12.3), and this result is consistent for both phylogenetic hypotheses (Figs. 12.2 and 12.3; Table 12.3). Given the various permutations tested here (18 in total), scales received a median likelihood support value of 0.93, compared with 0.04 for filaments, when using the H1 topology, and received almost identical support values, of 0.93 and 0.03, respectively, when using H2. However, it is evident that the assumed condition in pterosaurs has a significant effect on the support values at the dinosaurian node ($F = 26.2$, $p = 0.001$; Table 12.4, Fig. 12.4). If the pycnofibres in pterosaurs are assumed to be homologous to the filamentous structures/feathers seen in dinosaurs, median support for an ancestrally scaly dinosaur drops to 0.65 and support for a filamentous condition increases to 0.33 under H1 (Fig. 12.4c), and 0.63 and 0.36, respectively, under H2 (Fig. 12.4d). If, as argued previously, the condition in pterosaurs can be considered as an independent, potentially autapomorphic state, then the median value for a scaly dinosaurian ancestor increases (H1: $lik_{scales} = 0.94$, $lik_{filaments} = 0.02$, H2: $lik_{scales} = 0.93$, $lik_{filaments} = 0.02$). Of the various empirical parameters tested with the dataset, the choice of character evolution model has the greatest effect on the likelihood values obtained for the dinosaurian node (Partial $R^2 = 0.245$, $F = 56.9$, $p = 0.001$), followed by the aforementioned pterosaur condition (Partial $R^2 = 0.075$; Fig. 12.4) and, finally, the choice of branch scaling method (Partial $R^2 = 0.045$; Fig. 12.5). There are no significant differences in these results when topological differences are considered (Partial $R^2 = 0.001$; Table 12.4).

Much like the values obtained for the dinosaurian ancestor, estimates of other nodes of interest (e.g. Ornithischia, Saurischia) are affected by similar factors (e.g. the assumed state in pterosaurs; Table 12.4), with the exception that the choice of branch-scaling method does not significantly affect the ornithoscelidan (H2) results, and topological differences play a significant role in determining the ancestral condition of Theropoda. It is also worth noting that despite representing the sister-taxon of dinosaurs, pterosaurs have a significant systemic effect on ancestral conditions farther up the tree (i.e. at the nodes for Theropoda and Coelurosauria).

Hypothetical—In all hypothetical analyses (variants 1–4; Table 12.2), the discovery of a Late Triassic taxon with filaments (or feathers) would have a dramatic effect on the resultant dinosaurian nodal likelihoods (Fig. 12.6), with a stronger effect on the ancestral state than any other parameter (Partial $R^2 = 0.393$; Table 12.4). Whilst the empirical results indicate that the vast majority of analyses support a scaly dinosaurian ancestor (Fig. 12.6a), discovery of a single Late Triassic taxon bearing a filamentous or feathered pelage would greatly increase the uncertainty associated with the interpreted primitive condition in dinosaurs (Fig. 12.6b–d). Median support for scales ranges from 0.29 to 0.6 and for filaments from 0.4 to 0.7. The discovery of a Late Triassic 'feathered' taxon in each major branch of the dinosaur tree (i.e. one sauropodomorph, one theropod, and one ornithischian) would result in a complete reversal of the empirical results and would lend overwhelming support for a filament bearing or feathered dinosaurian ancestor (Fig. 12.6e).

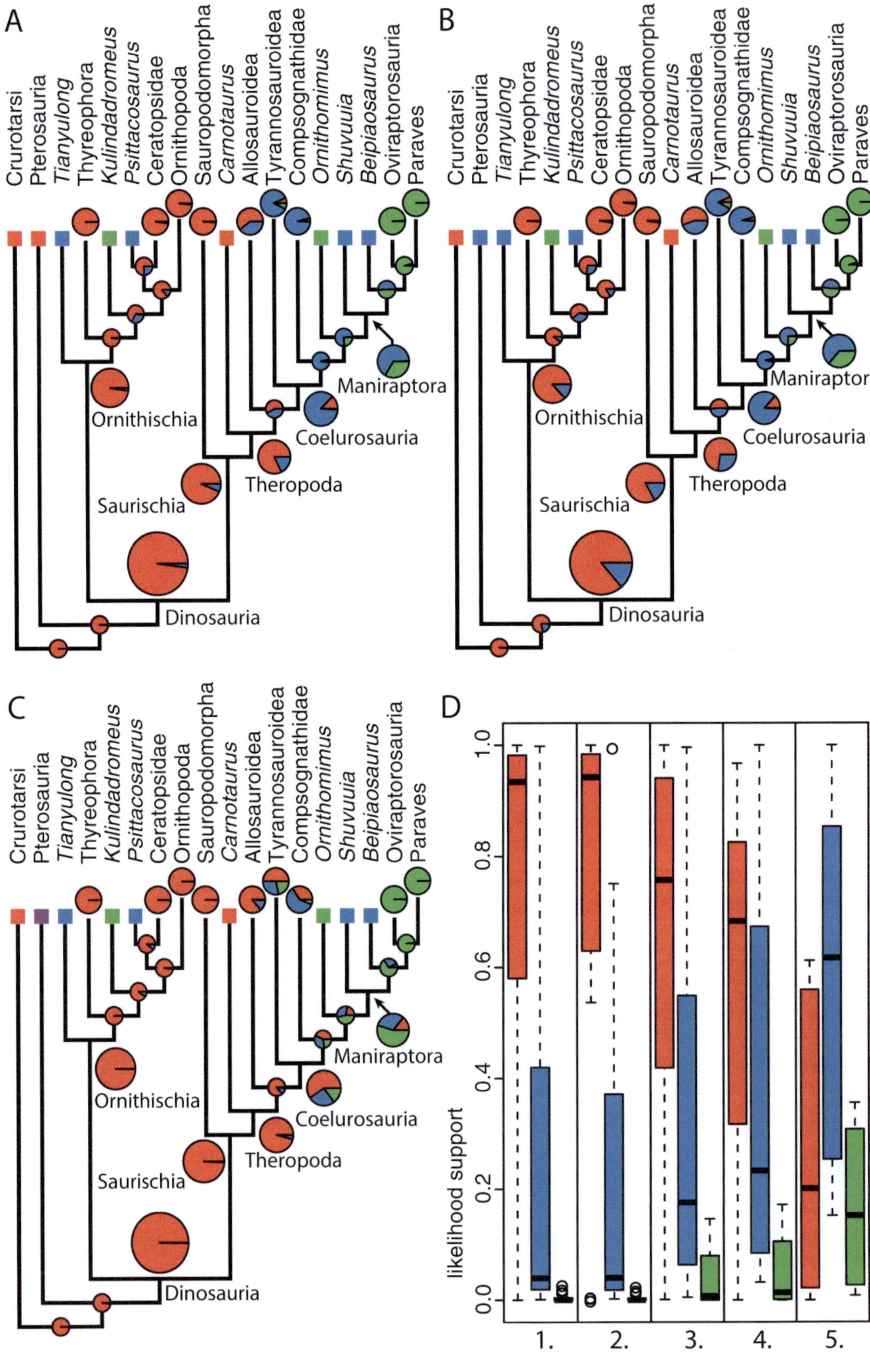

Fig. 12.2 Ancestral state reconstructions for ornithodiran integument based on the traditional dinosaurian relationships. Reconstructions along the 'equal' time-scaled phylogenetic tree (**a**–**c**) are based on the ORD

12.5 Discussion

12.5.1 Feather Evolution

In his treatise on the origin of feathers, Prum (1999) proposed a developmental model for feather evolution that predicts a stepwise evolutionary transition. Here, the archetypal bird feather evolved progressively from a simple, follicle-born filament (stage I), to a basally branched tuft of monofilaments (unbranched barbs: stage II), to branched bards (i.e. barbules, stage III), to differentiated barbules and the generation of a pennaceous vane (stage IV), and, finally, to the appearance of an asymmetrical feather (Prum 1999; Prum and Brush 2002). This progressive model is supported by qualitative observations, whereby early coelurosaurians possessed simple filaments (i.e. stage I: Chen et al. 1998) and later developmental stages are observed in later maniraptorans (Xu et al. 1999a), but no quantitative support has been provided for this scenario. Unfortunately, the fossil record of integument does not permit our integumentary characters to be coded to the level of resolution predicted by Prum's (1999) developmental model. However, the multistate character used here (three states) offers a basic framework with which to test whether feather evolution in dinosaurs occurred in a step-wise, ordered fashion or not. Our results support the notion that an ordered model of evolution provides a better fit to the available data; however, it is not strictly progressive (i.e. increasing in complexity). Models that did not allow reversals (i.e. PROG models) received extremely low support values (Table 12.3) and indicate that feather evolution is more complex than a progression from simple to more complex structures (Barrett et al. 2015; Bell et al. 2017). For example, there is at least one instance in which a filamentous (possibly feathered) covering was lost, at least partially, in tyrannosaurids (Bell et al. 2017). However, our results do support one prediction of Prum's (1999) model, in that there seems to be an order to these integumentary characters in which, whether through progression or reversals, character states are ordered: 0-1-2-1-0. Ordered models with reversals received the greatest support, sometimes exclusively compared with standard static (equal-rates) or random (unequal-rates) models. Despite general quantitative support for the ordered prediction of Prum's (1999) developmental model, our results cannot be used to support or reject the developmental origins of feathers, whether follicle-born (Prum 1999), epidermal-based (Sawyer et al. 2000), or from scales (Maderson and Alibardi 2000). Such a test would require the construction of more detailed integumentary characters and a broader phylogenetic framework. At this point, all we can say is that the evolution of feathers occurred in an ordered fashion, but that it was not strictly progressive.

12.5.2 Ancestral Dinosaur Integument

The minor updates and modifications made to our previous dataset and the inclusion of an extra outgroup (i.e. a scaled crurotarsan) did little to change the overall conclusions offered in Barrett et al. (2015: fig. 1) (Fig. 12.2). Modifications to traditional dinosaurian phylogenetic relationships (Baron et al. 2017), if anything, strengthen the results previously obtained for some of the deepest nodes within the tree (e.g. Dinosauria and Ornithischia). Our empirical results retain strong support for a scaly ancestral condition in dinosaurs with strong likelihood values that reject the notion that ornithischian and saurischian (or theropod under H2) filamentous structures share a common ancestry (Godefroit et al. 2014a; Saveliev and

Fig. 12.2 (continued) model in (**a**) and (**b**) and the 2R model in (**c**). Pterosaurs coded as plesiomorphically scaled (**a**), filament-covered (**b**), or independent (**c**). Character states are scales (state 0; red), filaments (state 1; blue), feathered (state 2; green), or alternate novel integument (state 3; purple). Ranges of probabilities for each state at particular nodes based on all permutations described in the text (**d**). Dinosauria = 1, Ornithischia = 2, Saurischia = 3, Theropoda = 4, Coelurosauria = 5

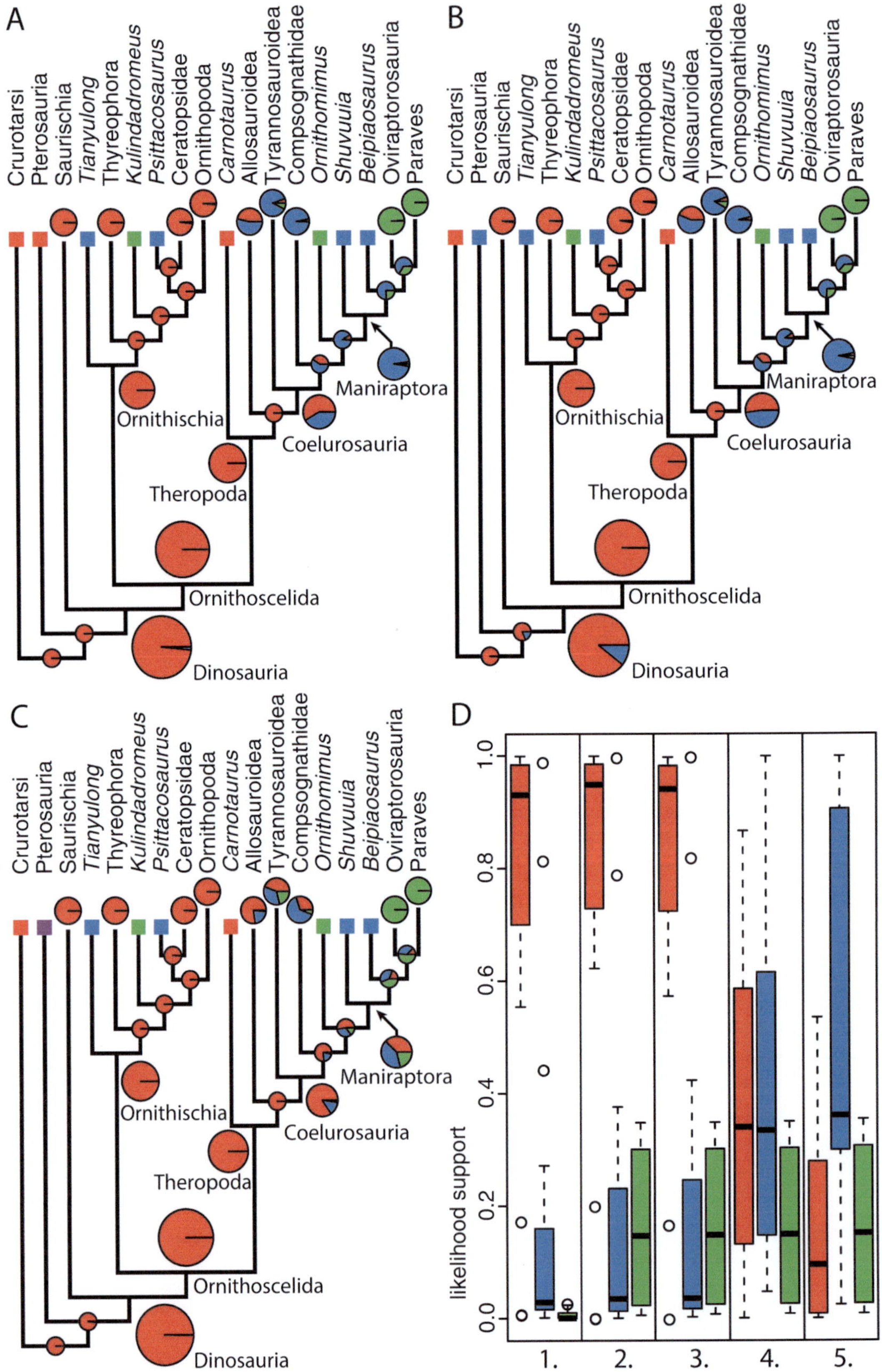

Fig. 12.3 Ancestral state reconstructions for ornithodiran integument based on the recent Baron et al. (2017) phylogenetic relationships for Dinosauria. Reconstructions along the 'equal' time-scaled phylogenetic tree (**a**–**c**) are

Alifanov 2014) and support the hypothesis that ornithischian filaments are not true feathers (Mayr et al. 2002, 2016; Barrett et al. 2015). It is not until the Theropoda node, where both filaments and scales receive similar support values, that the ancestral state of scales becomes more ambiguous, and it is only at Coelurosauria that filaments receive overall stronger support values.

Interestingly, contrary to the conclusions of Barrett et al. (2015), the outgroup condition (i.e. that of pterosaurs), although significant in its influence on the dinosaurian ancestral condition, recovered a very low partial R^2 value (0.08) indicating that it is not the principal driver of the ancestral state, especially when compared with other sources of variation, namely, the inclusion of hypothetically 'feathered' Late Triassic taxa and the choice of evolutionary model (Table 12.4). Even residuals (unexplained variance) have a much higher partial R^2 value (0.24) than that explained by variation in the pterosaur condition. This result is probably due to the inclusion of a second outgroup, that of scaled crurotarsans, which serves to stabilise the results in this region of the tree. This effect is particularly noteworthy if pterosaurs are given their own derived condition ('naked'), as the results obtained are virtually identical to those when pterosaurs are coded as scaly (Figs. 12.2 and 12.3). Despite this, it is worth noting that the primitive pterosaur condition remains equivocal, and pterosaurs may well have developed their integument along pathways that were independent of those seen in other ornithodirans. For example, most pterosaurs with soft tissue remains lack scales on their bodies and had naked leathery skin (Frey et al. 2003; Unwin 2006; Wang et al. 2007; Kellner et al. 2010). Although some taxa (specifically anurognathids) had pycnofibres, these are not common (see Sect. 12.2; Kellner et al. 2010). Moreover, remains of Late Triassic/Early Jurassic pterosaurs often lack evidence of pycnofibres, even in cases where these specimens are exceptionally preserved, with skin outlines or wing fibres (Dalla Vecchia 2009). Given the early diverging positions of these taxa within pterosaur phylogeny, it seems increasingly evident that pterosaurs likely lost scales apomophically and that pycnofibres cannot be assumed as the primitive condition for pterosaurs. They might have been acquired later in their evolutionary history, potentially more than once, though further exceptionally preserved specimens will be required to test these hypotheses. However, if pterosaur skin structures are not homologous with those in dinosaurs, it follows that Pterosauria is not an appropriate outgroup for assessing the primitive condition in dinosaurs, as their primitive condition might have been entirely different (i.e. naked, rather than feathered or scaled). Other archosaurs would thus be better outgroups, but unfortunately skin impressions for other avemetatarsalians or crurotarsans are lacking, with the exception of 'typical' reptilian scales in Crocodylomorpha. Skin impressions are also lacking in non-archosaurian archosauriforms, with few exceptions. *Chirotherium* footprints from the Triassic Period frequently include evidence of scales on the feet (e.g. King et al. 2005) and, more notably, *Longisquama* possesses a series of highly apomorphic 'scales' that extend dorsally, some of which are longer than the preserved body length (Sharov 1970; Reisz and Sues 2000; Buchwitz and Voigt 2012). It is due to these major uncertainties in determining the primitive skin condition in archosaurs that we have opted for a more conservative approach than that presented in Barrett et al. (2015), including two outgroups for dinosaurs: (1) the condition in living crocodylomorphs and (2) the addition of the

Fig. 12.3 (continued) based on the ORD model in (**a**) and (**b**) and the 2R model in (**c**). Pterosaurs coded as plesiomorphically scaled (**a**), filament-covered (**b**), or independent (**c**). Character states are scales (state 0; red), filaments (state 1; blue), feathered (state 2; green), or alternate novel integument (state 3; purple). Ranges of probabilities for each state at particular nodes based on all permutations described in the text (**d**). Dinosauria = 1, Ornithischia = 2, Ornithoscelida = 3, Theropoda = 4, Coelurosauria = 5

Table 12.4 Results of the two-way npMANOVAs

Effect	df	Sum of squares	Mean square	F-statistic	Partial R^2	p-Value
Node: Dinosauria						
Pterosaurs	2	2.0535	1.02673	26.235	0.07522	0.001
Branches	1	1.2321	1.23213	31.483	0.04513	0.001
Topology	1	0.0225	0.02249	0.575	0.00082	0.457
Model	3	6.6862	2.22873	56.948	0.24491	0.001
Hypothetical	4	10.7310	2.68275	68.549	0.39308	0.001
Residuals	168	6.5748	0.03914		0.24084	
Node: Ornithischia						
Pterosaurs	2	0.5895	0.2948	5.621	0.02612	0.006
Branches	1	6.2918	6.2918	119.980	0.27881	0.001
Topology	1	0.0540	0.0540	1.031	0.00240	0.314
Model	3	4.8199	1.6066	30.637	0.21359	0.001
Hypothetical	4	2.0010	0.5003	9.539	0.08867	0.001
Residuals	168	8.8100	0.0524		0.39041	
Node: Saurischia (H1 only)						
Pterosaurs	2	0.4083	0.20413	5.620	0.03388	0.007
Branches	1	0.5569	0.55688	15.333	0.04621	0.002
Model	3	3.6578	1.21925	33.571	0.30352	0.001
Hypothetical	4	4.5590	1.13975	31.382	0.37831	0.001
Residuals	79	2.8692	0.03632		0.23808	
Node: Ornithoscelida (H2 only)						
Pterosaurs	2	0.5596	0.27979	6.887	0.04228	0.004
Branches	1	0.0418	0.04181	1.029	0.00316	0.310
Model	3	2.5890	0.86300	21.243	0.19563	0.001
Hypothetical	4	6.8344	1.70860	42.058	0.51642	0.001
Residuals	79	3.2094	0.04062		0.24251	
Node: Theropoda						
Pterosaurs	2	0.2704	0.1352	2.936	0.01119	0.029
Branches	1	8.5113	8.5113	184.805	0.35206	0.001
Topology	1	0.2555	0.2555	5.547	0.01057	0.003
Model	3	3.3110	1.1037	23.964	0.13695	0.001
Hypothetical	4	4.0904	1.0226	22.204	0.16919	0.001
Residuals	168	7.7374	0.0461		0.32004	
Node: Coelurosauria						
Pterosaurs	2	0.1877	0.0938	3.062	0.00981	0.028
Branches	1	9.5613	9.5613	312.033	0.50005	0.001
Topology	1	0.0799	0.0799	2.609	0.00418	0.088
Model	3	2.9106	0.9702	31.663	0.15222	0.001
Hypothetical	4	1.2335	0.3084	10.064	0.06451	0.001
Residuals	168	5.1478	0.0306		0.26923	

Pterosaurs, the assumed condition in that group; Branches, the method for time-scaling the tree; Topology, whether the tradition or novel phylogenetic hypothesis for dinosaurs is considered; Model, the model of evolution assumed for character evolution; Hypothetical, the various ad hoc tests implemented in which filamentous Late Triassic taxa are artificially implemented

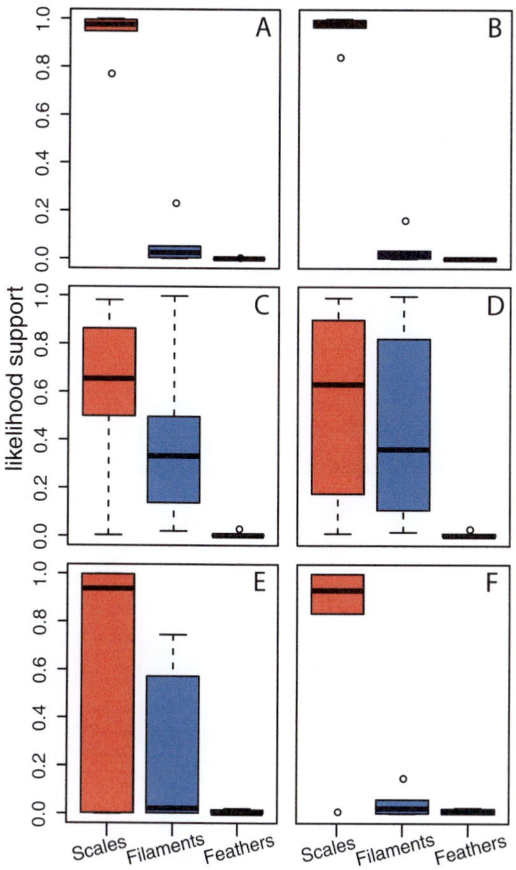

Fig. 12.4 Range of likelihood support values for the integumentary condition of the dinosaurian ancestor given various integumentary assumptions in pterosaurs. Pterosaurs are coded as either plesiomorphically scaled (**a**, **b**), filamented (**c**, **d**), or derived with respect to Dinosauria (**e**, **f**). Plots (**a**, **c and e**) are based on the traditional dinosaurian relationships and (**b**, **d and f**) are based on the recent results of Baron et al. (2017)

potentially apomorphic evolutionary trajectory in pterosaurs.

As is far too often the case, the fossil record is fraught with sampling problems. Ideally, our sampling of taxa with fossilised integumentary structures would span the duration of the Mesozoic, with evidence distributed evenly across all clades and major time intervals. The reality is that 70% of the available data come from the Late Cretaceous and the remaining 30% from the Jurassic, with almost all of the latter data points occurring within the Late Jurassic (Table 12.1). The sole Middle Jurassic occurrence is

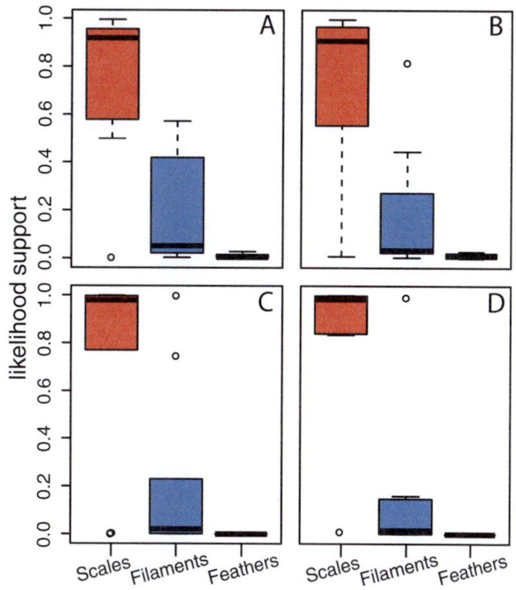

Fig. 12.5 Range of likelihood support values for the integumentary condition of the dinosaurian ancestor given two branch scaling approaches. In plots (**a**) and (**b**), zero-length branches are rescaled by adding a minimum branch length of 1 Ma ('mbl' approach), whereas plots (**c**) and (**d**) add 1 Ma to the root node and then distribute zero-length branches more evenly throughout the tree ('equal' approach). Plots (**a**) and (**c**) are based on the traditional dinosaurian relationships; (**b**) and (**d**) are based on the recent results of Baron et al. (2017)

Kulindadromeus, whose age is uncertain and may actually hail from the Late Jurassic (Godefroit et al. 2014a) and the only Early Jurassic occurrence concerns the scaled thyreophoran *Scelidosaurus*. No occurrences of dinosaur skin are known from the Triassic, presenting a serious limitation to assessing the deep evolutionary history of their integument. However, some evidence of filamentous integumentary structures is known from Early Jurassic theropod tracks that are either of tetanuran or neoceratosaurian affinities (Kundrát 2004). Unfortunately, given its unknown phylogenetic position, this occurrence could not be included in the present analysis. Nevertheless, Kundrát's (2004) observations provide evidence that at least some filamentous structures are older than our current record of feathered taxa would suggest (based on body fossils), and it is conceivable that filaments were present in some Late Triassic dinosaurs.

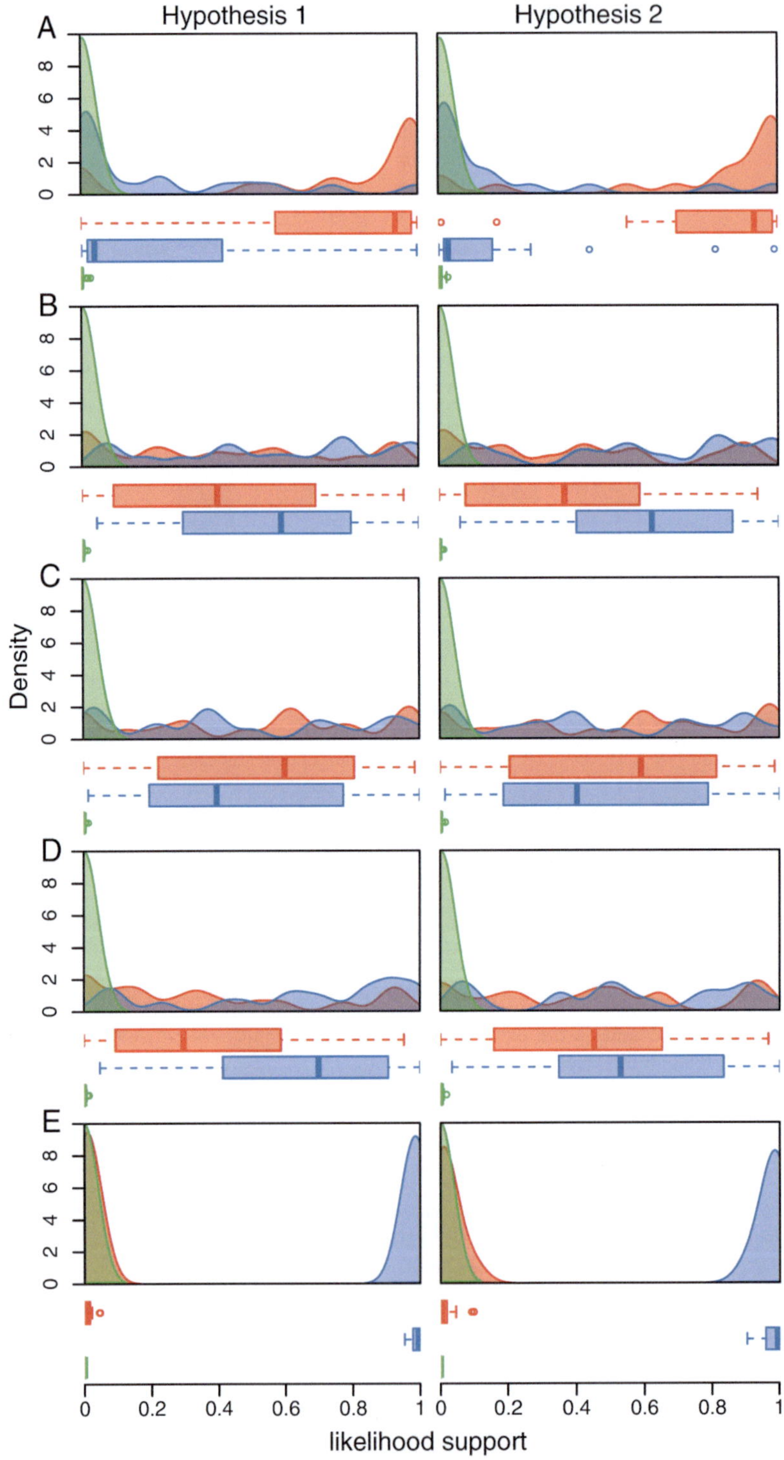

Fig. 12.6 Distribution and ranges of likelihood values for the integumentary condition of the dinosaurian ancestor given the four hypothetical variants outlines in the text (Fig. 12.1; Table 12.2). Variants applied to both the traditional dinosaurian relationships (Hypothesis 1) and the novel relationships presented by Baron et al. (2017; Hypothesis 2). Plots show density distributions and box plots based on the empirical results (**a**) and hypothetical variants 1 (**b**), 2 (**c**), 3 (**d**), and 4 (**e**). Colours follow those of Figs. 12.2 and 12.3

Accordingly, we ran a series of four hypothetical scenarios in which a filamented Late Triassic taxon was inserted into the dinosaurian phylogenetic tree. These hypothetical taxa were placed as sister-taxa to each of the three major dinosaurian clades (Ornithischia, Sauropodomorpha, and Theropoda: variants 1–3), as this placement should inflict the strongest effect on the ancestral condition recovered at the dinosaurian node. Despite a strong influence on the ancestral state of integument in Dinosauria (Fig. 12.6; Table 12.4), the one-off discovery of a Triassic 'basal' ornithischian, sauropodomorph, or theropod would not be enough to overturn the results of Barrett et al. (2015) in favour of a deep filamentous or feathered ancestry and would only serve to increase the ambiguity around the dinosaurian node. It is only if all three hypothetical feathered taxa are included (variant 4) that results flip, with overwhelming support for filaments at the dinosaur origin. The discovery of a Triassic Lagerstätte could supply evidence of such taxa, but on the basis of current evidence, the chances of finding a feathered dinosaur that occupies each, or all three, of these phylogenetic positions seem low, particularly with respect to sauropodomorphs, which currently show no evidence of integumentary structures other than scales, and, if Baron et al.'s (2017) hypothesis holds, then they are more distant relatives of other feathered taxa than currently thought. Taken together, our results demonstrate that only under very specific criteria can our current probabilistic framework be overturned and, thus, it seems rather unlikely and premature to state that filaments (feathers) represent the primitive dinosaurian condition; many more 'feathered' data points would be needed to support the latter hypothesis.

12.5.3 Taphonomic Considerations

Taphonomic experiments have revealed that preservational conditions can alter the morphology of feathers and filamentous integumentary structures in certain depositional settings, such as lacustrine shales (Foth 2012a, b; but see Saitta et al. 2017). It was also assumed that relatively coarse-grained fluvial systems are biased towards the preservation of scaly skin impressions and against that of filamentous integument (Zelenitsky et al. 2012). Rauhut et al. (2012) argued that the presence of both scales and filaments in some taxa (e.g. *Juravenator*, *Psittacosaurus*, *Anchiornis*) indicates that the apparent lack of filaments in animals that preserve impressions (e.g. ceratopsids, hadrosaurids) of scaly skin in these coarse-grained sediments could be the result of unspecified taphonomic processes. However, recent discoveries of several ornithomimid specimens from the Late Cretaceous of Alberta show that fluvial sediments, including channel sandstones, can preserve feathers (Zelenitsky et al. 2012; Van der Reest et al. 2016). Fine-grained mudstones or iron-rich concretions can preserve exquisite details of the integument in taxa with polygonal scales (e.g. Anderson et al. 1998; Evans and Reisz 2007) and as such are not necessarily biased against the preservation of filamentous structures interspersed between scales, for example. Furthermore, polygonal scales preserved in fluvial systems are often assumed to be impressions; however, the preservation of three-dimensional structure and internal complexity in skin associated with some well-preserved specimens suggests the possibility of more complex preservational scenarios that may include preservation of the integument in some form (Evans and Reisz 2007). Given these taphonomic considerations, integumentary soft tissues, including feathers, may be more readily preserved than previously assumed in fluviodeltaic rocks (Zelenitsky et al. 2012), and integumentary structures preserved in fluvial environments should be carefully assessed on an individual basis in order to determine the presence or absence of filamentous structures or feathers.

12.6 Conclusions

Evidence from extant taxa favours the deep homology of ectodermal placodes across Amniota (Sawyer et al. 2000; Musser et al.

2015; Di-Poï and Milinkovitch 2016), suggesting shared genetic and developmental mechanisms in the processes underlying the production of varied integumentary features, including scales, monofilaments, and feathers. This observation raises the possibility that the filamentous structures present in different ornithodiran clades (pterosaurs, theropods, and ornithischians) did not necessarily have a single evolutionary origin but could represent expressions of similar integument morphology based upon independent modifications to shared developmental processes. In addition, it is clear that these processes, which can be shown to proceed in an ordered fashion (e.g. Prum 1999; Prum and Brush 2002; this study), can also produce novel 'less complex' conditions, rather than simply providing a continuum of 'increasing complexity', such as the loss of scales from the body surface of both pterosaurs and coelurosaurian theropods. Phylogenetic optimisations of the integumentary features present in ornithodirans indicate that the homology of pterosaur pycnofibres and dinosaur filaments and feathers is poorly supported, and there is currently no evidence to determine if all of the filamentous structures produced by these taxa developed from true follicles or not (Sawyer et al. 2003a, b). As a result, pterosaurs are poor models for constraining the nature of the integument at the ancestral node for Dinosauria. In addition, other aspects of ornithodiran skin evolution remain unresolved but might be difficult to address using evidence from the fossil record. These issues include deducing when follicles appeared and determining when scales were lost in pterosaur and coelurosaur evolution.

Although the combined presence of filamentous structures in the majority of theropods and a minority of ornithischians makes it tempting to infer a covering of 'feathers' for the ancestral dinosaur on the basis of simple phylogenetic bracketing (e.g. Witmer 2009; Brusatte 2012; Godefroit et al. 2014a), this interpretation conflicts with empirical and modelled data on the phylogenetic distribution of archosaur integument types, irrespective of the preferred phylogenetic framework, the outgroup condition in pterosaurs, or whether any one major dinosaur lineage had as yet unknown Late Triassic feathered representatives. Although support for a feathered dinosaur ancestor remains a plausible hypothesis, it requires stronger empirical evidence, namely, the unambiguous documentation of filaments or feathers in multiple Late Triassic dinosaur lineages.

Acknowledgements We thank the editors (C. Foth and O. Rauhut) for inviting us to contribute to this volume, for their insightful comments, and for their patience. We thank P. Bell, R. Benson, R. Butler, M. Davis and M. Young for discussions related to feather evolution and dinosaur skin.

Appendix

Character coding data set used in the analyses

Taxon	Char1a	Char2a	Char1b	Char2b	Char1c	Char2c	Max	Min
Crurotarsi	0	0	0	0	0	0	247.2	0
Pterosauria	0	0	1	1	3	2	212	66
Allosaurus_fragilis	0	0	0	0	0	0	157.3	152.1
Albertosaurus_sarcophagus	0	0	0	0	0	0	72.1	66
Anchiornis_huxleyi	2	1	2	1	2	1	163.5	157.3
Archaeopteryx_lithographica	2	1	2	1	2	1	152.1	145
Auca_Maheuvo_titanosaur	0	0	0	0	0	0	83.6	72.1
Aurornis_xui	2	1	2	1	2	1	163.5	157.3
Barosaurus_lentus	0	0	0	0	0	0	157.3	152.1
Beipiaosaurus_inexpectus	1	1	1	1	1	1	129.4	125
Brachylophosaurus_canadensis	0	0	0	0	0	0	83.6	72.1
*Camarasaurus_*sp.	0	0	0	0	0	0	157.3	152.1

(continued)

Taxon	Char1a	Char2a	Char1b	Char2b	Char1c	Char2c	Max	Min
Carnotaurus_sastrei	0	0	0	0	0	0	83.6	72.1
Caudipteryx_dongi	2	1	2	1	2	1	129.4	125
Caudipteryx_zhoui	2	1	2	1	2	1	129.4	125
Centrosaurus_apertus	0	0	0	0	0	0	83.6	72.1
cf._*Stegosaurus*_sp.	0	0	0	0	0	0	157.3	152.1
Chasmosaurus_belli	0	0	0	0	0	0	83.6	72.1
Compsognathus_longipes	0	0	0	0	0	0	152.1	145
Concavenator_corcovatus	0	0	0	0	0	0	129.4	125
Confuciusornis_sanctus	2	1	2	1	2	1	129.4	125
Corythosaurus_casuarius	0	0	0	0	0	0	83.6	72.1
Corythosaurus_intermedius	0	0	0	0	0	0	83.6	72.1
Daspletosaurus_torosus	0	0	0	0	0	0	83.6	72.1
Dilong_paradoxus	2	1	2	1	2	1	129.4	125
*Diplodocus*_sp.	0	0	0	0	0	0	157.3	152.1
Dyoplosaurus_acutosquameus	0	0	0	0	0	0	83.6	72.1
Edmontosaurus_annectens	0	0	0	0	0	0	72.1	66
Edmontosaurus_regalis	0	0	0	0	0	0	83.6	72.1
Eosinopteryx_brevipenna	2	1	2	1	2	1	163.5	157.3
Epidexipteryx_hui	2	1	2	1	2	1	163.5	157.3
Euoplocephalus_tutus	0	0	0	0	0	0	83.6	72.1
Gigantspinosaurus_szechuanensis	0	0	0	0	0	0	163.5	157.3
Gorgosaurus_libratus	0	0	0	0	0	0	83.6	72.1
Gryposaurus_monumentensis	0	0	0	0	0	0	83.6	72.1
Gryposaurus_notabilis	0	0	0	0	0	0	83.6	72.1
Hesperosaurus_mjosi	0	0	0	0	0	0	157.3	152.1
Hypacrosaurus_altispinus	0	0	0	0	0	0	72.1	66
Jinfengopteryx_elegans	2	1	2	1	2	1	129.4	125
Juravenator_starkii	1	1	1	1	1	1	152.1	145
Kulindadromeus_zabaikalicus	2	1	2	1	2	1	169	144
Lambeosaurus_clavinitialis	0	0	0	0	0	0	83.6	72.1
Lambeosaurus_magnicristatus	0	0	0	0	0	0	83.6	72.1
Liaoningosaurus_paradoxus	0	0	0	0	0	0	129.4	125
Magnipaulia_laticaudus	0	0	0	0	0	0	83.6	72.1
Maiasaura_peeblesorum	0	0	0	0	0	0	83.6	72.1
Mamenchisaurus_youngi	0	0	0	0	0	0	163.5	152.1
Mantellisaurus_atherfieldensis	0	0	0	0	0	0	129.4	125
Microraptor_gui	2	1	2	1	2	1	125	113
Microraptor_zhaoianus	2	1	2	1	2	1	125	113
Ornithomimus_edmontonicus	2	1	2	1	2	1	83.6	66
Parasaurolophus_walkeri	0	0	0	0	0	0	83.6	72.1
Pedopenna_daohugouensis	2	1	2	1	2	1	163.5	157.3
Pelorosaurus_becklesii	0	0	0	0	0	0	139.8	132.9
Prosaurolophus_maximus	0	0	0	0	0	0	83.6	72.1
Protarchaeopteryx_robusta	2	1	2	1	2	1	129.4	125
*Psittacosaurus*_sp.	1	1	1	1	1	1	129.4	125
Rahonavis_ostromi	2	1	2	1	2	1	72.1	66
Saurolophus_angustirostris	0	0	0	0	0	0	72.1	66
Saurolophus_osborni	0	0	0	0	0	0	72.1	66

(continued)

Taxon	Char1a	Char2a	Char1b	Char2b	Char1c	Char2c	Max	Min
Scansoriopteryx_heilmanni	2	1	2	1	2	1	163.5	157.3
Scelidosaurus_harrisonii	0	0	0	0	0	0	199.3	190.8
Sciurumimus_albersodoefri	1	1	1	1	1	1	157.3	152.1
Scolosaurus_cutleri	0	0	0	0	0	0	83.6	72.1
Shuvuuia_deserti	1	1	1	1	1	1	83.6	72.1
Similicaudipteryx_yixianensis	2	1	2	1	2	1	129.4	125
Sinocalliopteryx_gigas	1	1	1	1	1	1	129.4	125
Sinornithosaurus_millenii	2	1	2	1	2	1	129.4	125
Sinosauropteryx_prima	1	1	1	1	1	1	129.4	125
Tarbosaurus_bataar	0	0	0	0	0	0	72.1	66
Tarchia_kielanae	0	0	0	0	0	0	72.1	66
Tehuelchesaurus_benitezii	0	0	0	0	0	0	157.3	152.1
*Tenontosaurus*_sp.	0	0	0	0	0	0	125	113
Tianyulong_confuciusi	1	1	1	1	1	1	163.5	157.3
Triceratops_horridus	0	0	0	0	0	0	72.1	66
Tyrannosaurus_rex	0	0	0	0	0	0	72.1	66
Velociraptor_mongoliensis	2	1	2	1	2	1	83.6	72.1
Xiaotingia_zhengi	1	1	1	1	1	1	163.5	157.3
Yutyrannus_huali	1	1	1	1	1	1	129.4	125

Codings are explained in the text. Char1 refers to a multistate treatment of the filamentous/feathered condition; Char2 lumps all non-scaly conditions into a single state (i.e. binary); a, b, and c represent different assumptions made for ancestral condition in pterosaurs. Maximal and minimal ages are in millions of years

References

Anderson BG, Lucas SG, Barrick RE, Heckert AB, Basabilvaso GT (1998) Dinosaur skin impressions and associated skeletal remains from the upper Campanian of southwestern New Mexico: new data on the integument morphology of hadrosaurs. J Vertebr Paleontol 18(4):739–745. https://doi.org/10.1080/02724634.1998.10011102

Andrews CW (1913) A descriptive catalogue of the marine reptiles of the Oxford Clay: based on the Leeds Collection in the British Museum (Natural History), London. Part 2. British Museum (Natural History), London

Arbour VM, Burns ME, Bell PR, Currie PJ (2014) Epidermal and dermal integumentary structures of ankylosaurian dinosaurs. J Morphol 275:39–50. https://doi.org/10.1002/jmor.20194

Avanzini M (2000) Synaptichnium tracks with skin impressions from the Anisian (middle Triassic) of the Southern Alps (Val di Non – Italy). Ichnos 7:243–251. https://doi.org/10.1080/10420940009380164

Bakker RT, Galton PM (1974) Dinosaur monophyly and a new class of vertebrates. Nature 248:168–172. https://doi.org/10.1038/248168a0

Baron MG, Norman DB, Barrett PM (2017) A new hypothesis of dinosaur relationships and early dinosaur evolution. Nature 543:501–506. https://doi.org/10.1038/nature21700

Barrett PM, Butler RJ, Edwards NP, Milner AR (2008) Pterosaur distribution in time and space: an atlas. Zitteliana B 28:61–107

Barrett PM, Evans DC, Campione NE (2015) Evolution of dinosaur epidermal structures. Biol Lett 11:20150229. https://doi.org/10.1098/rsbl.2015.0229

Bell PR (2012) Standardized terminology and potential taxonomic utility for hadrosaurid skin impressions: a case study for *Saurolophus* from Canada and Mongolia. PLoS One 7:e31295. https://doi.org/10.1371/journal.pone.0031295

Bell PR (2014) A review of hadrosaurid skin impressions. In: Eberth DA, Evans DC (eds) Hadrosaurs. Indiana University Press, Bloomington, pp 572–590

Bell PR, Fanti F, Currie PJ, Arbour VM (2014) A mummified duck-billed dinosaur with a soft-tissue cock's comb. Curr Biol 24:70–75. https://doi.org/10.1016/j.cub.2013.11.008

Bell PR, Campione NE, Persons WS IV, Currie PJ, Larson PL, Tanke DH, Bakker RT (2017) Tyrannosauroid integument reveals conflicting patterns of gigantism and feather evolution. Biol Lett 13:20170092. https://doi.org/10.1098/rsbl.2017.0092

Bennett SC (2015) New interpretation of the wings of the pterosaur *Rhamphorhynchus muensteri* based on the Zittel and Marsh specimens. J Paleontol 89:845–869. https://doi.org/10.1017/jpa.2015.68

Bernardi M, Petti FM, Avanzini M (2010) A webbed archosaur footprint from the Upper Jurassic (Carnian) of the Italian Southern Alps. In: Milàn J, Lucas SG,

Lockley MG, Spielmann JA (eds) Crocodile Tracks and Traces, vol 21. New Mexico Museum of Natural History and Science, Bulletin, pp 65–68

Bonaparte JF, Novas FE, Coria RA (1990) *Carnotaurus sastrei* Bonaparte, the horned, lightly built carnosaur from the middle Cretaceous of Patagonia. Los Angeles County Nat His Mus Contrib Sci 416:1–42

Brown B (1917) A complete skeleton of the horned dinosaur *Monoclonius*, and description of a second skeleton showing skin impressions. Bull Am Mus Nat Hist 37:281–306

Brusatte SL (2012) Dinosaur paleobiology. Wiley, Chichester

Brush AH (2000) Evolving a protofeather and feather diversity. Am Zool 40:631–639. https://doi.org/10.1093/icb/40.4.631

Buchwitz M, Voigt S (2012) The dorsal appendages of the Triassic reptile *Longisquama insignis*: reconsideration of a controversial integument type. Paläontol Z 86:313–331. https://doi.org/10.1007/s12542-012-0135-3

Butler RJ, Upchurch P, Norman DB (2008) The phylogeny of the ornithischian dinosaurs. J Syst Palaeontol 6:1–40. https://doi.org/10.1017/S1477201907002271

Chang C, Wu P, Baker RE, Maini PK, Alibardi L, Chuong C-M (2009) Reptile scale paradigm: Evo-Devo, pattern formation and regeneration. Int J Dev Biol 53:813–826. https://doi.org/10.1387/ijdb.072556cc

Charig AJ (1979) A new look at the dinosaurs. British Museum (Natural History), London

Chen PJ, Dong ZM, Zhen SN (1998) An exceptionally well-preserved theropod dinosaur from the Yixian Formation of China. Nature 391:147–152. https://doi.org/10.1038/34356

Chen CF, Foley J, Tang PC, Li A, Jiang TX, Wu P, Widelitz RB, Chuong CM (2015) Development, regeneration, and evolution of feathers. Annu Rev Anim Biosci 3:169–195. https://doi.org/10.1146/annurev-animal-022513-114127

Cheng X, Wang X, Jiang S, Kellner AWA (2015) Short note on a non-pterodactyloid pterosaur from Upper Jurassic deposits of Inner Mongolia, China. Hist Biol 27:749–754. https://doi.org/10.1080/08912963.2014.974038

Chiappe LM, Göhlich UB (2010) Anatomy of *Juravenator starki* (Theropoda: Coelurosauria) from the Late Jurassic of Germany. Neues Jahrb Geol Palaontol Abh 258:257–296. https://doi.org/10.1127/0077-7749/2010/0125

Christiansen NA, Tschopp E (2010) Exceptional stegosaur integument impressions from the Upper Jurassic Morrison Formation of Wyoming. Swiss J Geosci 103:163–171. https://doi.org/10.1007/s00015-010-0026-0

Coria RA, Chiappe LM (2007) Embryonic skin from the Late Cretaceous sauropods (Dinosauria) of Auca Mahuevo, Patagonia, Argentina. J Paleontol 81:1528–1532. https://doi.org/10.1666/05-150.1

Cruickshank ARI (1972) The proterosuchian thecodonts. In: Joysey KA, Kemp TS (eds) Studies in vertebrate evolution. Oliver & Boyd, Edinburgh, pp 89–119

Cunningham CW, Omland KE, Oakley TH (1998) Reconstructing ancestral character states: a critical reappraisal. Trends Ecol Evol 13:361–366. https://doi.org/10.1016/S0169-5347(98)01382-2

Czerkas SA (1992) Discovery of dermal spines reveals a new look for sauropod dinosaurs. Geology 20:1068–1070

Czerkas SA (1994) The history and interpretation of sauropod skin impressions. Gaia 10:173–182

Czerkas SA, Czerkas SJ (1997) The integument and life restoration of *Carnotaurus*. In: Wolberg DL, Stump E, Rosenberg G (eds) Dinofest International: Proceedings of a Symposium held at Arizona State University. Academy of Natural Sciences, Philadelphia, pp 155–158

Czerkas SA, Ji Q (2002) A new rhamphorhynchoid with a headcrest and complex integumentary structures. In: Czerkas SA (ed) Feathered dinosaurs and the origin of flight. The Dinosaur Museum, Blanding, pp 15–41

Dalla Vecchia FM (2009) Anatomy and systematics of the pterosaur *Carniadactylus* gen. n. Rosenfeldi (Dalla Vecchia, 1995). Riv Ital Paleontol Stratigr 115:159–188. https://doi.org/10.13130/2039-4942/6377

Davis M (2014) Census of dinosaur skin reveals lithology may not be the most important factor in increased preservation of hadrosaurid skin. Acta Palaeontol Pol 59:601–605. https://doi.org/10.4202/app.2012.0077

DePalma RA, Burnham DA, Martin LD, Larson PL, Bakker RT (2015) The first giant raptor (Theropoda: Dromaeosauridae) from the Hell Creek Formation. Paleontological Contributions 14. https://doi.org/10.17161/paleo.1808.18764

Dhouailly D (2009) A new scenario for the evolutionary origin of hair, feather, and avian scales. J Anat 214:587–606. https://doi.org/10.1111/j.1469-7580.2008.01041.x

Di-Poï N, Milinkovitch MC (2016) The anatomical placode in reptile scale morphogenesis indicates shared ancestry among skin appendages in amniotes. Sci Adv 2:e1600708. https://doi.org/10.1126/sciadv.1600708

Dodson P, Forster CA, Sampson SD (2004) Ceratopsidae. In: Weishampel DB, Dodson P, Osmólska H (eds) The Dinosauria, 2nd edn. University of California Press, Berkeley, pp 494–513

Ellenberger P (1974) Contribution à la classification des pistes de vertébrés du Trias. Les types du Stormberg d'Afrique du Sud. II. Le Stormberg Superieur. Palaeovertebrata, Memoire Extraordinaire, Montpellier

Evans DC, Reisz RR (2007) Anatomy and relationships of *Lambeosaurus magnicristatus*, a crested hadrosaurid dinosaur (Ornithischia) from the Dinosaur Park Formation, Alberta. J Vertebr Paleontol 27(2):373–393. https://doi.org/10.1671/0272-4634(2007)27[373

Felsenstein J (1985) Phylogenies and the comparative method. Am Nat 125:1–15. https://doi.org/10.1086/284325

Foster J, Hamblin A, Lockley M (2003) *Apatopus* trackway and other footprints from the Chinle Group of southern Utah: an update. Ichnos 10:165–167. https://doi.org/10.1080/10420940390258003

Foth C (2012a) On the identification of feather structures in stem-line representatives of birds: evidence from fossils and actuopalaeontology. Paläontol Z 86:91–102. https://doi.org/10.1007/s12542-011-0111-3

Foth C (2012b) The influence of taphonomy on fossil body plumages and the identification of feather structures in stem-lineage representatives of birds. In: Royo-Torres-R, Gascó F, Alcalá L (eds) 10th Annual Meeting of the European Association of Vertebrate Paleontologists. ¡Fundamental! 20:77–79

Foth C, Tischlinger H, Rauhut OWM (2014) New specimen of *Archaeopteryx* provides insights into the evolution of pennaceous feathers. Nature 511:79–82. https://doi.org/10.1038/nature13467

Foth C, Haug C, Haug JT, Tischlinger H, Rauhut OWM (2020) Two of a feather: a comparison of the preserved integument in the juvenile theropod dinosaurs *Sciurumimus* and *Juravenator* from the Kimmeridgian torleite formation of southern Germany. In: Foth C, Rauhut OWM (eds) The evolution of feathers – a palaeontological perspective. Springer, Heidelberg

Frey E, Martill DM (1998) Soft tissue preservation in a specimen of *Pterodactylus kochi* (Wagner) from the Upper Jurassic of Germany. Neues Jahrb Geol Palaontol Abh 210:421–441

Frey E, Tischlinger H, Buchy MC, Martill DM (2003) New specimens of Pterosauria (Reptilia) with soft parts with implications for pterosaurian anatomy and locomotion. In: Bufferaut E, Mazin J-M (eds) Evolution and palaeobiology of pterosaurs. Geological Society of London, Special Publications 217:233–266. https://doi.org/10.1144/GSL.SP.2003.217.01.14

Funston GF, Currie PJ (2016) A new caenagnathid (Dinosauria: Oviraptorosauria) from the Horseshoe Canyon Formation of Alberta, Canada, and a reevaluation of the relationships of Caenagnathidae. J Vertebr Paleontol 36:e1160910. https://doi.org/10.1080/02724634.2016.1160910

Gauthier J (1986) Saurischian monophyly and the origin of birds. In: Padian K (ed) The origin of birds and the evolution of flight. Memoirs of the California Academy of Science 8:1–55

Godefroit P, Cau A, Hu DY, Escuillie F, Wu WH, Dyke G (2013) A Jurassic avialan dinosaur from China resolves the early phylogenetic history of birds. Nature 498:359–362. https://doi.org/10.1038/nature12168

Godefroit P, Sinitsa SM, Dhouailly D, Bolotsky YL, Sizov AV, McNamara ME, Benton MJ, Spagna P (2014a) A Jurassic ornithischian dinosaur from Siberia with both feathers and scales. Science 345:451–455. https://doi.org/10.1126/science.1253351

Godefroit P, Sinitsa SM, Dhouailly D, Bolotsky YL, Sizov AV, McNamara ME, Benton MJ, Spagna P (2014b) Response to Comment on "A Jurassic ornithischian dinosaur from Siberia with both feathers and scales". Science 346:434c. https://doi.org/10.1126/science.1260146

Horner JR (1984) A "segmented" epidermal tail frill in a species of hadrosaurian dinosaur. J Paleontol 58:270–271

Hu DY, Hou LH, Zhang LJ, Xu X (2009) A pre-*Archaeopteryx* troodontid theropod from China with long feathers on the metatarsus. Nature 461:640–643. https://doi.org/10.1038/nature08322

Hu DY, Clark JA, Eliason CM, Qiu R, Li QG, Shawkey MD, Zhao C, D'Alba L, Jiang JK, Xu X (2018) A bony-crested Jurassic dinosaur with evidence of iridescent plumage highlights complexity in early paravian evolution. Nat Commun 9:217. https://doi.org/10.1038/s41467-017-02515-y

Ji Q, Yuan CX (2002) Discovery of two kinds of protofeathered pterosaurs in the Mesozoic Daohugou Biota in the Ningcheng region and its stratigraphic and biologic significances. Geological Review 48:221–224

Ji Q, Currie PJ, Norell MA, Ji SA (1998) Two feathered dinosaurs from northeastern China. Nature 393:753–761. https://doi.org/10.1038/31635

Ji Q, Ji SA, Lü JC, You HL (2005) First avialian bird from China (*Jinfengopteryx elegans* gen. et sp. nov.). Geol Bull China 24(3):197–210

Jiang SX, Wang XL (2011) Important features of *Gegepterus change* (Pterosauria: Archaeopterodactyloidea, Ctenochasmatidae) from a new specimen. Vertebrata PalAsiatica 49(2):172–184

Kellner AWA (1996) Reinterpretation of a remarkably well preserved pterosaur soft tissue from the Early Cretaceous of Brazil. J Vertebr Paleontol 16:718–722. https://doi.org/10.1080/02724634.1996.10011360

Kellner AWA, Wang XL, Tischlinger H, de Almeida Campos D, Hone DWE, Meng X (2010) The soft tissue of *Jeholopterus* (Pterosauria, Anurognathidae, Batrachognathinae) and the structure of the pterosaur wing membrane. Proc R Soc B 277:321–329. https://doi.org/10.1098/rspb.2009.0846

King MJ, Sarjeant WAS, Thompson DB, Tresise G (2005) A revised systematic ichnotaxonomy and review of the vertebrate ichnofamily Chirotheriidae from the British Triassic. Ichnos 12(4):241–299. https://doi.org/10.1080/10420940591009312

Klein H, Voigt S, Saber H, Schneider JW, Hminna A, Fischer J, Lagnaoui A, Brosig A (2011) First occurrence of a Middle Triassic tetrapod ichnofauna from the Argana Basin (western High Atlas, Morocco). Palaeogeogr Palaeoclimatol Palaeoecol 307 (1–4):218–231. https://doi.org/10.1016/j.palaeo.2011.05.021

Kundrát M (2004) When did theropods become feathered? Evidence for pre-*Archaeopteryx* feathery appendages. J Exp Zool B Mol Dev Evol 302:355–364. https://doi.org/10.1002/jez.b.20014

Lefèvre U, Cau A, Cincotta A, Hu DY, Chinsamy A, Escuillié F, Goderfroit P (2017) A new Jurassic theropod from China documents a transitional step in the macrostructure of feathers. Sci Nat 104:74. https://doi.org/10.1007/s00114-017-1496-y

Lefèvre U, Cau A, Hu D, Godefroit P (2020) Feather evolution in non-avialan paravians. In: Foth C, Rauhut OWM (eds) The evolution of feathers – a palaeontological perspective. Springer, Heidelberg

Lingham-Soliar T (2014) Comment on "A Jurassic ornithischian dinosaur from Siberia with both feathers and scales". Science 346:434b. https://doi.org/10.1126/science.1259983

Lü JC (2002) Soft tissue in an Early Cretaceous pterosaur from Liaoning Province, China. Memoir of the Fukui Prefectural Dinosaur Museum 1:19–28

Lü JC (2009) A baby pterodactyloid pterosaur from the Yixian Formation of Ningcheng, Inner Mongolia, China. Acta Geologica Sinica (English Edition) 83:1–8. https://doi.org/10.1111/j.1755-6724.2009.00001.x

Lü JC, Brusatte SL (2015) A large, short-armed, winged dromaeosaurid (Dinosauria: Theropoda) from the Early Cretaceous of China and its implications for feather evolution. Sci Rep 5:11775. https://doi.org/10.1038/srep11775

Lü JC, Ji Q, Wei XF, Liu YQ (2012) A new ctenochasmatoid pterosaur from the Early Cretaceous Yixian Formation of western Liaoning, China. Cretac Res 34:26–30. https://doi.org/10.1016/j.cretres.2011.09.010

Maddison WP (1991) Squared-change parsimony reconstructions of ancestral states for continuous-valued characters on a phylogenetic tree. Syst Zool 40:304–314

Maderson PFA, Alibardi L (2000) The development of the sauropsid integument: a contribution to the problem of the origin and evolution of feathers. Am Zool 40:513–529. https://doi.org/10.1093/icb/40.4.513

Maidment SCR (2010) Stegosauria: a historical review of the body fossil record and phylogenetic relationships. Swiss J Geosci 103:199–210. https://doi.org/10.1007/s00015-010-0023-3

Mantell GA (1852) On the structure of the *Iguanodon*, and on the fauna and flora of the Wealden Formation. Notices of the Meetings of the Royal Institution 1:141–146

Martill DM, Unwin DM (1989) Exceptionally well preserved pterosaur wing membrane from the Cretaceous of Brazil. Nature 340:138–140

Mayr G, Peters SD, Plodowski G, Vogel O (2002) Bristle-like integumentary structures at the tail of the horned dinosaur *Psittacosaurus*. Naturwissenschaften 89:361–365. https://doi.org/10.1007/s00114-002-0339-6

Mayr G, Pittman M, Saitta E, Kaye TG, Vinther J (2016) Structure and homology of *Psittacosaurus* tail bristles. Palaeontology 59:793–802. https://doi.org/10.1111/pala.12257

Mazin J-M, Hantzpergue P, Lafaurie G, Vignaud P (1995) Pterosaur trackways from the Tithonian-of-Crayssac (Quercy, France). Comptes Rendus de l'Académie des Sciences, Série IIA, Sciences de la Terre et des Planètes 321:417–424

McKellar RC, Chatterton BDE, Wolfe AP, Currie PJ (2011) A diverse assemblage of Late Cretaceous dinosaur and bird feathers from Canadian amber. Science 333:1619–1622. https://doi.org/10.1126/science.1203344

Morris WJ (1976) Hypsilophodont dinosaurs: a new species and comments on their systematics. In: Churcher CS (ed) Athlon: essays on palaeontology in honour of Loris Shano Russell. Toronto University Press, Toronto, pp 93–113

Musser JM, Wagner GP, Prum RO (2015) Nuclear β-catenin localization supports homology of feathers, avian scutate scales, and alligator scales in early development. Evol Dev 17:185–194. https://doi.org/10.1111/ede.12123

Norell MA, Xu X (2005) Feathered dinosaurs. Annu Rev Earth Planet Sci 33:277–299. https://doi.org/10.1146/annurev.earth.33.092203.122511

Norman DB (1985) The illustrated encyclopedia of dinosaurs. Salamander Books, London

O'Connor J (2020) The plumage of basal birds. In: Foth C, Rauhut OWM (eds) The evolution of feathers – a palaeontological perspective. Springer, Heidelberg

Oksanen J, Blanchet FG, Friendly M, Kindt R, Legendre P, McGlinn D, Minchin PR, O'Hara RB, Simpson GL, Solymos P, Stevens MHH, Szoecs E, Wagner H (2017) Vegan: Community Ecology Package, version 2.4, 2nd edn. CRAN

Ortega F, Escaso F, Sanz JL (2010) A bizarre, humped Carcharodontosauria (Theropoda) from the Lower Cretaceous of Spain. Nature 467:203–206. https://doi.org/10.1038/nature09181

Ostrom JH (1976) *Archaeopteryx* and the origin of birds. Biol J Linn Soc 8:91–182. https://doi.org/10.1111/j.1095-8312.1976.tb00244.x

Ouyang H, Ye Y (2002) The first mamenchisaurian skeleton with complete skull: *Mamenchisaurus youngi*. Sichuan Science and Technology Press, Chengu

Padian K (2008) The Early Jurassic pterosaur *Dorygnathus banthensis* (Theodori, 1830). Spec Pap Palaeontol 80:1–107

Padian K, Rayner JMV (1993) Structural fibres of the pterosaur wing: anatomy and aerodynamics. Naturwissenschaften 80:361–364

Padian K, Li C, Pchelnikova J (2010) The trackmaker of *Apatopus* (Late Triassic, North America): implications for the evolution of archosaur stance and gait. Palaeontology 53:175–189. https://doi.org/10.1111/j.1475-4983.2009.00924.x

Paradis E (2012) Analysis of phylogenetics and evolution with R, 2nd edn. Springer, New York

Paul GS (1988) Physiological, migratorial, climatological, geophysical, survival, and evolutionary implications of Cretaceous polar dinosaurs. J Paleontol 62:640–652

Pinegar RT, Loewen MA, Cloward KC, Hunter RJ, Weege CJ (2003) A juvenile allosaur with preserved integument from the basal Morrison Formation of central Wyoming. J Vertebr Paleontol 23(3 Suppl):87A–88A

Prieto-Márquez A (2010) Global phylogeny of Hadrosauridae (Dinosauria: Ornithopoda) using parsimony and Bayesian methods. Zool J Linnean Soc 159:435–502

Prieto-Márquez A, Chiappe LM, Joshi SH (2012) The lambeosaurine dinosaur *Magnapaulia laticaudus* from the Late Cretaceous of Baja California, Northwestern Mexico. PLoS One 7:e38207. https://doi.org/10.1371/journal.pone.0038207

Prum RO (1999) Development and evolutionary origin of feathers. J Exp Zool (Mol Dev Evol) 285:291–306

Prum RO, Brush AH (2002) The evolutionary origin and diversification of feathers. Q Rev Biol 77:261–295. https://doi.org/10.1086/341993

Rauhut OWM, Foth C, Tischlinger H, Norell MA (2012) Exceptionally preserved juvenile megalosauroid theropod dinosaur with filamentous integument from the Late Jurassic of Germany. Proc Natl Acad Sci U S A 109:11746–11751. https://doi.org/10.1073/pnas.1203238109

Reisz RR, Sues HD (2000) The 'feathers' of *Longisquama*. Nature 408:428. https://doi.org/10.1038/35044204

Saitta ET, Rogers C, Brooker RA, Abbott GD, Kumar S, O'Reilly SS, Donohoe P, Dutta S, Summons RE, Vinther J (2017) Low fossilization potential of keratin protein revealed by experimental taphonomy. Palaeontology 60(4):547–556. https://doi.org/10.1111/pala.12299

Saveliev SV, Alifanov VR (2014) A new type of skin derivatives in ornithischian dinosaurs from the Late Jurassic of Transbaikalia (Russia). Dokl Biol Sci 456:182–184. https://doi.org/10.1134/s0012496614030053

Sawyer RH, Knapp LW (2003) Avian skin development and the evolutionary origin of feathers. J Exp Zool B Mol Dev Evol 298B:57–72. https://doi.org/10.1002/jez.b.26

Sawyer RH, Glenn T, French JO, Mays B, Shames RB, Barnes GL Jr, Rhodes W, Ishikawa Y (2000) The expression of beta (β) keratins in the epidermal appendages of reptiles and birds. Am Zool 40:530–539. https://doi.org/10.1668/0003-1569(2000)040[0530:TEOBKI]2.0.CO;2

Sawyer RH, Salvatore BA, Potylicki TTF, French JO, Glenn TC, Knapp LW (2003a) Origin of feathers: Feather beta (β) keratins are expressed in discrete epidermal cell populations of embryonic scutate scales. J Exp Zool B Mol Dev Evol 295B:12–24. https://doi.org/10.1002/jez.b.5

Sawyer RH, Washington LD, Salvatore BA, Glenn TC, Knapp LW (2003b) Origin of archosaurian integumentary appendages: the bristles of the wild turkey beard express feather-type β keratins. J Exp Zool B Mol Dev Evol 297B:27–34. https://doi.org/10.1002/jez.b.17

Schluter D, Price T, Mooers AØ, Ludwig D (1997) Likelihood of ancestor states in adaptive radiation. Evolution 51:1699–1711. https://doi.org/10.1111/j.1558-5646.1997.tb05095.x

Schweitzer MH, Watt JA, Avci R, Knapp L, Chiappe L, Norell M, Marshall M (1999) Beta-keratin specific immunological reactivity in feather-like structures of the Cretaceous alvarezsaurid, *Shuvuuia deserti*. J Exp Zool (Mol Dev Evol) 285:146–157. https://doi.org/10.1002/(SICI)1097-010X(19990815)285:2<146::AID-JEZ7>3.0.CO;2-A

Sereno PC (2012) Taxonomy, morphology, masticatory function and phylogeny of heterodontosaurid dinosaurs. ZooKeys 226:1–225. https://doi.org/10.3897/zookeys.226.2840

Sharov AG (1970) An unusual reptile from the Lower Triassic of Fergana. Paleontol Zh 4:112–116

Sharov AG (1971) New flying reptiles from the Mesozoic of Kazakhstan and Kirgizstan. Trudy Palaontologicheskogo Instituta Akademia Nauk SSSR 130:103–113

Starrfelt J, Liow LH (2016) How many dinosaur species were there? Fossil bias and true richness estimated using a Poisson sampling model. Philos Trans R Soc B Biol Sci 371:20150219. https://doi.org/10.1098/rstb.2015.0219

Sullivan C, Wang Y, Hone DWE, Wang YQ, Xu X, Zhang FC (2014) The vertebrates of the Jurassic Daohugou Biota of northeastern China. J Vertebr Paleontol 34:243–280. https://doi.org/10.1080/02724634.2013.787316

Sullivan C, Xu X, O'Connor JK (2017) Complexities and novelties in the early evolution of avian flight, as seen in the Mesozoic Yanliao and Jehol Biotas of Northeast China. Palaeoworld 26:212–229. https://doi.org/10.1016/j.palwor.2016.12.001

Swofford DL, Maddison WP (1987) Reconstructing ancestral character states under Wagner parsimony. Math Biosci 87:199–229. https://doi.org/10.1016/0025-5564(87)90074-5

Thompson RS, Parish JC, Maidment SCR, Barrett PM (2012) Phylogeny of the ankylosaurian dinosaurs (Ornithischia: Thyreophora). J Syst Palaeontol 10:301–312. https://doi.org/10.1080/14772019.2011.569091

Thornley AL (1970) Epidermal remnants of *Proterosuchus vanhoepeni* (HTN). Palaeontol Afr 132:57–60

Unwin DM (2006) The Pterosaurs: from deep time. Pi, New York

Upchurch P, Barrett PM, Dodson P (2004) Sauropoda. In: Weishampel DB, Dodson P, Osmólska H (eds) The Dinosauria, 2nd edn. University of California Press, Berkeley, pp 259–322

Van der Reest AJ, Wolfe AP, Currie PJ (2016) A densely feathered ornithomimid (Dinosauria: Theropoda) from the Upper Cretaceous Dinosaur Park Formation, Alberta, Canada. Cretac Res 58:108–117. https://doi.org/10.1016/j.cretres.2015.10.004

Wang XL, Zhou ZH, Zhang FC, Xu X (2002) A nearly completely articulated rhamphorhynchoid pterosaur with exceptionally well-preserved wing membranes and "hairs" from Inner Mongolia, northeast China. Chin Sci Bull 47:226–230. https://doi.org/10.1360/02tb9054

Wang XL, Kellner AWA, Zhou ZH, de ACD (2007) A new pterosaur (Ctenochasmatidae, Archaeopterodactyloidea) from the Lower Cretaceous Yixian Formation of China. Cretac Res 28:245–260. https://doi.org/10.1016/j.cretres.2006.08.004

Wild R (1993) A juvenile specimen of *Eudimorphodon ranzii* Zambelli (Reptilia, Pterosauria) from the upper

Triassic (Norian) of Bergamo. Rivista del Museo Civico di Scienze Naturali 16:95–120

Witmer LM (2009) Dinosaurs: fuzzy origins for feathers. Nature 458:293–295. https://doi.org/10.1038/458293a

Witton MP (2016) Recreating an age of reptiles. Rhedphare, UK

Xing LD, McKellar RC, Wang M, Bai M, O'Connor JK, Benton MJ, Zhang JP, Wang Y, Tseng KW, Lockley MG, Li G, Zhang WW, Xu X (2016) Mummified precocial bird wings in mid-Cretaceous Burmese amber. Nat Commun 7:12089. https://doi.org/10.1038/ncomms12089

Xu X (2006) Feathered dinosaurs from China and the evolution of major avian characters. Integr Zool 1:4–11. https://doi.org/10.1111/j.1749-4877.2006.00004.x

Xu X (2020) Filamentous integuments in non-avialan theropods and their kin: advances and future perspectives for understanding the evolution of feathers. In: Foth C, Rauhut OWM (eds) The evolution of feathers – a palaeontological perspective. Springer, Heidelberg

Xu X, Guo Y (2009) The origin and early evolution of feathers: insights from recent paleontological and neontological data. Vertebrata PalAsiatica 47:311–329

Xu X, Wang XL, Wu XC (1999a) A dromaeosaurid dinosaur with a filamentous integument from the Yixian Formation of China. Nature 401:262–266. https://doi.org/10.1038/45769

Xu X, Tang ZL, Wang XL (1999b) A therizinosauroid dinosaur with integumentary structures from China. Nature 399:350–354. https://doi.org/10.1038/20670

Xu X, Zhou ZH, Wang XL (2000) The smallest known non-avian theropod dinosaur. Nature 408:705–708. https://doi.org/10.1038/35047056

Xu X, Zhou ZH, Wang XL, Kuang XW, Zhang FC, Du XK (2003) Four-winged dinosaurs from China. Nature 421:335–340. https://doi.org/10.1038/nature01342

Xu X, Norell MA, Kuang XW, Wang XL, Zhao Q, Jia CK (2004) Basal tyrannosauroids from China and evidence for protofeathers in tyrannosauroids. Nature 431:680–684. https://doi.org/10.1038/nature02855

Xu X, Zheng XT, You HL (2009) A new feather type in a nonavian theropod and the early evolution of feathers. Proc Natl Acad Sci U S A 106:832–834. https://doi.org/10.1073/pnas.0810055106

Xu X, Wang KB, Ma QY, Xing LD, Sullivan C, Hu DY, Cheng SQ, Wang S (2012) A gigantic feathered dinosaur from the Lower Cretaceous of China. Nature 484:92–95. https://doi.org/10.1038/nature10906

Xu X, Zhou ZH, Dudley R, Mackem S, Chuong CM, Erickson GM, Varricchio DJ (2014) An integrative approach to understanding bird origins. Science 346:1253293. https://doi.org/10.1126/science.1253293

Xu X, Zheng XT, Sullivan C, Wang XL, Xing LD, Wang Y, Zhang XM, O'Connor JK, Zhang FC, Pan YH (2015) A bizarre Jurassic maniraptoran theropod with preserved evidence of membranous wings. Nature 521:70–73. https://doi.org/10.1038/nature14423

Xu X, Currie P, Pittman M, Xing LD, Meng QJ, Lü JC, Hu DY, CY Y (2017) Mosaic evolution in an asymmetrically feathered troodontid dinosaur with transitional features. Nat Commun 8:14972. https://doi.org/10.1038/ncomms14972

Zelenitsky DK, Therrien F, Erickson GM, DeBuhr CL, Kobayashi Y, Eberth DA, Hadfield F (2012) Feathered non-avian dinosaurs from North America provide insight into wing origins. Science 338:510–514. https://doi.org/10.1126/science.1225376

Zhang FC, Zhou ZH, Xu X, Wang XL, Sullivan C (2008) A bizarre Jurassic maniraptoran from China with elongate ribbon-like feathers. Nature 455:1105–1108. https://doi.org/10.1038/nature07447

Zheng XT, You HL, Xu X, Dong ZM (2009) An Early Cretaceous heterodontosaurid dinosaur with filamentous integumentary structures. Nature 458:333–336. https://doi.org/10.1038/nature07856

MIX
Papier aus verantwortungsvollen Quellen
Paper from responsible sources
FSC® C105338

If you have any concerns about our products,
you can contact us on
ProductSafety@springernature.com

In case Publisher is established outside the EU,
the EU authorized representative is:
Springer Nature Customer Service Center GmbH
Europaplatz 3, 69115 Heidelberg, Germany

Printed by Libri Plureos GmbH
in Hamburg, Germany